Contemporary Trends and Issues in Science Education

Volume 64

The book series Contemporary Trends and Issues in Science Education provides a forum for innovative trends and issues impacting science education. Scholarship that focuses on advancing new visions, understanding, and is at the forefront of the field is found in this series. Authoritative works based on empirical research and/or conceptual theory from disciplines including historical, philosophical, psychological and sociological traditions are represented here. Our goal is to advance the field of science education by testing and pushing the prevailing sociocultural norms about teaching, learning, research and policy. Book proposals for this series may be submitted to the Publishing Editor: Claudia Acuna E-mail: Claudia.Acuna@springer.com

Antonio Joaquín Franco-Mariscal
Editor

Critical Thinking in Science Education and Teacher Training

Springer

Editor
Antonio Joaquín Franco-Mariscal
Facultad de Ciencias de la Educación
Universidad de Málaga
Málaga, Spain

ISSN 1878-0482 ISSN 1878-0784 (electronic)
Contemporary Trends and Issues in Science Education
ISBN 978-3-031-78577-1 ISBN 978-3-031-78578-8 (eBook)
https://doi.org/10.1007/978-3-031-78578-8

This Springer imprint is published by the registered company Springer Nature Switzerland AG
The registered company address is: Gewerbestrasse 11, 6330 Cham, Switzerland

Foreword

In the dynamic landscape of modern education, fostering critical thinking skills is widely recognized as a cornerstone of effective teaching and learning. The recent COVID-19 pandemic has provided us with the opportunity to examine whether our educational systems were successful in preparing critical thinkers/citizens who are equipped with the skills of making informed decisions. The experiment has proven that there is more that needs to be done in the future. The recent introduction of artificial intelligence (AI) in our everyday lives highlights the need to prepare our students and teachers to navigate the uncertainty. The innovation of AI and the implications from this new technology highlight the need for students, teachers, and citizens in general to be able to critically evaluate and reflect information presented to them and act accordingly. The field of science education holds a great responsibility in preparing students and teachers for this new challenging world. The ability to think critically is not only desirable, but also essential for navigating the complexities of both the natural and the real world.

But what exactly do we mean by critical thinking in the context of science education, and how can we harness its power to enrich the learning experiences of our students? At its essence, critical thinking in science education encompasses a multifaceted approach to inquiry and problem-solving that supports students to engage thoughtfully with scientific concepts and phenomena. It involves the application of reasoning and argumentation, evidence-based analysis and explanations, and creative problem-solving skills to explore and understand the intricacies of the natural world. Moreover, critical thinking in this context extends beyond the mere acquisition of knowledge; it involves the creation of a mindset characterized by curiosity, skepticism, intellectual flexibility, the ability to deal with uncertainty—a mindset that is essential for success in both scientific endeavors and everyday life.

In this edited volume, the authors offer a comprehensive exploration of critical thinking in science education, drawing upon their expertise to provide valuable insights and practical strategies for educators at all levels. Central to their approach is the recognition that critical thinking is not a standalone skill but rather an integrated and multifaceted process that permeates all aspects of scientific inquiry and has connections to everyday life. The authors advocate for a holistic approach to critical thinking development, one that is grounded in real-world problems and meaningful scientific inquiry. They argue that by contextualizing scientific concepts within the framework of everyday life, educators can enhance student engagement and foster deeper understanding of complex scientific phenomena. From exploring socio-scientific issues and uncertainty, to tackling socially relevant questions, the integration of real-world problems serves as a catalyst for critical thinking development, providing students with opportunities to apply their knowledge and skills in meaningful ways.

Building upon this foundation, the authors outline three key domains of critical thinking—knowledge, skills, and dispositions—that are essential for success in science education. Within these domains, they emphasize the importance of promoting scientific practices such as argumentation, inquiry, and modeling, which are fundamental to the advancement of scientific knowledge and understanding. By integrating these practices into the curriculum, educators can empower students to think critically about scientific concepts and phenomena, fostering a deeper appreciation for the scientific method and its implications for society.

Furthermore, the authors offer practical strategies and approaches for implementing critical thinking in the science classroom and beyond. From designing inquiry-based learning experiences to fostering a culture of scientific inquiry, these strategies provide educators with the tools and resources they need to cultivate critical thinking skills among their students effectively.

As readers engage with the contents of this book, they are invited to reflect on several overarching questions that can serve as organizers for their exploration of critical thinking in science education:

- What are the key characteristics of critical thinking in the context of science education, and how can they inform our approach to teaching and learning?
- How can we effectively integrate real-world problems and meaningful scientific inquiry into the science curriculum to enhance student engagement and critical thinking skills?
- What are the core domains of critical thinking in science education, and how can educators foster the development of these domains in their students?
- What practical strategies and approaches can educators use to implement critical thinking in the science classroom, and how can they adapt these strategies to meet the diverse needs of their students?
- How can critical thinking support us to prepare our students to become responsible citizens and at the same time remain hopeful for their futures?
- What is the role of critical thinking in supporting future citizens as they struggle with the idea of artificial intelligence?

In essence, this edited volume serves as a valuable resource for educators, researchers, and policymakers alike, offering insights and strategies for implementing and promoting critical thinking skills in science education. By embracing the principles and practices outlined within these pages, we can empower the next generation of scientists, thinkers, and problem solvers to navigate an increasingly complex world with clarity, curiosity, and confidence.

Maria Evagorou
University of Nicosia
Nicosia, Cyprus

The original version of the book has been revised: Volume number has been replaced. A correction to this book can be found at https://doi.org/10.1007/978-3-031-78578-8_18

Acknowledgements

The editor and authors wish to express their gratitude to the series editor, Dana Zeidler, and the two anonymous reviewers whose valuable feedback during the double-blind review process enhanced the quality of this edited volume. We would also like to extend our appreciation to José Manuel Hierrezuelo-Osorio for his contributions to the English translation of this volume.

Introduction

The complexity of modern society increasingly demands citizens with critical thinking skills for decision-making and problem-solving, as well as for their personal development and intellectual autonomy. This book explores this challenge within science education, offering a model for its development (ENCIC-CT), along with diverse experiences tailored for students at varying educational levels and in teacher training, each serving as an illustration.

Critical Thinking in Science Education and Teacher Training includes 17 chapters and represents an international collective effort involving researchers from two Spanish universities (Málaga, and Santiago de Compostela), one European university (Cyprus), and two Latin American universities (Chile, and Uruguay).

The edited volume begins with a Foreword by Maria Evagorou, underscoring the significance of critical thinking in society and its development through science education. Following this, the book is structured into six parts.

In Part I, Franco-Mariscal and collaborators delineate the theoretical framework and strategies for fostering critical thinking in science education, introducing the ENCIC-CT model. This model is named after the acronym of the *Science Education and Competences* [ENseñanza de las CIencias y Competencias] (ENCIC) research group (https://www.encic.es/) at the University of Malaga (Málaga, Spain), where it was developed. Grounded in daily-life problems and socio-scientific issues, it encompasses various domains (knowledge, skills, and dispositions) and dimensions of critical thinking. The model utilizes scientific practices of argumentation, inquiry, and modeling as key educational objectives within science education, employing various teaching strategies for their development. The versatility of the model offers multiple pathways to address critical thinking in the classroom, as many combinations as there are possible within its various layers. Different chapters tackle some of these options, facilitating a sequential reading experience.

Part II comprises five chapters that explore socio-scientific issues, teaching in context, and critical thinking. Chapter 2, authored by Franco-Mariscal and collaborators, provides insights into the ENCIC-CT model applied to a contemporary issue such as the energy problem. Puig and Rodríguez Losada address the domains of knowledge, skills, and dispositions of critical thinking in Chap. 3, focusing on the

COVID-19 vaccination issue. Chapter 4 sees Ramos-Solano and Cebrián-Robles investigating the naturalization of spaces in preschool from personal autonomy, decision-making, and modeling perspectives. The comprehensive analysis of the urban space care problem through controversy mapping is thoroughly examined with architecture students in Chap. 5, authored by España-Naveira and collaborators. Lastly, García-Pérez and collaborators conclude this part in Chap. 6, addressing the critical analysis of information conducted by high school biology students on the issue of pasteurization.

Part III is dedicated to argumentation and comprises three chapters. In Chap. 7, Cañero-Arias and collaborators explore this dimension of critical thinking through the issue of energy production and consumption with high school students. Chapter 8, authored by Imbert and several Uruguayan researchers, focuses on argumentation through brief debates involving preservice teachers. Meanwhile, Chap. 9, by Palma-Jiménez and collaborators, also with teachers in initial training, addresses the issue of breastfeeding in public places using the Kialo online debate platform.

Part IV, focusing on inquiry and critical thinking, consists of three chapters. In Chap. 10, Muñoz-Campos and Franco-Mariscal integrate inquiry and argumentation by analyzing the degree of argumentation development among Spanish secondary school students in an inquiry focusing on homemade yogurt production. Chapter 11, by Alarcón-Orozco and collaborators, examines critical thinking through scientific thinking in the design of inquiries by preservice preschool teachers. The study conducted by Otárola and several Chilean and Spanish collaborators (Chap. 12) explores a comprehensive approach to raising environmental awareness through a service-learning proposal based on inquiry, highlighting social engagement from a critical thinking perspective.

Part V focuses on the educational possibilities of role-playing games, technologies, and multimedia resources to develop critical thinking in science education. The role-playing by López-Fernández, presented in Chap. 13, analyzes critical analysis of information and decision-making among secondary students focusing on the socio-scientific issue of Russian gas. Chapter 14, authored by Moreno-Martínez and Franco-Mariscal, advocates for leveraging the Zapworks augmented reality tool to enhance argumentation and presents some educational proposal designs by preservice science teachers. In Chap. 15, the study by Cruz-Lorite and researchers examines the influence of the design and dramatization of digital storytelling by preservice preschool teachers on the development of critical thinking knowledge, skills, and dispositions.

Part VI focuses on research ideas and concluding remarks. Chapter 16 presents three innovative approaches to fostering critical thinking. In the first, Franco-Mariscal and collaborators propose a mobile application that uses role-playing to enhance argumentation skills. The second idea, by López-Fernández and collaborators, explores the use of movies and TV series to cultivate critical thinking in preservice science teachers. Lastly, the third research idea, presented by Franco-Mariscal and collaborators, introduces an activity specifically designed to develop metacognitive abilities. The edited volume concludes with Chap. 17, which presents overarching insights on critical thinking in science education, as well as specific conclusions regarding the various dimensions of the ENCIC-CT model.

The book in your hands is the result of the outcomes of the Spanish National Plan R+D+i Project entitled *Citizens with Critical Thinking: A Challenge for Teachers in Science Education* (https://pensamientocritico.encic.es/), reference PID2019-105765GA-I00, funded by the Ministry of Science and Innovation and the State Research Agency of the Government of Spain (MCIN/AEI/10.13039/501100011033).

This project was carried out between 2020 and 2024 by 33 researchers from the ENCIC Research Group at the University of Malaga (Málaga, Spain). Antonio Joaquín Franco-Mariscal, the editor of this book, served as the project's principal investigator. The research was conducted in accordance with the protocol approved by the Ethics Committee on Experimentation of the University of Malaga (CEUMA) with reference 31-2022.

Málaga, Spain Antonio Joaquín Franco-Mariscal

Contents

Contributors

María Marta Alarcón-Orozco Science Education, Universidad de Málaga, Málaga, Spain

Sair Aparicio National Public Education Administration (ANEP), Consejo de Formación en Educación (CFE), Montevideo, Uruguay

Ángel Blanco-López Science Education, Universidad de Málaga, Málaga, Spain

Claudia Cabrera Borges National Public Education Administration (ANEP), Consejo de Formación en Educación (CFE), Montevideo, Uruguay;
ORT University, Montevideo, Uruguay

María José Cano-Iglesias Science Education, Universidad de Málaga, Málaga, Spain

Silvia Carámbula National Public Education Administration (ANEP), Consejo de Formación en Educación (CFE), Montevideo, Uruguay

Joaquín Cañero-Arias Science Education, Universidad de Málaga, Málaga, Spain;
Colegio La Presentación, Málaga, Spain

Daniel Cebrián-Robles Science Education, Universidad de Málaga, Málaga, Spain

Isabel María Cruz-Lorite Department of Education, University of Nicosia, Nicosia, Cyprus

Paloma España-Naveira Art and Architecture Department, Universidad de Málaga, Málaga, Spain

Enrique España-Ramos Science Education, Universidad de Málaga, Málaga, Spain

Alicia Fagúndez National Public Education Administration (ANEP), Consejo de Formación en Educación (CFE), Montevideo, Uruguay

Ana Belén Ferrer-Baldomero Colegio La Presentación, Málaga, Spain

Antonio Joaquín Franco-Mariscal Science Education, Universidad de Málaga, Málaga, Spain

José Antonio García-Pérez Science Education, Universidad de Málaga, Málaga, Spain

Cristina García-Ruiz Science Education, Universidad de Málaga, Málaga, Spain

José Manuel Hierrezuelo-Osorio Science Education, Universidad de Málaga, Málaga, Spain

Daisy Imbert Romero National Public Education Administration (ANEP), Consejo de Formación en Educación (CFE), Montevideo, Uruguay;
ORT University, Montevideo, Uruguay

María del Mar López-Fernández Science Education, Universidad de Málaga, Málaga, Spain

Teresa Lupión-Cobos Science Education, Universidad de Málaga, Málaga, Spain

Gabriela Martínez National Public Education Administration (ANEP), Consejo de Formación en Educación (CFE), Montevideo, Uruguay

Manuel Martínez-Lobos Universidad de Playa Ancha, Valparaíso, Chile

Sylvia Moraga-Toledo Liceo Artístico Guillermo Gronemeyer Zamorano, Valparaíso, Chile

Eva Morales-Soler Art and Architecture Department, Universidad de Málaga, Málaga, Spain

Noelia Margarita Moreno-Martínez Personality, Psychological Assessment and Treatment, Universidad de Málaga, Málaga, Spain

Verónica Muñoz-Campos Science Education, Universidad de Málaga, Málaga, Spain;
I.E.S. Portada Alta, Málaga, Spain

Fabianna Otárola Benavides Universidad de Playa Ancha, Valparaíso, Chile

Miriam Palma-Jiménez Science Education, Universidad de Málaga, Málaga, Spain

Mariana Pérez Orlando National Public Education Administration (ANEP), Consejo de Formación en Educación (CFE), Montevideo, Uruguay

Marcela Pérez Salatto National Public Education Administration (ANEP), Consejo de Formación en Educación (CFE), Montevideo, Uruguay

Blanca Puig Universidade de Santiago de Compostela (USC), Santiago de Compostela, Spain

Alba Ramos-Solano Science Education, Universidad de Málaga, Málaga, Spain

Noela Rodríguez Losada Science Education, Universidad de Málaga, Málaga, Spain

Jessica Soto-Martínez Liceo Artístico Guillermo Gronemeyer Zamorano, Valparaíso, Chile

Juan Zagalaz Didactics of Musical Expression, Universidad de Málaga, Málaga, Spain

Part I
Theoretical Framework and Strategies for the Development of Critical Thinking in Science Education

Chapter 1
The ENCIC-CT Model for the Development of Critical Thinking

Antonio Joaquín Franco-Mariscal, María José Cano-Iglesias, Enrique España-Ramos, and Ángel Blanco-López

1.1 Theoretical Perspectives and Approaches for the Development of Critical Thinking

This book explores critical thinking (hereinafter, CT), a complex construct that is challenging to define precisely. The following sections offer insights that may help to clarify its meaning, examining how various scholars have discussed and developed this concept from philosophical foundations and cognitive psychology perspectives.

Although the formal development of the concept of CT took place in the twentieth century, it is important to recognize that similar ideas were being explored in ancient Greece, even though the term itself was not used. A significant precursor to the development of CT is found in Socrates' method. While the Greek philosopher did not use the term CT, he is credited as one of the earliest practitioners and promoters of a form of thinking we now associate with this skill. Socrates employed a technique of dialogue and questioning known as the Socratic method to foster critical reflection and analysis in his interlocutors. This method involved asking probing and persistent questions to challenge ideas, examine assumptions, and seek truth through rational discussion (Farnsworth, 2021).

A. J. Franco-Mariscal (✉) · M. J. Cano-Iglesias · E. España-Ramos · Á. Blanco-López
Science Education, Universidad de Málaga, Málaga, Spain
e-mail: anjoa@uma.es

M. J. Cano-Iglesias
e-mail: mjcano@uma.es

E. España-Ramos
e-mail: enrienri@uma.es

Á. Blanco-López
e-mail: ablancol@uma.es

A. J. Franco-Mariscal (ed.), *Critical Thinking in Science Education and Teacher Training*, Contemporary Trends and Issues in Science Education 64,
https://doi.org/10.1007/978-3-031-78578-8_1

In the seventeenth century, the philosopher and mathematician Descartes (1641) made significant contributions to CT that have profoundly impacted both philosophy and scientific methodology. Key contributions include: (a) the method of systematic doubt, which involves rigorously questioning all beliefs that are not absolutely certain, thereby highlighting the importance of self-reflection and certainty in evaluating our beliefs; (b) the Cartesian method, encompassing evidence, analysis, synthesis, and revision—essential components of CT; and (c) rationalism, which asserts that reason and logic are fundamental tools for acquiring knowledge.

Similarly, Kant's contributions to CT are crucial as they promote a rigorous and systematic examination of the foundations and limits of knowledge, morality, and reason, offering tools for a more profound and reflective analysis of our beliefs and practices. His key contributions include: (a) *The Critique of Pure Reason*, where Kant (1781) investigated the boundaries and capacities of human reason, introducing the idea that reason has inherent limits on what it can know; (b) transcendental epistemology, which explores how cognitive structures shape our understanding of the world; (c) the categorical imperative as a fundamental moral principle that guides action, providing a framework for critically and objectively assessing moral decisions; (d) individual autonomy in making rational and moral decisions independently; and (e) *The Critique of Practical Reason*, where Kant (1788) examines morality and practical reason, offering a critical perspective on how rational principles are applied to ethical behavior.

In the twentieth century, philosophers and educators like Dewey and Russell played a pivotal role in shaping the concept of CT. Dewey (1910) advocated for education as a process of reflective and CT, which involves carefully analyzing information, experiences, and ideas before drawing conclusions. This reflective thinking requires actively questioning, analyzing, and evaluating beliefs and assumptions. It is an approach that fosters CT through experimentation and the resolution of real and relevant problems. Dewey also argued that CT should be taught in schools, as education's role is to prepare individuals to actively engage in democracy and make informed decisions, which necessitates CT.

Russell's (1973) contributions to CT center around logic and argumentation, which are essential in shaping the CT construct. His contributions include: (a) the advancement of mathematical logic and reasoning, which influenced the development of formal reasoning systems and the clarification of complex arguments; (b) philosophical analysis, an approach aimed at clarifying concepts and arguments through rigorous logical and linguistic examination, essential for enhancing precision and clarity in reasoning and communication; (c) skepticism and critique of unsupported claims; (d) the defense of intellectual freedom and the right to challenge established ideas; and (e) social and political engagement on issues such as pacifism, civil rights, and education, demonstrating that CT should be applied to real-world societal issues.

Paul and Elder (2005) are considered the leading proponents of CT theory, and their contributions have influenced its understanding and teaching. Their work focuses on identifying the intellectual skills necessary for CT and the importance of explicit instruction in fostering these skills. According to their perspective, CT entails a mode of thinking where individuals enhance the quality of their cognition by engaging with the underlying structures of thought and subjecting them to rigorous intellectual standards. As a result, a critical and practiced thinker formulates problems and questions with clarity and precision; assesses and synthesizes relevant information; employs abstract ideas to interpret data effectively, and reaches conclusions and solutions based on relevant criteria. Moreover, such thinkers exhibit an open mind; recognizes and evaluates assumptions, implications, and practical consequences as needed; and communicates their insights when devising solutions to complex problems (Paul & Elder, 2005).

Another prominent author is Facione (1990), whose models and approaches have influenced the understanding and teaching of CT, emphasizing cognitive skills and dispositions or critical attitudes. This author argues that CT development is closely related to its practice, to action. According to Facione (1990), the essential cognitive skills of CT include interpretation, analysis, evaluation, inference, explanation, and self-regulation.

Interpretation consists of "comprehending and expressing the meaning or significance of a wide variety of experiences, situations, data, events, judgments, conventions, beliefs, rules, procedures, or criteria." (Facione, 1990, p. 5). It includes the subskills of categorization, decoding significance, and clarifying meaning.

Analysis is "to identify the intended and actual inferential relationships among statements, questions, concepts, descriptions, or other forms of representation intended to express belief, judgment, experiences, information, or opinions" (p. 5). This skill includes examining ideas, detecting arguments, and analyzing arguments.

Evaluation is "to assess the credibility of statements or other representations which are accounts or descriptions of a person's perception, experience, situation, judgment, belief, or opinion; and to assess the logical strength of the actual or intended inferential relationships among statements, descriptions, questions or other forms of representation" (p. 6).

Inference means "to identify and secure elements needed to draw reasonable conclusions; to form conjectures and hypotheses; to consider relevant information and to reduce the consequences flowing from data, statements, principles, evidence, judgments, beliefs, opinions, concepts, descriptions, questions, or other forms of representation" (p. 6). As sub-skills, it includes querying evidence, conjecturing alternatives, and drawing conclusions.

Explanation is understood as the ability to present the results of one's own reasoning reflectively and coherently. This involves being able to provide someone with a comprehensive view of the entire landscape, both "to state and to justify that reasoning in terms of the evidential, conceptual, methodological, criteriological, and contextual considerations upon which one's results were based; and to present one's reasoning in the form of cogent arguments" (p. 6). The sub-skills of explanation include describing methods and results, justifying procedures, proposing and defending with good reasons one's causal and conceptual explanations of events or viewpoints, and presenting complete and well-reasoned arguments in the context of seeking the greatest possible understanding.

Self-regulation means "self-consciously to monitor one's cognitive activities, the elements used in those activities, and the results educed, particularly by applying skills in analysis, and evaluation to one's own inferential judgments with a view toward questioning, confirming, validating, or correcting either one's reasoning or one's results" (p. 7). Self-examination and self-correction are its sub-skills. Some authors refer to this skill as metacognition.

Facione (1990) asserts that critical thinkers can be characterized not only by their cognitive abilities but also by their approach to and way of living life. These dispositions or critical attitudes encompass: curiosity about a broad range of issues, a commitment to staying well-informed; alertness for opportunities to apply CT; confidence in reasoned inquiry processes; self-confidence in one's capacity for reasoning; openness to diverse worldviews; flexibility in considering alternatives and viewpoints; empathy for others' perspectives; impartiality in evaluating reasoning; honesty in confronting personal biases, prejudices, stereotypes, or egocentric tendencies; caution in deferring, making, or modifying judgments; and a readiness to reconsider and revise perspectives where honest reflection suggests that change is warranted.

Ennis (1987) is another pioneering author in the study of CT. His works addressed the definition and assessment of CT and have influenced the development of strategies for teaching and measuring this skill. According to Ennis (1987), "CT is a reasonable reflective thinking that is focused on deciding what to believe or do".

Lipman (2003) advocates for an education that transcends mere knowledge transmission, instead emphasizing CT as a set of higher-order cognitive skills. This author conceives of CT as part of multidimensional thinking, alongside creative thinking and careful thinking. According to Halpern (2006), CT is central to problem-solving, inference formulation, and decision-making.

The concept of CT has extended to education in general, where it currently represents a focal point of interest by encompassing cross-cutting competencies essential for personal, social, and professional life (Vázquez & Manassero, 2020). With the exponential growth of information creation each year, CT skills are increasingly indispensable for individuals to become more adaptable, flexible, and capable of dealing with this rapidly evolving information. Drawing upon the integration of existing frameworks with recent conceptualizations of CT, Dwyer et al. (2014) propose an integrated CT framework for the twenty-first century. This framework identifies memory/knowledge and understanding as fundamental processes required for successfully applying CT (i.e., analysis, evaluation, and inference). Moreover, it also includes reflective judgment, as well as the self-regulatory functions of metacognition, which ultimately dictate how each thinking process will be carried out.

As emerges from the various perspectives provided, CT is a complex construct involving different skills and dispositions, making it difficult to find a consensus definition in the literature. Furthermore, it is essential to analyze how it is being understood within the specific context of science education.

1.2 Critical Thinking in Science Education

Currently, there is unanimous agreement regarding the importance of CT for citizens and its consideration as one of the major goals of science education (Osborne, 2014). Nowadays, science and technology play a significant role in the challenges confronting our society, many of which are interconnected with the Sustainable Development Goals (SDGs) established by the United Nations (2015) to safeguard the planet, eradicate poverty, and ensure prosperity for all individuals. Moreover, science and technology are also present in many daily life problems. It is evident that, without critical reflection on these issues, decisions in economic, social, political spheres, among others, could lead to irreparable consequences.

Therefore, it is necessary to educate CT skills in students across all disciplines and educational levels, empowering them to make decisions in society and take action (Colucci-Gray & Gray, 2022; Hodson, 2003; Simonneaux, 2014). Unfortunately, for many years, science education has overlooked the development of CT skills (Halpern & Danna, 2023), despite the current demand for science education to play an active role in addressing issues related to energy, the environment, food, or health, among others. Consequently, integrating CT into science education poses a formidable challenge (Puig & Jiménez-Aleixandre, 2022), prompting an increasing focus on this subject within the specialized literature of science education (Bailin, 2002; Jiménez-Aleixandre & Puig, 2022; Torres & Solbes, 2016).

Next, we discuss some contributions on how to understand CT and its importance in science education. Osborne (2014) suggests that criticism and questioning are essential for the practice of science, and without them, scientific knowledge could not be constructed.

Bailin (2002) suggests that many efforts to promote CT in science education have been hindered by certain misconceptions about the nature of CT, "in particular the characterization of CT in terms of processes or skills and the separation of CT and knowledge" (p. 361). This author argues that CT always emerges in response to a task, question, problematic situation, or challenge. This encompasses activities such as problem-solving, theorizing, evaluating theories, conducting research, and engaging in creative tasks, all of which are inherently context-specific. Addressing these challenges critically involves drawing upon a complex range of knowledge that is context-dependent. Given the contextual nature of CT, its application in science education necessitates a focus on tasks, problems, and issues within the scientific curriculum that call for, or stimulate, CT.

Solbes and Torres (2012) define CT as a necessity for individuals to develop their own way of thinking, enabling them to engage with social situations and play an active role in scientific and cultural decisions. For these authors, CT development requires individuals acquiring a set of skills to address and discuss socio-scientific issues, alongside understanding science as a human activity that is related to technology, environment, and society. Among the skills highlighted by Solbes and Torres (2012) are questioning argument validity, transcending dominant discourses, identifying argumentative fallacies, comprehensively analyzing socio-scientific issues, making informed decisions, among others.

Blanco-López et al. (2015) in a study on the key aspects of scientific competence for citizenship, found that critical attitude/thinking was the aspect with the highest level of agreement and consensus among a group of Spanish experts in science from different spheres and fields: scholar-scientists and engineers, researchers and private sector scientists, philosophers of science, science educators, and science communicators. In addition to CT, four other aspects were emphasized: individual responsibility; ability to reason, analyze, interpret, and construct an argument in relation to scientific phenomena and knowledge; ability to search for, analyze, synthesize, and communicate information and; ability to work as part of a team. These authors conclude that:

> the five key aspects identified in this study should be considered jointly in the context of school science education, since they are interrelated skills that citizens will require when tackling important issues and making decisions in various spheres of their life (personal, social, professional, etc.) (p. 164).

According to Vieira and Tenreiro (2016), CT shares common attributes with scientific literacy. Its core components encompass knowledge (including great ideas and scientific explanations, history of science, Nature Of Science, STS relations), skills/processes/abilities (analyse and assess information, evidence and arguments; formulate and test hypotheses, conjectures; draw and assess conclusions; make and assess generalisations; make and assess judgements; argue; communicate; assess courses of action), dispositions/attitudes (open-mindedness; impartialness; intellectual integrity; perseverance; respect for evidence; appreciation and enthusiasm for science) and standards/criteria (rigour, accuracy, clarity, consistency, validity, control of variables).

Vázquez and Manassero (2018) synthesize CT into four dimensions: creativity, reasoning and argumentation, complex processes (problem-solving and decision-making), and evaluation and judgment. Each dimension encompasses various skills that extend beyond mere cognitive abilities. Moreover, contexts, attitudes, values, and emotions are acknowledged to play crucial roles in CT. In this regard, it is important to highlight that research has shown that science education is influenced by feelings, values, and ideals, which can act as facilitators or obstacles to learning (Membiela et al., 2022).

Jiménez-Aleixandre and Puig (2022) indicate that CT comprises a series of components related to deliberate judgment (argumentation), and other components linked to citizenship education. This latter aspect involves the ability to develop an independent opinion, challenge socially and culturally established ideas, and analyze and critique discourses that justify social inequalities (Fairclough, 1995). According to these authors, CT implies not only thinking reasonably but also "independently." In essence, it involves developing an informed opinion that integrates scientific reasoning and values, as well as independent CT for social action.

García-Carmona (2023) examines the differences and relationships between CT and scientific thinking, two key intellectual processes essential for the comprehensive scientific education of citizens, which are sometimes ambiguously employed in specific literature on science education. As a result, some studies interchangeably refer to either type of thinking to represent the same cognitive and metacognitive abilities, frequently leaving their differences and similarities unclear. This study concludes that while they differ in terms of the purposes of their application and some skills or processes, they also share common elements. Moreover, they form a symbiotic relationship metaphorically, meaning each gains coherence or develops optimally when nourished or enriched by the other.

Furthermore, the difficulty associated with the CT construct extends beyond its definition or the skills it encompasses. Educators also encounter obstacles when attempting to bring it into the classroom, as they often lack clear examples to serve as guidance and suitable tools to evaluate the development of different skills (Hierrezuelo-Osorio et al., 2022).

In this regard, Vila et al. (2023) developed two tools, namely the Operational Map of CT (MOPC) and the Scheme for Designing CT Activities (EDAPC) respectively, to assist in designing classroom activities that enhance CT. According to the MOPC, a student must engage a set of cognitive skills while also adhering to dispositions that facilitate thoughtful action. Additionally, the student must be capable of self-regulating the emotions and values evoked by the context. These dimensions are applied to a specific context that demands the activation of relevant knowledge. Skills, dispositions, values/emotions, and knowledge enable the development of metacognitive competencies in the thinker, which are indispensable attributes for CT. The EDAPC outlines the following sequence: (1) Explicitly or implicitly define the scientific content to be addressed in alignment with the curriculum, (2) specify the CT elements (in accordance with the MOPC) to be developed, (3) select a relevant context for students that correlates with the scientific content to be explored, (4) craft the activity framework, establish scaffolding, bolster argumentation, present the activity in a multimodal format, and incorporate communicative and collaborative dimensions, and (5) integrate metacognitive questions aimed at fostering self-regulation throughout the activity.

In summary, it is essential not only to have approaches to the CT construct in science education, but also to provide examples of proposals for addressing it in the classroom at all educational levels. This book aims to fulfill precisely that objective.

1.3 The ENCIC-CT Model for the Development of Critical Thinking

Although CT can be developed through science education by addressing various types of problem (Bailin, 2002), several authors emphasize the significance of creating learning environments based on contexts where science and technology serve as driving forces of learning. These settings aim to foster not only knowledge but also skills, attitudes, and values (Chen & Xiao, 2021; Owens et al., 2021). Moreover, such learning contexts facilitate a deeper understanding of science and technology, helping students to forge connections with their everyday lives.

There is agreement on the importance of developing CT within context of daily life problems where science and technology play a significant role (Cebrián-Robles et al., 2021a; Solbes & Torres, 2012), particularly those involving controversies where citizens need to make decisions that affect them personally and socially (Evagorou et al., 2012). As a result, robust lines of research have emerged within the science education focusing on Socio-Scientific Issues (SSI) (Zeidler et al., 2019).

From this standpoint, building on Solbes and Torres's (2012) characterization of the competencies needed for fostering CT within the SSIs, Blanco-López et al. (2017) identified eight dimensions for their development in science education: vision of science, knowledge, critical analysis of information, comprehensive analysis of the problem, argumentation, personal autonomy, decision-making, and communication. Progress in CT development is closely tied to advancements in each dimension, which can be nurtured by addressing them individually, yet it's imperative to include issues where all dimensions are integrated simultaneously.

Different research conducted by members of the Science Education and Competences Research Group [ENseñanza de las CIencias y Competencias] (ENCIC) (https://www.encic.es/) support this approach to science education, which has been shown to be effective in teaching both chemistry (López-Fernández et al., 2022) and engineering (Cano-Iglesias, 2024). Recently, García-Carmona's (2023) literature review on CT considers it to be an approach that is clearly contextualized in science education.

However, in the research project from which this book originates, a thorough review of this approach has been undertaken, incorporating additional aspects not covered previously. These include dispositions towards CT, explicit references to scientific practices of argumentation, inquiry, and modeling, a restructuring of the considered dimensions into three domains (knowledge, skills, and dispositions), as well as strategies for their development.

From our perspective, considering all these aspects is essential for achieving a deeper and clearer understanding of the term CT within the context of science education. Our view of CT goes beyond isolated thinking skills, encompassing the integrated development of cognitive dimensions and dispositions, which must be systematically addressed in science education through real-world problems and SSIs. These relevant, authentic contexts enable students not only to analyze, evaluate, and synthesize information but also to apply these thought processes in making informed, well-reasoned decisions. Furthermore, to effectively promote CT in science education, it is crucial to utilize core scientific practices such as argumentation, inquiry, and modeling. These practices, combined with effective teaching strategies, provide students with opportunities to challenge assumptions, interpret data, infer, explain phenomena, make predictions, evaluate evidence, and consider multiple perspectives in problem-solving. This reflective and rigorous process helps students develop CT

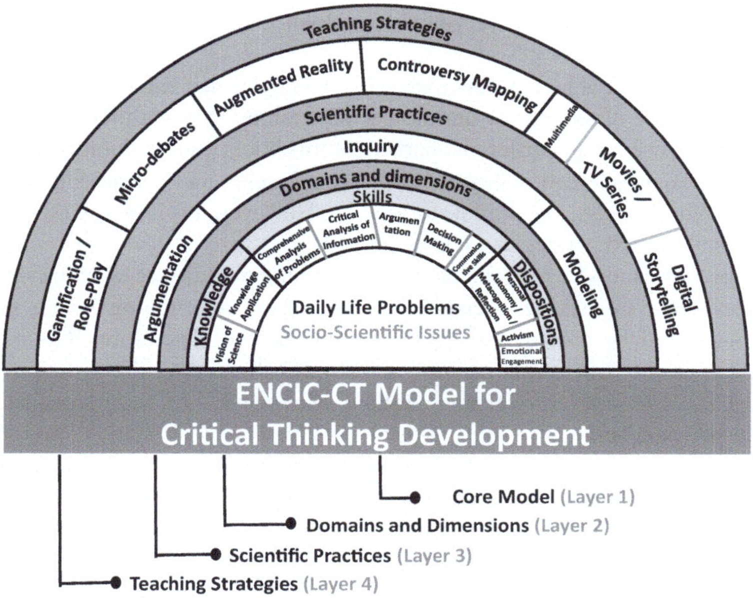

Fig. 1.1 ENCIC-CT model for the development of CT through the analysis of daily life problems

that is not only analytical but also applicable to a wide range of situations both within and beyond the scientific domain.

Figure 1.1 illustrates the ENCIC-CT model, showcasing our understanding of CT development through daily life problems in science education. Named after the research group where it originated, this model consists of four key components: daily life problems/SSIs, domains and dimensions, scientific practices, and teaching strategies. Each component is depicted as a layer, representing various levels of CT conceptualization and its practical application into science and technology education.

The first layer represents the core of the model, which involves the use of daily life problems and SSIs (Bencze et al., 2020). This serves as a strong starting point for developing CT in science education (Solbes & Torres, 2012).

The second layer displays the dimensions of CT grouped into three domains: knowledge, skills, and dispositions, as included in other CT proposals such as Vieira and Tenreiro (2016), Jiménez-Aleixandre and Puig (2022), Halpern and Danna (2023), and Vila et al. (2023). It is understood that this second layer can be applicable across any teaching area.

Now, in science education, we consider that developing these dimensions involves employing scientific practices such as argumentation (Erduran & Jiménez-Aleixandre, 2007; Kuhn, 2019), inquiry (European Commission, 2015; National Research Council, 2012), and modeling (Justi & Gilbert, 2002). These practices, which are considered major goals of science education (National Research Council,

2012), are tailored to address everyday life problems. They constitute the third layer of the model.

The fourth and final layer refers to possible strategies that can be used to develop CT in the science classroom or during teacher training. Among them, the Fig. 1.1 displays those used in the different chapters of this book: gamification/role-playing, micro-debates, augmented reality, controversy mapping, and multimedia resources such as movies/TV series or digital storytelling. However, it is important to note that these are not the sole strategies to be employed.

The use of this model for developing instructional modules or teaching–learning sequences involves making decisions at each layer, determining which specific components will be considered in the teaching design. Therefore, it is a versatile model, as the different concentric semicircles illustrating the different layers in Fig. 1.1 can be moved over one another, allowing CT to be approached from different perspectives.

The following sections provide detailed descriptions of the different layers and elements included in this model, drawing on the literature and studies conducted by the ENCIC research group.

1.4 First Layer of the ENCIC-CT Model: Daily Life Problems/Socio-Scientific Issues

In order to effectively develop citizens' CT, it is essential to approach people's everyday lives and identify socially relevant issues where they must make significant decisions before taking action. These problems should be authentic, relatable, current, open-ended, and controversial, suitable for the classroom, with information accessible to students and relevant to the content we wish to teach regarding science and technology. Furthermore, they should encompass social, ethical, economic, environmental, political aspects, etc., and enable the expression of attitudes, values, and emotions. These real-life problems constitute the central focus of the ENCIC-CT model and can affect us at three levels: as individuals (e.g., food or energy consumption), as members of a local community (e.g., water treatment or the location of a power plant), or as global citizens (e.g., global warming or biodiversity loss).

Daily life problems provide useful contexts for learning (Lupión et al., 2017) and for the development of CT, as they offer real-world scenarios that increase the likelihood of applying values and attitudes commonly used in everyday decision-making (Sadler & Zeidler, 2005). Among these problems, SSIs (Zeidler et al., 2019) are widely used approaches in science education. The SSI approach primarily focuses the scientific component while also acknowledging the social dimensions of the issues. Scientific education underscores the importance of equipping students with the ability to make informed decisions about SSIs, as their implications can significantly impact society. Hence, the importance of having scientifically and technologically literate students capable of fully exercising their rights and participating in decision-making in democratic societies. This training should empower them to adopt responsible, well-reasoned positions based on scientific and technological knowledge (Yacoubian, 2018).

1.4.1 Socio-scientific Issues

The SSIs refer to real, complex, and unresolved problems that are intrinsically linked to the intersection between science, technology, and society. These SSIs are controversial and can be approached from multiple perspectives (social, ethical, economic, environmental, legal, political, etc.) (Chen & Xiao, 2021; Sadler & Zeidler, 2005; Simonneaux, 2001; Zohar & Nemet, 2002). Acevedo (2006) asserts that controversies surrounding techno-scientific issues serve as clear demonstrations of the relationship between values education and science education. Consequently, SSIs require students to engage critically with science-related scenarios, enabling them to propose solutions and participate in debates on genuine problems (Jiménez-Aleixandre et al., 2000), thus fostering scientific literacy and democratic participation in society (Yacoubian & Khishfe, 2018).

Some SSIs are related to sustainable energy, food, biotechnology, etc. On the other hand, certain SDGs outlined in the Agenda 2030 (United Nations, 2015) address science and societal concerns, thus they could be considered to identify relevant problems for CT development. Health (SDG 3), energy (SDG 7), climate change (SDG 13), resources (SDG 14), etc., are some examples of these contexts focused on relevant citizenship issues. Likewise, historical episodes within the realm of science offer opportunities to explore the construction of scientific knowledge within authentic socio-cultural contexts. This allows for the explicit exploration of some features of Nature of Science (NOS) (Acevedo & García-Carmona, 2017), contributing to a holistic, functional, and realistic understanding of science (Manassero & Vázquez, 2019).

1.4.2 Some Examples of SSIs

In order to illustrate the potential of these issues for CT development in science education, below are a few examples.

- *Banning Single-Use Plastics*

Plastics are the most widely used material in object manufacturing, known for their durability. However, they pose an environmental problem due to their slow decomposition, persisting as waste in natural habitats, especially in oceans and seas (Elias, 2018). Additionally, their production necessitates the use of fossil fuels, exacerbating environmental concerns. Moreover, plastics and their additives are toxic to many living organisms, including humans (Eriksen & Al, 2014). In response to these issues, the United Nations set forth regulations to curb the excessive use of single-use plastics (such as cotton swabs, disposable cutlery, etc.), plastic bags, and overreliance on plastic packaging by 2022, specifying measures through a European Union directive (Koch & Barber, 2019). Addressing the environmental challenges posed by plastics demands concerted efforts across society, including education, which serves as a catalyst for instigating changes in citizens' attitudes and behaviors by fostering awareness of the far-reaching impacts of their actions on the environment (Marcén & Molina, 2006).

The problem of plastics is relevant and affects students' daily lives, as they frequently use single-use plastics. This issue offers an opportunity to explore scientific concepts such as plastic degradation, its physicochemical properties, and its environmental impact. Additionally, it allows for the examination of various dimensions of the issue, including social factors related to convenience, economic considerations, legislative aspects, and health implications. A study conducted by López-Fernández et al. (2021) with Spanish secondary students demonstrated advancements in understanding plastic chemistry and observed attitude shifts among some students regarding this issue following their participation in a role-playing game.

- *Genetically Modified Food*

Genetically modified foods are products altered using genetic engineering techniques, allowing them to acquire desirable traits in crops, such as disease resistance, tolerance to adverse conditions, or increased nutritional content (España-Ramos & Rueda, 2023). However, this practice has sparked intense debate in society due to concerns raised regarding food safety, human health, and environmental impact.

Regarding food safety, there are concerns about the potential long-term effects of consuming genetically modified foods, including the possibility of toxicity or allergies (España-Ramos, 2008). Another concern is the environmental impact of genetically modified crops. Speculation exists regarding the potential for genetically modified plants to crossbreed with wild species, potentially resulting in new herbicide-resistant species. This could complicate pest management efforts and negatively impact biodiversity. Similarly, there is apprehension regarding the potential spread of modified genes in the environment, which could have adverse effects on natural ecosystems (Wolfenbarger & Phifer, 2000). The debate continues, and it is essential to continue researching the long-term effects of genetically modified foods on both health and the environment.

- *Secure and Sustainable Energy*

The issues surrounding energy and resource usage are gaining increasing attention from the public and are part of SDG 7 (United Nations, 2015), which aims to ensure access to affordable, reliable, sustainable, and modern energy for all citizens. Recent examples of these SSIs include the adoption of sustainable energy sources as alternatives to fossil fuels, efforts to reduce energy consumption (degrowth), Europe's gas dependency following the Ukraine war, and the controversial proposal by the European Commission to classify nuclear energy as green energy (Alderman & Pronczuk, 2022), at least until 2045. This decision has sparked significant controversy as it appears to overlook the serious issue of radioactive waste generated by nuclear plants or the profound safety risks, as evidenced by various accidents in nuclear power plants or the conflict in Ukraine, with battles occurring around Zaporiyia, Europe's largest nuclear plant. The contributions that education can make to sustainability are paramount (Scalabrino et al., 2022), and as a result, there are increasingly more educational proposals addressing this issue (Cruz-Lorite et al., 2023). The energy debate is ongoing and has become a social issue that requires solutions from all sectors.

- *Sustainable Mobility*

Sustainable mobility is an issue that involves energy, environmental, health, economic, technological, and other aspects. A relevant everyday problem could be the choice of a car (Moreno-Fontiveros et al., 2022) or the purchase of a bicycle (Cano-Iglesias, 2024). These problems offer rich opportunities for argumentation and decision-making as they are influenced by various factors, with advertising playing a prominent role (Grosick et al., 2013).

The choice of a bicycle is a common situation in our daily lives, given the wide array of options available on the market, each crafted from various materials and boasting unique features. Nowadays, increasing environmental awareness (Hadjichambis et al., 2020) and the urban expansion have led many people to opt for the bicycle as an eco-friendly mode of transportation (Parkin, 2008), in addition to its use for recreational purposes. Purchasing a bicycle involves several aspects that students must consider in their decision-making process, such as selecting a model tailored to their individual interests (type, intended use), the social, environmental, and global impact of its purchase and use, or the influence of advertising messages (product, price, promotion, brand, etc.) (Blanco & Forero, 2017). This issue provides an opportunity for students to engage in discussions supported by technological and chemical evidence, delving into their understanding of metallic material properties, mechanical attributes, and associated manufacturing processes (Cano-Iglesias, 2024).

1.5 Second Layer of the ENCIC-CT Model: Domains and Dimensions of the Critical Thinking

The ENCIC-CT model groups the dimensions of CT into three domains: knowledge, skills, and dispositions. As inferred from the theoretical framework, initially CT was primarily focused on skill development (Ennis, 1987; Facione, 1990; Halpern, 2006; Paul & Elder, 2005). Other proposals highlighted the significance of knowledge (Bailin, 2002) and dispositions (Facione, 1990; Dwyer et al., 2014). However, more recent proposals (Halpern & Danna, 2023; Jiménez-Aleixandre & Puig, 2022; Vieira & Tenreiro, 2016; Vila et al., 2023) recognize the importance of integrating all three mentioned domains.

The second layer of the model involves selecting the domains and dimensions of CT to be addressed in dealing with the selected everyday life problem. It is important to note that although these dimensions are presented separately, some may possess attributes from more than one domain due to potential overlap.

1.5.1 Knowledge Domain

A consensus characteristic of CT is its requirement for specific contextual knowledge to evaluate particular knowledge or beliefs (Anderman et al., 2012). Even in environments conducive to CT development, a lack of knowledge could compromise its progress. Specifically, we consider that CT development in the context of daily life problems or SSIs requires:

- *Scientific knowledge application and knowledge about science.* Students should be able to use/apply scientific knowledge (Blanco-López et al., 2015), being well-informed about the topics addressed, not limiting themselves to dominant discourses (Solbes & Torres, 2012), and understanding alternative perspectives (Bailin, 2002).
- *Vision of science/Epistemic knowledge about science.* Understood as a human activity with multiple relationships with technology, society, and the environment. There are numerous SSI that involve situations where scientific knowledge that has not yet been consolidated is demanded. These issues, situated on the current frontier of science, serve as valuable contexts for challenging the perception of "neutrality" and "objectivity" traditionally attributed to science (Khishfe, 2014). Kolstø (2001) suggests that these issues can help promote a more complex view of science, where cases of "settled science" coexist with cases of "science in progress".

1.5.2 Skills Domain

As previously mentioned, there are various proposals of skills that underpin CT (Facione, 1990; Jiménez-Aleixandre & Puig, 2022; Vieira & Tenreiro, 2016). In the case of addressing everyday life problems, we consider that students must deploy the skills proposed by Facione (1990) across the following five dimensions:

- *Comprehensive analysis of the problem.* The problems of daily life cannot be compartmentalized like scientific disciplines; instead, they necessitate a holistic approach, considering their complexity and taking into account scientific, technical, ethical, cultural, philosophical, social, environmental, economic, and other aspects. They demand a thoughtful analysis and personal decision-making, wherein scientific knowledge, epistemological beliefs, skills, attitudes, values, disposition for action, etc., play an important role (Wu & Tsai, 2011).

- *Critical analysis of information.* Students need to be able to assess the credibility of different sources of information, considering underlying interests (Blanco-López et al., 2015). Osborne and Pimentel (2023) argue "that the current science curricula are failing to educate students to be competent outsiders to science" (p. 1). Currently, we are confronted with an excess of information (infodemic), especially on social networks. Within this flood of data, there is a significant amount of false information pretending to be scientific. Therefore, it poses a challenge to scientific literacy to prepare students to evaluate the claims on social networks.
- *Argumentation.* The development of CT requires that students be able to construct solid arguments while questioning the validity of claims, rejecting evidence-lacking conclusions, and identifying argumentative fallacies (Jiménez-Aleixandre, 2010). In the realm of science, argumentation stands as an important scientific practice (Kuhn, 2019), which is further described later on.
- *Decision-making.* Involves making rational choices and informed judgments as integral components of the arguments used to solve problems. In decision-making within SSI contexts, factors related to personal preferences, beliefs, and environmental influences come into play (Muñetón et al., 2017). This is where CT becomes relevant (analyzing, reflecting, selecting information, etc.) to ensure the most appropriate decision for the presented problem. The use of SSIs offers benefits to students such as greater mastery and retention of the subject matter, enhanced ability to articulate positions, and higher-quality decision-making (Sadler & Zeidler, 2005).
- *Communicative skills*: Being able to communicate decisions using appropriate language, according to the context and goals or intentions, constitutes an important component of CT skills, in line with the skill of explanation as described by Facione (1990).

1.5.3 Dispositions Domain

Besides skills, Facione (1990) highlighted the importance of dispositions in CT development, a component also recognized by authors such as Halpern and Danna (2023), Jiménez-Aleixandre and Puig (2022) or Vieira and Tenreiro (2016). The ENCIC-CT model includes several dispositions and critical attitudes related to:

- *Personal autonomy/Metacognition/Reflection.* Students should be able to develop an independent opinion, maintaining an open mind towards divergent world-views, flexibility in considering alternatives and opinions, impartiality in evaluating reasoning, honesty in confronting their own predispositions, prejudices, stereotypes, or egocentric tendencies, prudence in postponing, making, or altering judgments, and willingness to reconsider and revise views, whether their own or others', where honest reflection suggests that change is justified (Facione, 1990). Moreover, they must recognize their own limitations (Osborne & Pimentel, 2023). This dimension encompasses personal autonomy and facilitates decision-making

(Eggert et al., 2013). Metacognition serves as a powerful tool across various situations to enhance CT, such as in solving scientific problems or evaluating scientific evidence and detecting cognitive biases, that is, in misinterpretations of reality (Vila et al., 2023).
- *Social engagement/Activism.* It is important to recognize that CT encompasses not only reflection but also action (see examples of scientists persecuted for their contributions in the work of Solbes & Torres, 2015). As we have pointed out, the CT development involves actively engagement in its practice, as it is intrinsically action-oriented (Facione, 1990; Jiménez-Aleixandre & Puig 2022). Therefore, we advocate that addressing relevant issues should lead to responsible decision-making, leading to tangible actions that contribute to problem-solving. Hodson (2003) proposed that to effectively address current social and environmental challenges, we need a generation of critically informed individuals with scientific and political knowledge who move into action rather than simply expressing opinions and discussing these issues. From this standpoint, the approach of socio-political activism is closely linked to CT, within the framework of a critical scientific literacy that becomes a driving force for socio-political action.
- *Emotional engagement.* CT can be driven or inhibited by emotions, as they influence how individuals approach decision-making and problem-solving. They provide a subjective perspective that complements rational analysis, thus they can either promote or hinder reflection, analysis, and decision-making (Elster, 1994; Kang & Keinonen, 2018). Furthermore, emotions can help establish an emotional connection with the problem, which may increase empathy and understanding across different perspectives. They can stimulate creativity and new ideas (Abdullah et al., 2022). Different researches suggest that positive emotions among learners are associated with engagement in challenging projects and creative problem solving, while negative emotions have been linked to poor academic performance and school attrition (Bellocchi, 2015; Pekrun & Linnenbrink-García, 2014).

The dimensions of CT presented are closely interconnected with each other, with all of them being equally important. Additionally, the order in which they appear in Fig. 1.1 does not imply a sequence in which they must be developed, nor does it imply addressing all dimensions simultaneously in every activity. For instance, we can utilize knowledge followed by critical analysis of information to argue about science or technology, leading to decision-making and ultimately communication. It is essential to emphasize that fostering CT in students is associated with progress across all dimensions and, most importantly, with their integration.

1.6 Third Layer of the ENCIC-CT Model: Scientific Practices

Science education has long benefited from robust research areas focusing on what are known as scientific practices (Next Generation Science Standards, 2013; National Research Council, 2012), which prominently include argumentation, inquiry, and modeling. These practices, in the context of everyday life problems, are suitable for promoting the development of CT as they involve acquiring and applying knowledge, utilizing skills related to analysis, explanation, inference, decision-making, etc., and in their development, important dispositions such as personal autonomy, emotional engagement, or activism are crucial. The development of scientific practices in science education is framed within an approach that views learning as a process of socialization within scientific culture (Driver et al., 2000), involving students' participation in the characteristic practices of the scientific community. This entails constructing scientific knowledge and understanding why it is constructed, examined, and evaluated in a particular way (Reiser et al., 2012).

1.6.1 Argumentation Scientific Practice

At this point, it is important to distinguish between the argumentation skill, integrated as a dimension of CT within the skill domain of the ENCIC-CT model, and the argumentation as scientific practice outlined in the third layer. The argumentation skill refers to students' general ability to construct effective arguments, while the scientific practice focuses on how this skill is applied in the context of scientific research.

Argumentation is indispensable in the construction of scientific knowledge, as it facilitates the exchange and examination of ideas and concerns. Through this process, more significant and comprehensible conclusions about phenomena are crafted. Argumentation is the expression of rational judgment in which the verbal articulation, in social contexts, reveals the reasons for accepting or refuting a viewpoint or set of ideas (Hahn & Oaksford, 2012). This skill enables individuals to move beyond their intuition to justify their ideas and conclusions to themselves and others (Mercier & Sperber, 2011).

Arguing about daily life problems involves informal reasoning processes. This type of reasoning is applied outside formal contexts, where problems may not be well-defined, premises may not be explicitly stated, and conclusions may not be delimited (Cruz-Lorite et al., 2023; Wu, 2013). Various authors indicate different types of informal reasoning according to the domains to which arguments are oriented (e.g., scientific-technological, economic, social, etc.) (Cano-Iglesias, 2024; Christenson et al., 2012; Ozturk & Yilmaz-Tuzun, 2017) or based on the considerations they rely on (rationalistic, emotional, or intuitive) (Sadler & Zeidler, 2005). The use and prevalence of these types of reasoning can be influenced by the problem being addressed or prior knowledge. For instance, Christenson et al. (2012) found that in SSIs related to global warming or nuclear energy, students used reasoning primarily based on attitudinal, affective, or value-based evidence.

On the other hand, different studies reveal that the main challenges students encounter in making reasoned decisions about daily life problems are associated with a lack of scientific knowledge to argue effectively (Cebrián-Robles et al., 2021b; Henderson et al., 2018; Rodríguez-Mora et al., 2022), or with the frequent use of beliefs or values in decision-making (von Winterfeldt, 2013). This fact is concerning in science education, which aims to teach making reasoned decisions based on scientific evidence (Bravo & Jiménez-Aleixandre, 2018), on social attitudes and values (Siribunnam et al., 2014), and on the use of justifications grounded in scientific knowledge.

Therefore, argumentation constitutes an important aspect for decision-making (Bravo & Jiménez-Aleixandre, 2018) because science education involves both learning science, which concerns epistemological content, and learning to do science, which considers it necessary to develop the inquiry skills inherent to science, including the development and evaluation of arguments (Duschl & Osborne, 2002). Doing science entails proposing and discussing ideas, evaluating alternatives, and choosing among different explanations (Jiménez-Aleixandre, 2010). Moreover, it requires the construction and critique of arguments (Osborne et al., 2016).

1.6.2 Inquiry Scientific Practice

Another important scientific practice is inquiry, also known in science education as Inquiry-Based Learning (IBL) or Inquiry-Based Science Education (IBSE). The term inquiry has various meanings in the educational context (Barrow, 2006). Sometimes, it is understood as a cognitive skill –the ability to inquire and investigate scientifically- that students need to develop. In other cases, it refers to the understanding of the methods used by scientists, that is, the nature of scientific inquiry. A third interpretation refers to various teaching strategies aimed at fostering students' inquiry skills, facilitating their understanding of scientific inquiry, and mastering scientific concepts.

A widely accepted definition is proposed by the National Research Council (2000), which describes inquiry as a multifaceted activity involving observation, question formulation, information gathering from various sources, as well as planning and designing investigations. Moreover, it entails reviewing ideas based on available experimental evidence, using tools to acquire, analyze, and interpret data, formulating responses, explanations, and predictions, and ultimately, communicating findings. The report "Science Education for Responsible Citizenship" (European Commission, 2015) defines it as:

> [...] *A complex process of sense-making and constructing coherent conceptual models where students formulate questions, investigate to find answers, build new understandings,*

meanings and knowledge, communicate their learning to others and apply their learning productively in unfamiliar situations. (European Commission, 2015, p. 68).

There is a wide variety of methodological approaches to inquiry in the classroom (Rönnebeck et al., 2016). The 5E learning cycle approach by Bybee et al. (2006) is one of the most commonly used, organizing teaching into five stages: Engage, Explore, Explain, Elaborate, and Evaluate. In the National Research Council proposal (1996), these stages, understood as characteristics of inquiry-based teaching, are formulated as: (1) Raising science-oriented questions that allow active student participation, (2) gathering evidence by students to develop and evaluate their own explanations for the questions posed, (3) developing explanations to address the questions based on the evidence gathered, (4) evaluating the explanations, which may incorporate alternative explanations reflecting scientific understanding, and (5) communicating and justifying the proposed explanations.

Various authors modify these proposals in different ways.

Pedaste et al. (2015) propose a nonlinear model comprising five phases and nine interrelated sub-phases, connected through multiple pathways. This configuration represents the diversity of possible implementations in the classroom. Discussion, communication, and reflection are proposed as overarching elements of the model, largely determining the quality of the inquiry carried out by students. Bevins and Price's (2016) inquiry approach includes conceptual, procedural, epistemic, and personal aspects related to affective and motivational factors.

Drawing on an analysis of various inquiry proposals, Franco-Mariscal (2015a) developed an approach that identifies the inherent dimensions of this practice, aimed at fostering the development of scientific competencies through contextualized situations. According to this approach, scientific competency includes seven dimensions: research formulation, information management, research planning and design, data collection and processing, data analysis and concluding, results communication, and attitude or critical reflection and teamwork. This approach highlights some aspects that are often not sufficiently considered in science education, such as information management, results communication, critical reflection, and teamwork. These dimensions, in turn, would be linked to CT.

It is necessary to consider the role that inquiry can play in addressing everyday life problems. In broad terms, students as citizens interact with science based on information from the media, and increasingly, from social networks, regarding research, scientific advancements, or SSIs (Kolstø, 2001). Therefore, individuals are not required to gather data in the manner of scientists. Instead, to comprehend and engage with problems effectively, they must analyze the data presented to them within news and information sources. They need to consider the potential biases and limitations inherent in such sources, and based on this assessment, evaluate their credibility and make informed decisions. Nevertheless, in the context of daily life problems, it is possible to involve students in citizen inquiries (Moraga-Toledo et al., 2024). In these inquiries, students can assume the roles of concerned and critical citizens, collectively addressing SSIs that directly impact them (e.g. the nutritional habits of peers or families, research on recreational possibilities for children in the community, identification of sources of pollution near the school, etc.). They gather information, collect data from reality, propose solutions, and, if possible, implement or at least disseminate them, even on a small scale. For more complex issues, a comprehensive inquiry cycle could be undertaken, but without gathering data from reality, which would be replaced by data provided by scientists.

1.6.3 Modeling Scientific Practice

Modeling is considered an essential component of scientific activity (Justi & Gilbert, 2002), and teaching through models is a key area of research in both science education and CT (Clement & Rea-Ramírez, 2008; Gilbert & Justi, 2016). The development and use of models are considered significant scientific practices (National Research Council, 2012), and it is recognized that they should be part of students' foundation in basic science education (Oliva et al., 2015).

Modeling is understood by some authors as a necessary competence to construct and enhance models of physical objects, processes, or phenomena. However, providing a precise definition is challenging due to the fragmented nature of the theoretical frameworks that address it (Nicolaou & Constantinou, 2014). Nevertheless, it is possible to establish common characteristics (Justi & Gilbert, 2002). Modeling could be understood as an integrated set of knowledge, metaknowledge, skills, and epistemic values necessary for carrying out the task of modeling comprehensively. This encompasses not only learning the models of school science but also working with them, elaborating on them, revising them, as well as discussing and opining about them, understanding their value, usefulness, approximate and evolving nature, and their limitations (Oliva et al., 2015). Consequently, modeling represents a complex task that demands a wide range of skills. In addition to knowledge of the specific domain, it involves numerous strategies, skills, and certain epistemological commitments (Halloun, 2007; Harrison & Treagust, 2000; Justi & Gilbert, 2002; Prins et al., 2009), as well as attitudinal aspects such as the willingness to model (Ammoneit et al., 2024).

Modeling involves processes closely related to the components of the scientific research cycle: posing problems, formulating hypotheses, seeking information, developing new ideas and explanations, etc. This would also require tasks such as interpreting, manipulating, and expressing phenomena through a variety of signs, whether propositional or iconic (Perales, 2006). Furthermore, models are laden with values that define and justify the "rules of the game" of a scientific discipline. These values pertain to perceiving the rationality of models, recognizing their role in the development of hypotheses, explanations, and scientific arguments, as well as appreciating their utility and their approximate, limited, and evolving nature.

Modeling has been commonly used in science education within the context of scientific problems, and there has even been questioning about whether it is possible to integrate approaches from everyday life contexts with modeling in educational practice. This involves considering if the situations that enable modeling are the same as those that facilitate the application of knowledge in everyday contexts (Izquierdo, 2013). However, according to Ke et al. (2021), modeling serves as a valuable tool for analyzing SSIs, as it enables the display of relevant aspects of these matters through multiple models, aiding in understanding their complexity, exploring their causes and effects, and facilitating the proposal of potential solutions. According to these authors, multiple models contribute to improving the understanding of the phenomena of interest and can be mechanistic, systemic, mathematical, or socio-scientific models. Mechanistic models explain phenomena, allowing students to adjust their perception of reality. Systemic models describe variables and interactions to predict behaviors and properties. Mathematical models utilize empirical data to reach quantitative conclusions about causal relationships. Socio-scientific models consider social contexts to explain and predict outcomes, distinguishing themselves from purely scientific models.

In summary, the use of models and modeling helps students enhance their reasoning skills and epistemological beliefs about the nature and purpose of scientific models (Soulios & Psillos, 2016), and thus, CT.

1.7 Fourth Layer of the ENCIC-CT Model: Strategies for Developing Critical Thinking

Different research conducted by the ENCIC research group has highlighted some strategies with the potential to develop CT in students at various educational levels. Among them, gamification and role-playing, micro-debates, controversy mapping, and various technologies like augmented reality, as well as multimedia resources such as movies, TV series, and digital storytelling, stand out.

In initial teacher training, in addition to using these strategies by putting students in the role of science learners, the analysis or design of educational proposals is also employed, in which they adopt the role of teaching learners. This section describes these strategies.

1.7.1 Gamification and Role-Playing

Gamification entails applying game strategies and mechanics in non-gaming contexts, such as education, to enhance learning outcomes (Deterding et al., 2011). Among its advantages, it stands out for fostering motivation by presenting scientific knowledge in a "wrapper" (the game) that captures students' attention and encourages them to dedicate time to the proposed activities. The use of games promotes active, cooperative, and competitive learning, creativity, and imagination (Kapp, 2012). Additionally, they help students relate to their peers, thus working on dispositions such as emotions, self-control, or concentration on the task, aspects that enhance the classroom climate.

The use of educational games can contribute to developing skills and positive perceptions towards science, which will impact improving motivation regarding the learning of scientific content (Franco-Mariscal et al., 2016). The literature includes a wide variety of games: puzzles (Granath & Russell, 1999), cards (Martí & Rubio, 2014), word formation (Helser, 2003), board games (Girón & Franco-Mariscal, 2023), daily life contexts such as a house (Franco-Mariscal et al., 2018), a car (Franco-Mariscal, 2015b), or food (Franco-Mariscal, 2018), football competitions (Franco-Mariscal, 2014), role-playing (Franco-Mariscal et al., 2023), etc.

This book will focus on role-playing, a simulation game where each participant assumes a specific role in the context of a simulated issue (Cruz-Lorite et al., 2023), following certain rules and interacting with other participants. This strategy not only promotes students' CT but also allows teachers to design and implement a range of tasks to encourage its development. Additionally, these games offer an alternative approach with the potential to boost motivation and provide meaningful learning experiences (Franco-Mariscal et al., 2023). Therefore, an educational component is added to the gaming aspect.

The main feature of role-playing is their simulation of specific situations through gameplay and various characters, which stem from a real-life problem (although it may be more or less fictionalized). Thus, they incorporate a theatrical element that allows participants to draw on science and technology to develop their roles.

Role-playing games are highly suitable for addressing SSIs in the classroom (Simonneaux, 2001), as they start from an issue and a scenario involving various roles representing different perspectives. Their development requires establishing instructions and game rules both for the staging and for the subsequent synthesis and conclusions phase. In their design, it is also necessary to create a dossier with news and other documents related to the issue to assist students in preparing for the different roles.

This activity prioritizes the process over the content, with the active involvement and participation of students for its successful execution (Rashid & Qaisar, 2017). Despite the educational benefits offered by role-playing, their use in science classrooms remains limited (McSharry & Jones, 2000) due to their complexity, requiring substantial preparatory work by teachers and a strong focus on student engagement.

According to España-Ramos (2023), role-playing enable the achievement of these objectives:

- Identifying problems, searching and selecting information, and proposing solutions and actions.
- Bringing orality into the classroom (Simonneaux, 2001), as it requires dialogue among participants, both during the preparation of the game and in its staging.
- Enhancing the capacity to argue by connecting explanations with evidence. Engaging with different arguments can aid in clarifying one's own ideas.
- Facilitating the expression of different perspectives on an issue and the criteria on which they are based, as well as the possibility of experiencing changes in

opinion. This helps identify the criteria supporting one's own position and those of other positions (Simonneaux, 2008).
- Making decisions responsibly and based on a solid foundation (López-Fernández et al., 2021).
- Highlighting attitudes, values, and emotions regarding the problem presented (España-Ramos, 2008).
- Facilitating the comprehension of scientific and technical content pertinent to the issue, infusing this knowledge with significance by demonstrating its practical relevance in addressing real-life challenges, and engaging students in situations where they must persuade others with differing viewpoints using reasoned arguments.
- Developing collaborative work. Role-playing is a collective activity that demands interaction and collaboration, both in the preparation phase and during enactment. Therefore, it is important for roles to be prepared by teams of several students.

As evident, these objectives are closely intertwined with the dimensions of CT in the ENCIC-CT model, which are also associated with creative thinking and careful thinking (Lipman, 2003), as well as a range of key competences, particularly scientific competence (Blanco-López et al., 2015).

1.7.2 Micro-debates

Debate is considered a suitable activity for developing CT from everyday life problems, as it is based on reasoning, argumentation, decision-making, and communication (Fang et al., 2019; Simonneaux, 2001). The debate involves students evaluating and identifying data and information, reflecting on various opinions, defending their position in a reasoned manner, and making decisions based on quality evidence (Wang et al., 2017). All of this facilitates the presentation of different perspectives. Thus, debate becomes an effective strategy for promoting CT as its dimensions are involved in a large number of tasks. Additionally, debate enhances the acquisition of competences and knowledge, while fostering autonomy and participation, thereby contributing to comprehensive student education by developing knowledge and attitudes in a cross-cutting and interdisciplinary manner (Erduran & Jiménez-Aleixandre, 2007). Studies such as Simonneaux (2001) indicate that debates and role-playing promote changes in students' opinions, whereas these changes do not occur in other activities such as visits to exhibitions. Martini et al.'s study (2021) reported that approximately 50% of students changed their opinions after a class debate on nuclear energy.

Next, the Micro-debate strategy is described, a short-duration debate specifically designed to develop CT (Cano-Iglesias, 2024). This activity aims to instruct and enhance students in argumentation skills, critical analysis of information, and decision-making, either through active participation in a debate on an SSI or as an audience member. The participants in this activity are three students (presenter and

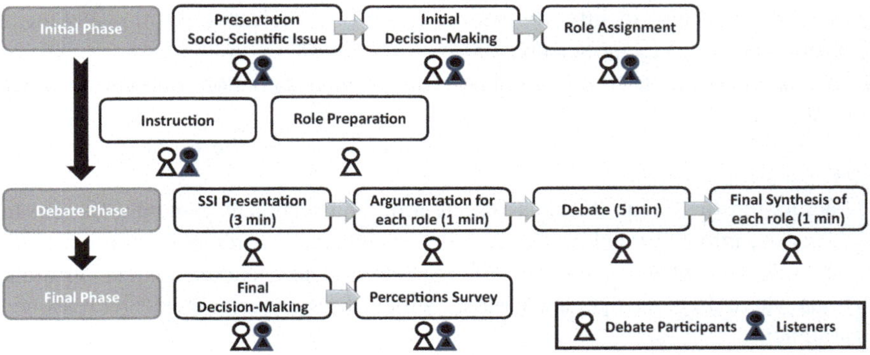

Fig. 1.2 Structure of the micro-debate activity

two debaters, one in favor and one against the issue) and the rest of the students who act as listeners. The Micro-debate activity is structured into three phases (Fig. 1.2).

- *Initial phase*: The SSI is presented through a question, and students are asked to make an initial reasoned decision without accessing information. In this phase, the presenter and debaters for and against the issue are assigned, with one week given for information gathering. All three roles must complete a form with at least five arguments to be used in the debate. In addition, debaters must include counterarguments for refutation, and the presenter must include a fake news item. Information can be sourced from any written or audiovisual medium with scientific rigor, and sources must be cited.

Between the initial phase and the debate phase, students receive a three-hour instruction in which different authors' understandings of CT and argumentation, critical analysis of information, and decision-making are presented. This training includes the ENCIC-CT model and the argumentation models of Toulmin (2003), Jiménez-Aleixandre's adaptation (2010), and Osborne et al.'s (2016) learning progression. Additionally, students engage in some straightforward tasks involving critical analysis of information and argumentation.

- *Debate phase*: Involves a three-minute presentation of the SSI by the presenter, supported by a digital slideshow. Following this, each debater is given an initial minute to present their arguments. Subsequently, there is a five-minute debate period, and finally, each debater has one minute for a concluding statement to persuade the audience of their position.
- *Final phase*: After the debate, all students, including the listeners, complete a questionnaire that includes a final decision-making aimed at assessing potential changes in stance resulting from the influence of the arguments presented during the debate. The questionnaire also investigates their perceptions of the activity's development.

Sufficient time should be allocated for this activity to ensure that all students have the opportunity to participate as debaters. Additionally, the teacher should assign each group of students a specific day for the activity and inform them of the assigned issue at least one week in advance.

1.7.3 Controversy Mapping

The Cartography of Controversy is a didactic methodology derived from Actor-Network Theory (Latour, 2007; Simonneaux, 2020; Venturini, 2010), providing to be an effective pedagogical approach for exploring and visualizing SSIs. Its capacity to unravel complex scenarios, such as the current model of meat production and consumption, addressing COVID-19, urban care, and controversies related to water use (España-Naveira et al., 2023), among others, makes it particularly valuable.

This methodology posits that understanding the formation of social phenomena requires more than examining individual actors; it is crucial to analyze the networks of actors that emerge in their various interactions (Venturini & Latour, 2009). In this context, participants in the controversy are termed actants, defined as any entity, whether human or not (including people, ideas, animals, objects, etc.), capable of influencing the controversy (Latour, 1996). Among the tools within the Cartography of Controversy, controversy mapping stands out as the most popular, capable of analyzing the complexity of SSIs (Christodoulou et al., 2021). These maps offer a visualization of controversy complexity by depicting actants (grouped according to their affinity into zones known as poles) and the favorable or unfavorable relationships established among them.

Cabello-Garrido et al. (2023) employed controversy mapping to explore and visualize the viewpoints and ideas of preservice science teachers regarding controversies associated with the meat production and consumption model. They then compared these initial controversy maps with those crafted by the research team, which were informed by a more detailed socio-epistemological analysis. The findings suggest that this approach is a useful way to engage students in addressing common questions, providing a framework for expanding their inquiry and understanding. It offers a platform from which they can develop their CT skills to take action for change.

Actants involved in an SSI can be introduced into the classroom through strategies that facilitate the simulation of arguments and counterarguments, such as role-play (España-Ramos, 2023). In these games, debate assumes a central position (Colucci-Gray & Gray, 2022), fostering the exploration of academic content by incorporating multiple viewpoints and argumentative perspectives (Simonneaux, 2008).

1.7.4 Augmented Reality

Different reports predict an increase in the use and integration of devices and applications in education as innovative quality factors. The Horizon Report (2021), compiled by experts globally to identify and describe emerging and disruptive technologies that could impact teaching, learning, and research, already highlighted in its ninth edition Mobile Learning, game-based learning, and augmented reality (AR) and virtual reality (Michael & Chen, 2006). In the educational settings, the use of digital technologies represents an innovation for both teachers and learners, enhancing educational standards. This transformation endeavors to reshape the curriculum, methodologies, and modes of knowledge dissemination, thereby offering new alternatives for teaching and learning.

This book focuses on the potential of AR, which is receiving considerable attention in education (Cortés et al., 2021; Geroimenko, 2020). AR enables the creation of mixed reality teaching and learning environments where real and virtual elements are combined within an interactive real-time setting (Moreno-Martínez & Franco-Mariscal, 2020).

AR in education involves integrating three-dimensional virtual elements or objects, generated by computers or additional digital information, into the learning context in the classroom, using devices such as mobile phones, tablets or AR glasses. Its objective is to complement, reinforce, enhance, amplify, and enrich learning experiences. By providing students with more stimuli, AR not only facilitates content learning but also cultivates students' creativity and sparks their curiosity to investigate and explore, thus empowering them to construct knowledge. This tool is highly motivating and holds immense potential for transforming learning (Abdullah et al., 2022).

AR applied to science education becomes an innovative, inclusive, formative, versatile, open, interactive, and dynamic methodology that enhances teaching and learning processes. Within this context, there is a wide range of AR tools available to create useful scenarios for teaching physics, chemistry, biology, and geology. By adding 3D objects and layers of virtual information, these tools enrich the information and stimuli provided by the surrounding physical environment. Moreover, these tools are not exclusive to the scientific field and allow for the creation of scenarios in other educational areas.

Some applications, such as AR Viewer or Sketchfab, enable the projection and integration of any three-dimensional model downloaded from 3D object galleries like 3D Warehouse or Archive 3D in various formats. Additionally, they support the inclusion of 3D objects previously crafted using graphic design and 3D modeling software such as 3DC or Tinkercad into real-world settings. Moreover, tools like Object Viewer Merge, Merge Things, Explorer Merge, or Mr. Body necessitate the printing and assembly of the Merge cube for visualizing AR scenarios. Noteworthy tools for science education encompass Unite AR, Hope, Augmented Class!, Zapworks, ARLOOPA, and Jigspace (Moreno-Martínez & Franco-Mariscal, 2023).

AR can promote the development of CT by providing practical experiences that encourage problem-solving, facilitate exploration and discovery, promote collaboration and communication, and teach users to critically evaluate information.

In relation to problem-solving, AR allows students to confront challenges within these environments to evaluate possible solutions and make decisions regarding complex issues. Similarly, AR offers opportunities for exploration and discovery in both virtual and real-world settings. Students can investigate and discover new concepts, relationships, and phenomena, which promotes CT by questioning, analyzing, and synthesizing information. Hence, AR, in addition to incorporating gamification-based learning, promotes discovery-based learning.

AR also facilitates collaboration among students by enabling interaction and cooperative work in shared virtual environments. This fosters CT by involving them in collaborative discussions, debates, and analysis to reach shared solutions and conclusions. Furthermore, AR environments empower students to interact with and access information within these scenarios, which they must analyze and evaluate for its quality and reliability. This promotes CT by teaching users to question and analyze information before accepting it as valid.

In summary, AR opens up new perspectives for fostering CT within science education, leveraging the broad opportunities afforded by the design and implementation of associated technologies. These technologies introduce novel methods of accessing and processing information. Likewise, AR proves highly advantageous for educators, facilitating content creation through a collaborative approach based on the principles of the makerspace philosophy (Vuorikari et al., 2019).

1.7.5 Multimedia Resources

Multimedia integrates various forms of media, including text, video, audio, graphics, and animation, into the learning environment, offering a potent new educational tool (Asthana, 2008). The utilization of multimedia resources with daily life problems could aid students in developing several thinking skills. Problem-based learning fosters CT and problem-solving abilities, empowering students to construct their knowledge actively (Hoffmann & Ritchie, 1997; Neo & Neo, 2001).

This book explores the potential of movies and television through excerpts from movies or series, which allow for the illustration of knowledge, stimulate debate, contextualize information, and offer insights into the scientific approach to various issues. Additionally, the exploration of digital storytelling is emphasized. Through students' creation and recording, contextualized within the resolution of scientific problems and complemented by music and sound effects, digital storytelling can effectively enhance CT. Moreover, it serves as a catalyst for promoting digital competence and nurturing creativity (Zagalaz, 2021).

1.7.5.1 Movies and Television Series

Science and technology permeate various modes of communication, including radio, television, news, magazines, music, and cinema (Gough, 1993). Particularly, cinema, television, and the internet are extensively utilized by the general population, owing to their broad accessibility, consumption, and influential capabilities (Government of Spain, 2021).

In general, science fiction movies depict issues related to science and technology that prompt viewers to question their feasibility in real life (Rose, 2003). Consequently, they represent promising strategies for facilitating the learning of scientific content portrayed in their scenes (Petit et al., 2021) and for fostering problem-solving challenges, thereby promoting CT. Furthermore, they offer a perspective on science that may diverge from reality (Dudo et al., 2011), which can contribute to its enhancement.

Several authors have explored the advantages of incorporating fictional resources such as films, TV programs, series, stories, or novels into science education (Fraknoi, 2002; Franco-Mariscal, 2021; Hamalosmanoglu et al., 2020; Koutnikova, 2017). These resources can facilitate the creation of mental images that correlate with underlying scientific theories, aid in comprehending abstract concepts, offer visual engagement, foster enjoyment, enhance the practical application of learned content, and stimulate interest in science, among other benefits (Allday, 2003; Barnett & Kafka, 2007).

The literature provides examples of how physics can be taught by presenting challenging situations of interest using sequences from TV series or movies. For instance, concepts of kinematics can be illustrated using scenes from Game of Thrones (Franco-Mariscal, 2021), principles of dynamics can be demonstrated through Prison Break (Franco-Mariscal, 2009), and optics can be explored using Peppa Pig (Franco-Mariscal, 2016). In the field of chemistry, Mojica (2019) demonstrated the use of a diverse range of video clips from movies, TV series, and YouTube to enhance the teaching and learning of this subject. Likewise, various biology-related topics are explored in films such as Jurassic Park, The Martian, GATTACA, Twenty Thousand Leagues Under the Seas, Planet of the Apes, Frankenstein, and The Walking Dead, all of which feature plots with biological origin (Barnett & Kafka, 2007). The work of Lawrence and Malinconico (2008) demonstrates how to teach geology using films such as Dante's Peak, Deep Impact, or The Day After Tomorrow. Ultimately, fiction often presents concepts, situations, or ideas related to real science that prompt viewers to contemplate and question whether these scenarios could occur in real life (Rose, 2003), thus helping to enhance CT.

1.7.5.2 Digital Storytelling

Digital storytelling is an effective strategy for developing CT, as it involves narrating a story in audio format to facilitate learning and instruction (Lambert, 2009; Xu et al., 2011). Instead of reading, listeners experience the narrative through audio playback devices. Digital storytelling often incorporates sound effects and music to enrich the listening experience and captivate the audience. Stories can be delivered by a single narrator or a collaborative team. This strategy enjoys popularity across various entertainment platforms, including radio programs, podcasts, mobile applications, and audio streaming services.

Narrative is a key feature of this multimedia resource (Negrete, 2003; Solomon, 2006) as it not only sparks students' interest in science by connecting it to everyday life problems (Franco-Mariscal, 2007), but also enhances scientific communication by conveying information in a precise, engaging, and imaginative manner. Additionally, it opens up opportunities for dramatization.

Digital storytelling can significantly enhance CT when students are actively involved in both designing and recording the story. By participating in the creation of digital stories, students are required to analyze, evaluate, and synthesize the scientific information they wish to convey, which fosters a deeper understanding of the science. To facilitate the design process, it is useful to highlight that scientific explanations can be structured similarly to traditional narratives. This means that the scientific content should feature protagonists (such as electrons or genes) who go through a sequence of events leading to an outcome (the phenomenon being studied) (Ogborn et al., 1996). This structural similarity not only aids in organizing the scientific information but also makes the learning experience more engaging and accessible for students.

Knowledge is present in students' understanding and application of scientific and technological concepts to develop the storyline of digital storytelling. Their vision of science is also reflected within this domain (Hadzigeorgiou, 2017).

Critical analysis of information stands out as a key skill, as students must analyze elements related to the plot, character development, conflicts, and scientific knowledge associated with the narrative to create a coherent and compelling scientific storyline. Additionally, during the process of searching for information for the story, students must assess the quality and reliability of their information sources.

The dispositions are also highly present in the creation and recording of digital storytelling, particularly science communication and emotional engagement. When crafting a digital storytelling piece about a scientific topic, students must effectively communicate their ideas through narration and audio production. This entails carefully selecting appropriate scientific terminology and considering the target audience. Creativity plays a vital role at this stage, allowing students to infuse their narratives with imaginative elements.

Emotions are expressed in various forms throughout the process. During recording, students may experience a range of feelings, including excitement, nervousness, satisfaction, frustration, or joy, driven by the creative challenges and character interpretations involved. Additionally, digital storytelling aims to evoke emotions in listeners, prompting students to skillfully convey these emotions through their narration, intonation, and vocal expression to elicit a heartfelt response from the audience. Moreover, personal autonomy/metacognition/reflection are essential aspects because, after completing the digital storytelling, students can reflect on their creation process, identify areas for improvement, and assess the effectiveness of their work in terms of how they communicate science and how it has impacted their peers.

Finally, digital storytelling is beneficial for developing CT in science education as it enables students to select, analyze, evaluate, synthesize, and communicate scientific information in a creative and meaningful way. Additionally, it requires students to establish connections between science and everyday life, and to reflect on the implications and applications of the knowledge gained.

Acknowledgements This work is part of the Spanish National Plan R+D+i Project entitled "Citizens with critical thinking: A challenge for teachers in science education", reference PID2019-105765GA-I00, funded by MCIN/AEI/10.13039/501100011033.

References

Abdullah, N., Baskaran, V. L., Mustafa, Z., Ali, S. R., & Zaini, S. H. (2022). Augmented reality: the effect in students' achievement, satisfaction and interest in science education. *International Journal of Learning, Teaching and Educational Research, 21*(5), 326–350. https://doi.org/10.26803/ijlter.21.5.17

Acevedo, J. A. (2006). Relevancia de los factores no-epistémicos en la percepción pública de los asuntos tecnocientíficos [The relevance of non-epistemic factors in the public perception of technoscientific issues]. *Revista Eureka sobre Enseñanza y Divulgación de las Ciencias, 3*(3), 370–391. https://doi.org/10.25267/Rev_Eureka_ensen_divulg_cienc.2006.v3.i3.03

Acevedo, J. A., & García-Carmona, A. (2017). *Controversias en la historia de la ciencia y cultura científica* [Controversies in the history of science and scientific culture]. Los Libros de la Catarata. https://oei.int/publicaciones/controversias-en-la-historia-de-la-ciencia-y-cultura-cientifica

Alderman, L., & Pronczuk, M. (2022, January 1). Europe plans to say nuclear power and natural gas are green investments. *The New York Times.*

Allday, J. (2003). Science in science fiction. *Physics Education, 38*(1), 27–30.

Ammoneit, R., Göhner, M. F., Bielik, T., & Krell, M. (2024). Why most definitions of modeling competence in science education fall short: Analyzing the relevance of volition for modelling. *Science Education, 108*(2), 443–466. https://doi.org/10.1002/sce.21841

Anderman, E. M., Sinatra, G. L., & Gray, D. L. (2012). The challenges of teaching and learning about science in the twenty-first century: Exploring the abilities and constraints of adolescent learners. *Studies in Science Education, 48*(1), 89–117. https://doi.org/10.1080/03057267.2012.655038

Asthana, A. (2008). Multimedia in education. In B. Furht (Eds.), *Encyclopedia of multimedia.* Springer. https://doi.org/10.1007/978-0-387-78414-4_140

Bailin, S. (2002). Critical thinking and science education. *Science & Education, 11*(4), 361–375. https://doi.org/10.1023/A:1016042608621

Barnett, M., & Kafka, A. (2007). Using science fiction movie scenes to support critical analysis of science. *Journal of College Science Teaching, 36*(4), 31–35.

Barrow, L. H. (2006). A brief history of inquiry: From Dewey to Standards. *Journal of Science Teacher Education, 17*(3), 265–278. https://doi.org/10.1007/s10972-006-9008-5

Bellocchi, A. (2015). Methods for sociological inquiry on emotion in educational settings. *Emotion Review, 7*(2), 151–156. https://doi.org/10.1177/1754073914554775

Bencze, L., Pouliot, C., Pedretti, E., Simonneaux, L., Simonneaux, J., & Zeidler, D. L. (2020). SAQ, SSI and STSE education: Defending and extending "science-in-context." *Cultural Studies of Science Education, 15*(S1), 825–851. https://doi.org/10.1007/s11422-019-09962-7

Bevins, S., & Price, G. (2016). Reconceptualising inquiry in science education. *International Journal of Science Education, 38*(1), 17–29. https://doi.org/10.1080/09500693.2015.1124300

Blanco, H. A., & Forero, D. L. (2017). *Comportamiento del consumidor frente a la decisión de compra en artículos deportivos relacionados con el uso de bicicleta en la ciudad de Bogotá.* [Consumer behavior regarding purchasing decisions for sports items related to bicycle use in the city of Bogotá.][Doctoral Thesis, University Externado of Colombia].

Blanco-López, A., España-Ramos, E., & Franco-Mariscal, A. J. (2017). Estrategias didácticas para el desarrollo del pensamiento crítico en el aula de ciencias [Teaching strategies for the development of critical thinking in the teaching of science]. *Ápice, Revista de Educación Científica, 1*(1), 107–115. https://doi.org/10.17979/arec.2017.1.1.2004

Blanco-López, A., España-Ramos, E., González-García, F. J., & Franco-Mariscal, A. J. (2015). Key aspects of scientific competence for citizenships: A Delphi study of the expert community in Spain. *Journal of Research in Science Teaching, 52*(2), 164–198. https://doi.org/10.1002/tea.21188

Bravo, B., & Jiménez-Aleixandre, M. P. (2018). Developing an initial learning progression for the use of evidence in decision-making contexts. *International Journal of Science and Mathematics Education, 16*(4), 619–638. https://doi.org/10.1007/s10763-017-9803-9

Bybee, R., Taylor, J., Gardner, A., Van Scotter, P., Powell, J., Westbrook, A., & Landes, N. (2006). *The BSCS 5E instructional model: Origins and effectiveness.* A Report Prepared for the Office of Science Education National Institutes of Health.

Cabello-Garrido, A., Cebrián-Robles, D., España-Ramos, E., González-García, F. J., Cruz-Lorite, M. I., España-Naveira, P., & Blanco-López, A. (2023). The dominant model of meat production and consumption as a socially acute question for activist education. *Cultural Studies of Science Education, 18*, 911–935. https://doi.org/10.1007/s11422-023-10188-x

Cano-Iglesias, M. J. (2024). *Argumentación y toma de decisiones como habilidades del pensamiento crítico: Una investigación con estudiantes de ingenierías industriales y del Máster en Profesorado de especialidades científico-tecnológicas* [Argumentation and decision making as critical thinking skills: An investigation with students of industrial engineering and of the Master's Degree in Teaching Science and Technology Specialities] [Doctoral Thesis, University of Málaga, Spain]. https://hdl.handle.net/10630/30996

Cebrián-Robles, D., Franco-Mariscal, A. J., Lupión, T., Acebal, M. C., & Blanco-López, A. (2021a) (Coords.), *Enseñanza de las ciencias y problemas relevantes de la ciudadanía* [Science education and relevant citizenship issues]. Graó.

Cebrián-Robles, D., Franco-Mariscal, A. J., & Blanco-López, A. (2021b). Secuencia de tareas para enseñar argumentación en ciencias a profesorado en formación inicial a través de CoRubric. Ejemplificación en una actividad sobre una central salina [Task sequence for teaching argumentation in science to pre-service teachers through CoRubric. Exemplification in an activity on a saline power plant]. *Didáctica de las Ciencias Experimentales y Sociales, 40*, 149–168.

Clement, J. J., & Rea-Ramírez, M. A. (2008). *Model based learning and instruction in science.* Springer. https://doi.org/10.1007/978-1-4020-6494-4

Chen, L., & Xiao, S. (2021). Perceptions, challenges and coping strategies of science teachers in teaching socioscientific issues: A systematic review. *Educational Research Review, 32*, 100377. https://doi.org/10.1016/j.edurev.2020.100377

Cortés, F., Frattini, G., Krapp, L. F., Martínez, H., Moreno, D. M., Roldán, M., Salomón, J., Stemkoski, L., Traeger, S., Dal Peraro, M., & Abriata, L. A. (2021). MoleculARweb: A web site for chemistry and structural biology education through interactive augmented reality out of the box in commodity devices. *Journal of Chemical Education, 98*(7), 2243–2255. https://doi.org/10.1021/acs.jchemed.1c00179

Christenson, N., Chang-Rundgren, S. N., & Höglund, H. O. (2012). Using the SEE-SEP model to analyze upper secondary students' use of supporting reasons in arguing socioscientific issues.

Journal of Science Education and Technology, 21, 342–352. https://doi.org/10.1007/s10956-011-9328-x

Christodoulou, A., Levinson, R., Davies, P., Grace, M., Nicholl, J., & Rietdijk, W. (2021). The use of Cartography of Controversy within socioscientific issues-based education: Students' mapping of the badger-cattle controversy in England. *International Journal of Science Education, 43*(15), 2479–2500. https://doi.org/10.1080/09500693.2021.1970852

Colucci-Gray, L., & Gray, D. (2022). Critical thinking in the flesh: Movement and metaphors in a world in flux. In B. Puig, & M. P. Jimenez-Aleixandre (Eds.), *Critical thinking in biology and environmental education: Facing challenges in a post-truth world* (pp. 21–39). Springer. https://doi.org/10.1007/978-3-030-92006-7_2

Cruz-Lorite, I. M., Cebrián-Robles, D., Acebal, M. C., & Evagorou, M. (2023). Analysis of the informal reasoning modes of preservice primary teachers when arguing about a socio-scientific issue on nuclear power during a role play. *Sustainability, 15*, 4291. https://doi.org/10.3390/su15054291

Descartes, R. (1641). *Meditationes de Prima Philosophia* [G. Heffernan, Trans., 1990]. University of Notre Dame Press.

Deterding, S., Dixon, D., Khaled, R., & Nacke, L. (2011). From game design elements to gamefulness: Defining gamification. In *Proceedings of the 15th international academic MindTrek conference: Envisioning future media environments*, MindTrek 2011 (Vol. 11, pp. 9–15). https://doi.org/10.1145/2181037.2181040

Dewey, J. (1910). *What is thought? In how we think* (1st ed., pp. 1–13). D.C. Health.

Driver, R., Newton, P., & Osborne, J. (2000). Establishing the norms of scientific argumentation in classroom. *Science Education, 84*(3), 287–312. https://doi.org/10.1002/(SICI)1098-237X(200005)84:3%3c287::AID-SCE1%3e3.0.CO;2-A

Dudo, A., Brossard, D., Shanahan, J., Scheufele, D. A., Morgan, M., & Signorielli, N. (2011). Science on television in the 21st century: Recent trends in portrayals and their contributions to public attitudes toward science. *Communication Research, 38*(6), 754–777. https://doi.org/10.1177/0093650210384988

Duschl, R. A., & Osborne, J. (2002). Supporting and promoting argumentation discourse in science education. *Studies in Science Education, 38*(1), 39–72. https://doi.org/10.1080/03057260208560187

Dwyer, C. P., Hogan, M. J., & Stewart, I. (2014). An integrated critical thinking framework for the 21st century. *Thinking Skills and Creativity, 12*, 43–52. https://doi.org/10.1016/j.tsc.2013.12.004

Eggert, S., Ostermeyer, F., Hasselhorn, M., & Bögeholz, S. (2013). Socioscientific decision making in the science classroom: The effect of embedded metacognitive instructions on students' learning outcomes. *Education Research International, 2013*, 309894. https://doi.org/10.1155/2013/309894

Elias, S. A. (2018). Plastics in the ocean. In D. A. Dellasala & M. I. Goldstein (Eds.), *Encyclopaedia of the antropocene* (pp. 133–149). Elsevier.

Elster, J. (1994). Rationality, emotions, and social norms. *Synthese, 98*(1), 21–49.

Ennis, R. H. (1987). A taxonomy of critical thinking dispositions and abilities. In J. B. Baron, & R. J. Sternberg (Eds.), *Teaching thinking skills: Theory and practice*. Freeman.

Erduran, S., & Jiménez-Aleixandre, M. P. (2007). *Argumentation in science education. Perspectives from classroom-based research*. Springer. https://doi.org/10.1007/978-1-4020-6670-2

Eriksen, M., & Al., E. (2014). Plastic pollution in the world's oceans: More than 5 trillion plastic pieces weighing over 250,000 tons afloat at sea. *PLOS ONE, 9*(12), E111913. https://doi.org/10.1371/journal.pone.0111913

España-Naveira, P., Cruz-Lorite, I. M., Cebrián-Robles, D., Cabello-Garrido, A., España-Ramos, E., González-García, F. J., & Blanco-López, A. (2023). Enfoque de cartografía de controversias para abordar cuestiones socialmente vivas desde la enseñanza de la ciencia y la tecnología [Controversy mapping approach to address socially relevant issues in science and technology education]. *Revista Eureka sobre Enseñanza y Divulgación de las Ciencias, 20*(3), 310101–310121. https://doi.org/10.25267/Rev_Eureka_ensen_divulg_cienc.2023.v20.i3.3101

España-Ramos, E. (2008). *Conocimiento, actitudes, creencias y valores en los argumentos sobre un tema socio-científico relacionado con los alimentos* [Knowledge, attitudes, beliefs, and values in arguments about a socio-scientific issue related to food] [Doctoral Thesis, University of Málaga, Spain]. http://hdl.handle.net/10630/2551

España-Ramos, E. (2023). El juego de rol como estrategia didáctica en el aula de ciencias [The role-playing as a didactic strategy in the science classroom]. In A. J. Franco-Mariscal, J. M. Hierrezuelo-Osorio, M. J. Cano-Iglesias, & A. Blanco-López (Coords.), *El juego de rol como estrategia para desarrollar habilidades de pensamiento crítico. Aplicado al aula de las ciencias* [Role-playing game as a strategy to develop critical thinking skills. Applied in the science classroom] (pp. 21–30). Pirámide.

España-Ramos, E., & Rueda, J. A. (2023). Juego de rol sobre los alimentos transgénicos [Role-playing about genetically modified foods]. In A. J. Franco-Mariscal, J. M. Hierrezuelo, M. J. Cano-Iglesias, & A. Blanco (Coords.), *El juego de rol como estrategia para desarrollar habilidades de pensamiento crítico. Aplicado al aula de las ciencias* [Role-playing game as a strategy to develop critical thinking skills. Applied in the science classroom] (pp. 45–56). Pirámide.

European Commission (2015). *Science education for responsible citizenship*. Publications Office of the European Union.

Evagorou, M., Jiménez-Aleixandre, M. P., & Osborne, J. (2012). Should we kill the grey squirrels? A study exploring students' justifications and decision making. *International Journal of Science Education, 34*(3), 401–428. https://doi.org/10.1080/09500693.2011.619211

Facione, P. A. (1990). *Critical thinking: A statement of expert consensus for purposes of educational assessment and instruction*. California Academic Press.

Fairclough, N. (1995). *Critical discourse analysis*. Longman.

Fang, S. C., Hsu, Y. S., & Lin, S. S. (2019). Conceptualizing socio-scientific decision making from a review of research in science education. *International Journal Science and Mathematics Education, 17*, 427–448. https://doi.org/10.1007/s10763-018-9890-2

Farnsworth, W. (2021). *The Socratic method: A practitioner's handbook hardcover*. Godine, Publisher.

Fraknoi, A. (2002). Teaching astronomy with science fiction: A resource guide. *Astronomy Education Review, 1*(2), 112–119.

Franco-Mariscal, A. J. (2007). *Enseñando Física y Química con ideas quijotescas* [Teaching Physics and Chemistry with Quixotic Ideas]. Spanish Government. Ministry of Education. CIDE.

Franco-Mariscal, A. J. (2009). Aprende física com Prison Break. [Learn Physics with Prison Break]. *Alambique, Didáctica de las Ciencias Experimentales, 60*, 82–94.

Franco-Mariscal, A. J. (2014). Diseño y evaluación del juego didáctico "Química con el mundial de Brasil 2014" [Design and evaluation of the educational game "Chemistry in 2014 Brazil World Cup"]. *Educación Química, 25*(E1), 276–283. https://doi.org/10.1016/S0187-893X(14)70568-3

Franco-Mariscal, A. J. (2015a). Competencias científicas en la enseñanza y el aprendizaje por investigación. Un estudio de caso sobre corrosión de metales en secundaria [Scientific competencies in teaching and learning through research. A case study on metal corrosion in secondary education]. *Enseñanza de las Ciencias, 33*(2), 231–252. https://doi.org/10.5565/rev/ensciencias.1645

Franco-Mariscal, A. J. (2015b). Exploring the everyday context of chemical elements: Discovering the elements of car components. *Journal of Chemical Education, 92*(10), 1672–1677. https://doi.org/10.1021/acs.jchemed.5b00164

Franco-Mariscal, A. J. (2016). Óptica con Peppa Pig [Optics with Peppa Pig]. *Alambique, Didáctica De Las Ciencias Experimentales, 83*, 57–63.

Franco-Mariscal, A. J. (2018). Discovering the chemical elements in food. *Journal of Chemical Education, 95*(3), 403–409. https://doi.org/10.1021/acs.jchemed.7b00218

Franco-Mariscal, A. J. (2021). Enseñar física con juego de tronos. Estudio del movimiento [Teaching Physics with Game of Thrones: A Study of Motion]. *Revista Mexicana de Física E, 18*(1), 63–68. https://doi.org/10.31349/RevMexFisE.18.63

Franco-Mariscal, A. J., España-Ramos, E., & Blanco-López, A. (2018). Teaching students about chemical elements using daily-life contexts. In O. E. Finlayson, E. McLoughlin, S. Erduran, & P. Childs, P. (Eds.), *Electronic proceedings of the ESERA 2017 conference. Research, practice and collaboration in science education*, Part 5/5 (co-Eds.) (pp. 710–718). Dublin City University.

Franco-Mariscal, A. J., Hierrezuelo-Osorio, J. M., Cano-Iglesias, M. J., & Blanco-López, A. (2023) (Coords.). *El juego de rol como estrategia para desarrollar habilidades de pensamiento crítico. Aplicado al aula de las ciencias* [Role-playing game as a strategy to develop critical thinking skills. Applied in the science classroom]. Pirámide.

Franco-Mariscal, A. J., Oliva-Martínez, J. M., Blanco-López, A., & España-Ramos, E. (2016). A game-based approach to learning the idea of chemical elements and their periodic classification. *Journal of Chemical Education, 93*(7), 1173–1190. https://doi.org/10.1021/acs.jchemed.5b00846

García-Carmona, A. (2023). Scientific thinking and critical thinking in science education. *Science & Education*. https://doi.org/10.1007/s11191-023-00460-5

Geroimenko, V. (2020). *Augmented reality in education. A new technology for teaching and learning*. Springer. https://doi.org/10.1007/978-3-030-42156-4

Gilbert, J. K., & Justi, R. (2016). *Modelling-based teaching in science education*. Springer. https://doi.org/10.1007/978-3-319-29039-3

Girón, J. R., & Franco-Mariscal, A. J. (2023). "*Atomizados*": An educational game for learning atomic structure. A case study with grade-9 students with difficulties learning chemistry. *Journal of Chemical Education, 100*(8), 3114–3123. https://doi.org/10.1021/acs.jchemed.2c00614

Government of Spain. Ministry of Culture and Sport (2021). *Anuario de estadísticas culturales* [Cultural Statistics Yearbook.] Secretaría General Técnica.

Gough, N. (1993). *Laboratories in fiction: Science education and popular media*. Deakin University.

Granath, P. L., & Russell, J. V. (1999). Using games to teach chemistry. 1. The old Prof card game. *Journal of Chemical Education, 76*(4), 485−486. https://doi.org/10.1021/ed076p485

Grosick, T. L., Talbert, C., Myers, M. J., & Angelo, R. (2013). Assessing the landscape: Body image values and attitudes among middle school boys and girls. *American Journal of Health Education, 44*(1), 41–52. https://doi.org/10.1080/19325037.2012.749682

Hadjichambis, A. C., Reis, P., Paraskeva-Hadjichambi, D., Činčera, J., Boeve-de Pauw, J., Gericke, N., & Knippels, M. C. (Eds.) (2020). *Conceptualizing environmental citizenship for 21st century education*. Springer. https://doi.org/10.1007/978-3-030-0249-1

Hadzigeorgiou, Y. (2017). Teaching the nature of science through storytelling: Some empirical evidence from a grade 9 classroom. *SFU Educational Review, 10*(2). https://doi.org/10.21810/sfuer.v10i2.318

Hahn, U., & Oaksford, M. (2012). *Rational argument*. Oxford University Press.

Halloun, I. (2007). Mediated modeling in science education. *Science & Education, 16*(7–8), 653–697. https://doi.org/10.1007/s11191-006-9004-3

Halpern, D. (2006). *Halpern critical thinking assessment using everyday situations: background and scoring standards* (2nd report) [Unpublished manuscript]. Claremont McKenna College.

Halpern, D. F., & Danna, D. S. (2023). *Though and knowledge. An introduction to critical thinking* (Vol. 2, 6th ed.). Routledge.

Hamalosmanoglu, M., Kizilay, E., & Saylan Kirmizigül, A. (2020). The effects of using animated films in the environmental education course on prospective teachers' behavior towards environmental problems and their attitude towards solid waste and recycling. *International Online Journal of Education and Teaching, 7*(3), 1178–1187.

Harrison, A. G., & Treagust, D. F. (2000). A typology of school science models. *International Journal of Science Education, 22*(9), 1011–1026. https://doi.org/10.1080/095006900416884

Helser, T. L. (2003). Elemental zoo. *Journal of Chemical Education, 80*(4), 409. https://doi.org/10.1021/ed080p409

Henderson, J. B., McNeill, K. L., González-Howard, M., Close, K., & Evans, M. (2018). Key challenges and future directions for educational research on scientific argumentation. *Journal of Research in Science Teaching, 55*(1), 5–18. https://doi.org/10.1002/tea.21412

Hierrezuelo-Osorio, J. M., Franco-Mariscal, A. J., & Blanco-López, A. (2022). Uso de dilemas socio-científicos para el desarrollo de habilidades de pensamiento crítico en docentes en formación inicial. Percepciones del profesorado [Use of socio-scientific dilemmas for the development of critical thinking skills in pre-service teachers. Perceptions of the teachers]. *Revista Interuniversitaria de Formación del Profesorado, 97*(36.1), 99–122.

Hodson, D. (2003). Time for action: Science education for an alternative future. *International Journal of Science Education, 25*(6), 645–670. https://doi.org/10.1080/09500690305021

Hoffmann, B., & Ritchie, D. (1997). Using multimedia to overcome the problems with problem based learning. *Instructional Science, 25*(2), 97–115. https://doi.org/10.1023/A:1002967414942

Horizon Report (2021). *Educause horizon report 2021: Teaching and learning*. Educause. https://library.educause.edu/-/media/files/library/2021/4/2021hrteachinglearning.pdf

Izquierdo, M. (2013). Consideraciones acerca de la diferencia entre 'contexto del alumno' y 'contexto de modelización científica escolar' y de las dificultades que de ella se derivan [Considerations on the Difference between 'student Context' and 'school Scientific Modeling Context' and the Challenges Arising from it]. *Seminario de Doctorado: Perspectivas sobre el contexto en la educación científica: Aproximaciones teóricas e implicaciones para la práctica educativa* [Doctoral Seminar: Perspectives on Context in Science Education: Theoretical Approaches and Implications for Educational Practice] (pp. 20–31). UAB/LIEC.

Jiménez-Aleixandre, M. P. (2010). *10 Ideas clave. Competencias en argumentación y uso de pruebas* [10 Key Ideas. Competencies in Argumentation and Use of Evidence]. Graó.

Jiménez-Aleixandre, M. P., Bugallo, A., & Duschl, R. A. (2000). "Doing the lesson" or "Doing science": Arguments in high school genetics. *Science Education, 84*(6), 757–792. https://doi.org/10.1002/1098-237X(200011)84:6%3c757::AID-SCE5%3e3.0.CO;2-F

Jiménez-Aleixandre, M. P., & Puig, B. (2022). Educating critical citizens to face post-truth: The time is now. In B. Puig & M. P. Jiménez-Aleixandre (Eds.), *Critical thinking in biology and environmental education. Facing challenges in a post-truth world* (pp. 3–19). Springer. https://doi.org/10.1007/978-3-030-92006-7

Justi, R., & Gilbert, J. K. (2002). Modelling teachers´ views on the nature of modelling, and implications for the education of modellers. *International Journal of Science Education, 24*(4), 369–387. https://doi.org/10.1080/09500690110110142

Kang, J., & Keinonen, T. (2018). The effect of student-centered approaches on students' interest and achievement in science: Relevant topic-based, open and guided inquiry-based, and discussion-based approaches. *Research in Science Education, 48*(4), 865–885. https://doi.org/10.1007/s11165-016-9590-2

Kant, I. (1781). *Critique of pure reason* (F. M. Müller, Trans., 1922). The McMillan Company.

Kant, I. (1788). *Critique of practical reason* (M. Gregor, & A. Reath, Trans., 1997). Cambridge University Press.

Kapp, K. (2012). *The gamification of learning and instruction*. Pfeiffer.

Ke, L., Sadler, T. D., Zangori, L., & Friedrichsen, P. J. (2021). Developing and using multiple models to promote scientific literacy in the context of socio-scientific issues. *Science & Education, 30*, 589–607. https://doi.org/10.1007/s11191-021-00206-1

Khishfe, R. (2014). Nature of Science and Decision-Making. *International Journal of Science Education, 34*(1), 67–100. https://doi.org/10.1080/09500693.2011.559490.

Koch, B. S., & Barber, M. M. (2019). Basuras marinas; impacto, actualidad y las acciones para mitigar sus consecuencias [Marine Litter: Impact, Current Status, and Actions to Mitigate its Consequences]. *Revista de Marina, 968*, 30–39.

Kolstø, S. (2001). Scientific literacy for citizenship: Tools for dealing with the science dimension of controversial socio-scientific issues. *Science Education, 85*(3), 291–310. https://doi.org/10.1002/sce.1011

Koutnikova, M. (2017). The application of comics in science education. *Acta Educationis Generalis, 7*(3), 88–98.
Kuhn, D. (2019). Critical thinking as discourse. *Human Development, 62*(3), 146–164. https://doi.org/10.1159/000500171
Lambert, J. (2009). *Digital storytelling cookbook.* Digital Diner Press.
Latour, B. (1996). On actor-network theory: A few clarifications. *Soziale Welt: Zeitschrift Fuer Sozialwissenschaftliche Forschung und Praxis, 47*(4), 369–381.
Latour, B. (2007). La cartographie des controverses. *Technology Review*, 82–83.
Lawrence, K. T., & Malinconico, L. L. (2008). Geology in the movies: Using Hollywood films as a teaching tool in introductory geosciences courses. *AGU Fall Meeting Abstracts* (Vol. 2008).
Lipman, M. (2003). *Thinking in education* (2nd ed.). Cambridge. https://doi.org/10.1017/CBO9780511840272
López-Fernández, M. M., González, F., & Franco-Mariscal, A. J. (2021). Should we ban single-use plastics? A role-playing game to argue and make decisions in a grade-8 school chemistry class. *Journal of Chemical Education, 98*(12), 3947 3956. https://doi.org/10.1021/acs.jchemed.2c00223
López-Fernández, M. M., González-García, F., & Franco-Mariscal, A. J. (2022). How can socio-scientific issues help develop critical thinking in chemistry education? A reflection on the problem of plastics. *Journal of Chemical Education, 99*(10), 3435–3442. https://doi.org/10.1021/acs.jchemed.2c00223
Lupión, T., López, R., & Blanco-López, A. (2017). What do science teachers think about developing scientific competences through context-based teaching? A case study. *International Journal of Science Education, 39*(7), 937–963. https://doi.org/10.1080/09500693.2017.1310412
Manassero, M. A., & Vázquez, A. (2019). Conceptualización y taxonomía para estructurar los conocimientos acerca de la ciencia. [Conceptualization and taxonomy to structure the knowledge about science.] *Revista Eureka sobre Enseñanza y Divulgación de las Ciencias, 16*(3), 3104. https://doi.org/10.25267/Rev_Eureka_ensen_divulg_cienc.2019.v16.i3.3104
Marcén, C., & Molina, P. J. (2006). *La persistencia de las opiniones de los escolares sobre el Medio Ambiente. Una particular visión retrospectiva desde 1980 a 2005* [The Persistence of Students' Opinions on the Environment: A Retrospective View from 1980 to 2005]. MMA.
Martí, V., & Rubio, J. (2014). ChemMend: A card game to introduce and explore the periodic table while engaging students' interest. *Journal of Chemical Education, 91*(6), 868–871. https://doi.org/10.1021/ed300733w
Martini, M., Widodo, W., Qosyim, A., Mahdiannur, M. A., & Jatmiko, B. (2021). Improving undergraduate science education students' argumentation skills through debates on socioscientific issues. *Jurnal Pendidikan IPA Indonesia, 10*(3), 428–438. https://doi.org/10.15294/jpii.v10i3.30050
McSharry, G., & Jones, S. (2000). Role-play in science teaching and learning. *School Science Review, 82*(298), 73–82.
Membiela, P., Vidal, M., Fragueiro, S., Lorenzo, M., García-Rodeja, I., Aznar, V., Bugallo, A., & González, A. (2022). Motivation for science learning as an antecedent of emotions and engagement in preservice elementary teachers. *Science Education, 106*(1), 119–141. https://doi.org/10.1002/sce.21686
Mercier, H., & Sperber, D. (2011). Why do humans reason? Arguments for an argumentative theory. *Behavioral and Brain Sciences, 34*, 57–74. https://doi.org/10.1017/S0140525X10000968
Michael, D., & Chen, S. (2006). *Serious games. Games that educate, train and inform.* Course Technology, CENGAGE Learning.
Mojica, E. R. E. (2019). Chemtertainment: Using video clips from movies, television series, and YouTube to enhance the teaching and learning experience of an introductory chemistry lecture class. In J. Parr (Ed.), *Videos in chemistry education: Applications of interactive tools* (pp. 21–34). ACS Symposium Series; American Chemical Society.
Moraga-Toledo, S., García-Ruiz, C., Lupión-Cobos, T., & Blanco-López, A. (2024). Opportunities and challenges for inquiry-based science education. In A. Marzabal & C. Merino (Eds.),

Rethinking science education in Latin-America: Diversity and equity for Latin American students in science education (pp. 225–244). Springer. https://doi.org/10.1007/978-3-031-52830-9
Moreno-Fontiveros, G., Cebrián-Robles, D., Blanco-López, A., & España-Ramos, E. (2022). Decisiones de estudiantes de 14/15 años en una propuesta didáctica sobre la compra de un coche [Fourteen/fifteen-year-old students' decisions in a teaching proposal on the buying of a car]. *Enseñanza de las Ciencias, 40*(1), 199–219. https://doi.org/10.5565/rev/ensciencias.3292
Moreno-Martínez, N. M., & Franco-Mariscal, A. J. (2020). Programa formativo de realidad aumentada y realidad virtual en la enseñanza de las ciencias en la educación superior [Training program on augmented reality and virtual reality in science education at the higher education]. In A. Alías, D. Cebrián-Robles, F. J. Ruiz, & I. Caraballo (Coords.), *Tecnologías para la formación de profesionales en educación* [Technologies for Professional Education Training] (pp. 232–256). Dykinson.
Moreno-Martínez, N. M., & Franco-Mariscal, A. J. (2023). Posibilidades didácticas de la herramienta de realidad aumentada ZapWorks en la enseñanza de las ciencias. Una experiencia con estudiantes de un Máster en Profesorado [Educational possibilities of the augmented reality tool ZapWorks in science education. An experience with Master's Degree in Teaching students]. *Tecnología, Ciencia y Educación, 24*, 91–118. https://doi.org/10.51302/tce.2023.2808
Muñetón, G., Ruiz, A. F., & Loaiza, O. L. (2017). Toma de decisiones: Explicaciones desde la ciencia aplicada del comportamiento. [Make a choice. Explanations from applied behavioral science.] *Espacios, 38*(13), 1–12.
National Research Council (NRC). (1996). *National science education standards*. The National Academy Press.
National Research Council (NRC). (2000). *Inquiry and the national science education standards*. National Academy Press.
National Research Council (NRC). (2012). *A framework for K-12 science education: Practices crosscutting concepts, and core ideas*. National Academy Press.
Negrete, A. (2003). Science via fictional narratives: Communicating science through literary forms. *Ludus Vitalis, 10*, 197–204.
Neo, M., & Neo, K. T. (2001). Innovative teaching: Using multimedia in a problem-based learning environment. *Journal of Educational Technology & Society, 4*(4), 19–31. https://doi.org/10.12944/CWE.6.1.28
Next Generation Science Standards (NGSS) (2013). *Next generation science standards: For states, by states*. Achieve, Inc. on behalf of the twenty-six states and partners that collaborated on the NGSS.
Nicolaou, C. T., & Constantinou, C. P. (2014). Assessment of the modeling competence: A systematic review and synthesis of empirical research. *Educational Research Review, 13*, 52–73. https://doi.org/10.1016/J.EDUREV.2014.10.001
Ogborn, J. G., Kress, G., Martins, I., & McGillicuddy, K. (1996). *Explaining science in the classroom*. Open University Press.
Oliva, J. M., Aragón, M. M., & Cuesta, J. (2015). The competence of modelling in learning chemical change: A study with secondary school students. *International Journal of Science and Mathematics Education, 13*(4), 751–791. https://doi.org/10.1007/s10763-014-9583-4
Osborne, J. (2014). Teaching critical thinking. New directions in science education? *School Science Review, 352*, 53–62.
Osborne, J., Henderson, J. B., MacPherson, A., Szu, E., Wild, A., & Yao, S. (2016). The development and validation of a learning progression for argumentation in science. *Journal Research in Science Teaching, 53*(6), 821–846. https://doi.org/10.1002/tea.21316
Osborne, J., & Pimentel, D. (2023). Science education in an age of misinformation. *Science Education, 107*(3), 553–571. https://doi.org/10.1002/sce.21790
Owens, D. C., Sadler, T. D., & Friedrichsen, P. (2021). Teaching practices for enactment of socio-scientific issues instruction: An instrumental case study of an experienced biology teacher. *Research in Science Education, 51*(2), 375–398. https://doi.org/10.1007/S11165-018-9799-3

Ozturk, N., & Yilmaz-Tuzun, O. (2017). Preservice science teachers' epistemological beliefs and infor mal reasoning regarding socio-scientific issues. *Research in Science Education, 47*, 1275–1304. https://doi.org/10.1007/s11165-016-9548-4

Parkin, J. (2008). *Bicycles as a sustainable transport mode: A review*. Energies Ed.

Paul, R., & Elder, L. (2005). *A guide for educator to critical thinking competency standards*. Foundation for Critical Thinking.

Pedaste, M., Mäeots, M., Siiman, L. A., De Jong, T., Van Riesen, S. A., Kamp, E. T., & Tsourlidaki, E. (2015). Phases of inquiry-based learning: Definitions and the inquiry cycle. *Educational Research Review, 14*, 47–61. https://doi.org/10.1016/j.edurev.2015.02.003

Pekrun, R., & Linnenbrink-García, L. (Eds.) (2014). *International handbook of emotions in education* (1st ed.). Routledge. https://bit.ly/3tQ8z5h

Perales, F. J. (2006). Pasado, presente y ¿futuro? de los libros de texto [Past, Present, and Future of Textbooks]. *Alambique, Didáctica de las Ciencias Experimentales, 48*, 57–63.

Petit, M. P., Solbes, J., & Torres, N. Y. (2021). El cine de ciencia ficción para desarrollar cuestiones sociocientíficas y el pensamiento crítico [Science Fiction Cinema for Developing Socioscientific Issues and Critical Thinking]. *Praxis & Saber, 12*(29), 52–73.

Prins, G. T., Bulte, A. M., Van Driel, J. H., & Pilot, A. (2009). Students' involvement in authentic modelling practices as contexts in chemistry education. *Research in Science Education, 39*(5), 681–700. https://doi.org/10.1007/s11165-008-9099-4

Puig, B. & Jiménez-Aleixandre, M. P. (Eds.) (2022). *Critical thinking in biology and environmental education. Facing challenges in a post-truth world*. Springer. https://doi.org/10.1007/978-3-030-92006-7

Rashid, S., & Qaisar, S. (2017). Role-play: A productive teaching strategy to promote critical thinking. *Bulletin of Education and Research, 39*(2), 197–213.

Reiser, B., Berland, L. K., & Keynon, L. (2012). Engaging students in the scientific practices of explanation and argumentation. *Science and Children, 49*(8), 8–13.

Rodríguez-Mora, F., Cebrián-Robles, D., & Blanco-López, A. (2022). An assessment using rubrics and the Rasch model of 14/15-year-old students' difficulties in arguing about bottled water consumption. *Research in Science Education, 52*, 1075–1091. https://doi.org/10.1007/s11165-020-09985-z

Rönnebeck, S., Bernholt, S., & Ropohl, M. (2016). Searching for a common ground—A literature review of empirical research on scientific inquiry activities. *Studies in Science Education, 52*(2), 161–197. https://doi.org/10.1080/03057267.2016.1206351

Rose, C. (2003). How to teach biology using the movie science of cloning people, resurrecting the dead, and combining flies and humans. *Public Understanding of Science, 12*(3), 289–296. https://doi.org/10.1177/0963662503123007

Russell, B. (1973). *The problems of philosophy*. Oxford University Press.

Sadler, T. D., & Zeidler, D. (2005). Patterns of informal reasoning in the context of socioscientific decision making. *Journal of Research in Science Teaching, 42*(1), 112–138. https://doi.org/10.1002/tea.20042

Scalabrino, C., Navarrete, A., & Oliva, J. M. (2022). A theoretical framework to address education for sustainability for an earlier transition to a just, low carbon and circular economy. *Environmental Education Research, 28*(5), 735–766. https://doi.org/10.1080/13504622.2022.2031899

Simonneaux, J. (2020). Outils didactiques pour réaliser la démarche d'enquête sur une QSV: Cartographier les controverses sur une QSV. https://t.ly/oPa

Simonneaux, L. (2001). Role-play or debate to promote students' argumentation and justification on an issue in animal transgenesis. *International Journal Science Education, 23*(9), 903–927. https://doi.org/10.1080/09500690010016076

Simonneaux, L. (2008). Argumentation in socio-scientific contexts. In S. Erduran (Ed.), *Argumentation in science education: perspectives from classroom-based research* (pp.179–199). Springer. https://doi.org/10.1007/978-1-4020-6670-2_9

Simonneaux, L. (2014). From promoting the techno-sciences to activism. In L. Bencze, & S. Alsop (Eds.), *Activist science and technology education* (pp. 98–112). Springer. https://doi.org/10.1007/978-94-007-4360-1

Siribunnam, S., Nuangchalerm, P., & Jansawang, N. (2014). Socio-scientific decision making in the science classroom. *Online Submission, 5*(4), 1777–1782.

Solbes, J., & Torres, N. Y. (2012). Análisis de las competencias de pensamiento crítico desde el aborde de las cuestiones sociocientíficas: un estudio en el ámbito universitario [Analysis of Critical Thinking Competencies from the Approach of Socio-scientific Issues: A Study in the University Setting]. *Didáctica de las Ciencias Experimentales y Sociales, 26*, 247–269. https://doi.org/10.7203/dces.26.1928

Solbes, J. & Torres, N. Y. (2015). Alternativas para reflexionar sobre aspectos críticos de la ciencia en el aula [Alternatives to Reflect on Critical Aspects of Science in the Classroom]. *Revista Científica, 22*(2), 31–44. https://doi.org/10.14483/10.14483/udistrital.jour.RC.2015.22.a3

Solomon, J. (2006). Meta-scientific criticisms, curriculum innovation, and the propagation of scientific culture. In I. Westbury, & G. Milburn (Eds.), *Rethinking schooling. Twenty-five years of the journal of curriculum studies* (pp. 265–280). Routledge.

Soulios, I., & Psillos, D. (2016). Enhancing student teachers' epistemological beliefs about models and conceptual understanding through a model-based inquiry process. *International Journal of Science Education, 38*(7), 1212–1233. https://doi.org/10.1080/09500693.2016.1186304

Torres, N., & Solbes, J. (2016). Contribuciones de una intervención didáctica usando cuestiones socio científicas para desarrollar el pensamiento crítico [Contributions of a Didactic Intervention Using Socioscientific Issues to Develop Critical Thinking]. *Enseñanza de las Ciencias, 34*(2), 43–65. https://doi.org/10.5565/rev/ensciencias.1638

Toulmin, S. E. (2003). *The uses of argument* (3rd ed.). Cambridge University Press.

United Nations. (2015). *Transforming our world: The 2030 agenda for sustainable development (No. A/RES/70/1)*. United Nations. https://t.ly/yfJ7

Vázquez, A., & Manassero, M. A. (2018). Una taxonomía de las destrezas de pensamiento: una herramienta clave para la alfabetización científica [A Taxonomy of Thinking Skills: A Key Tool for Scientific Literacy]. *Revista Tecné, Episteme y Didaxis*, extra, 1–7. VIII Congreso Internacional de formación de Profesores de Ciencias para la Construcción de Sociedades Sustentables.

Vázquez, A., & Manassero, M. A. (2020). Pensamiento científico y pensamiento crítico: competencias transversales para aprender [Scientific Thinking and Critical Thinking: Cross-Cutting Competencies for Learning]. In A. Vilches (Coord.), *Veinte años de avances y nuevos desafíos en la Educación CTS para el logro de Objetivos de Desarrollo Sostenible.* VII Seminario Iberoamericano CTS (pp. 519–522). CTS.

Venturini, T. (2010). Diving in magma: How to explore controversies with actor-network theory. *Public Understanding of Science, 19*(3), 258–273. https://doi.org/10.1177/0963662509102694

Venturini, T., & Latour, B. (2009). The social fabric: Digital footprints and quali-quantitative methods. In *Proceedings of future en seine* (pp.87–101). Editions Future en Seine.

Vieira, R. M., & Tenreiro, C. (2016). Fostering scientific literacy and critical thinking in elementary science education. *International Journal of Science and Mathematics Education, 14*, 659–680. https://doi.org/10.1007/s10763-014-9605-2

Vila, L., Márquez, C., & Oliveras, B. (2023). Una propuesta para el diseño de actividades que desarrollen el pensamiento crítico en el aula de ciencias [A didactic proposal for the development of critical thinking in the high school science classroom]. *Revista Eureka sobre Enseñanza y Divulgación de las Ciencias, 20*(1), 130201. https://doi.org/10.25267/Rev_Eureka_ensen_divulg_cienc.2023.v20.i1.1302

von Winterfeldt, D. (2013). Bridging the gap between science and decision making. *Proceedings of the National Academy of Sciences of the United States of America, 110*(Supp. 3), 14055–14061. https://doi.org/10.1073/pnas.1213532110

Vuorikari, R., Ferrari, A., & Punie, Y. (2019). *Makers paces for Education and Training: Exploring future implications for Europe*. Publications Office of the European Union. https://publications.jrc.ec.europa.eu/repository/handle/JRC117481

Wang, H. H., Chen, H. T., Lin, H. S., Huang, Y. N., & Hong, Z. R. (2017). Longitudinal study of a cooperation-driven, socio-scientific issue intervention on promoting students' critical thinking and self-regulation in learning science. *International Journal Science Education, 39*(15), 2002–2026. https://doi.org/10.1080/09500693.2017.1357087

Wolfenbarger, L., & Phifer, P. (2000). The ecological risks and benefits of genetically engineered plants. *Science, 290*, 2088–2093. https://doi.org/10.1126/science.290.5499.2088

Wu, Y., & Tsai, C. (2011). High school students' informal reasoning regarding a socio-scientific issue, with relation to scientific epistemological beliefs and cognitive structures. *International Journal of Science Education, 33*(3), 371–400. https://doi.org/10.1080/09500690903505661

Wu, Y.-T. (2013). University students' knowledge structures and informal reasoning on the use of genetically modified foods: multi-dimensional analyses. *Research in Science Education, 43*, 1873–1890. https://doi.org/10.1007/s11165-012-9343-9

Xu, Y., Park, H., & Baek, Y. (2011). A new approach toward digital storytelling: An activity focused on writing self-efficacy in a virtual learning environment. *Educational Technology and Society, 14*(4), 181–191.

Yacoubian, H. A., & Khishfe, R. (2018). Argumentation, critical thinking, nature of science and socioscientific issues: A dialogue between two researchers. *International Journal of Science Education, 40*(7), 796–807. https://doi.org/10.1080/09500693.2018.1449986

Zagalaz, J. (2021). *Creation of audiobooks as an educational resource: Second cycle of action in an experience based on action research*. University of Murcia.

Zeidler, D. L., Herman, B. C., & Sadler, T. D. (2019). New directions in socioscientific issues research. *Disciplinary and Interdisciplinary Science Education Research, 1*(1), 1–9. https://doi.org/10.1186/s43031-019-0008-7

Zohar, A., & Nemet, F. (2002). Fostering students' knowledge and argumentation skills through dilemmas in human genetics. *Journal of Research in Science Teaching, 39*(1), 35–62. https://doi.org/10.1002/tea.10008

Part II
Socio-scientific Issues, Teaching in Context, and Critical Thinking

Chapter 2
Preservice Science Teachers' Demonstration of Critical Thinking Dimensions in Addressing the Socio-scientific Issue of Implementing an Artificial Moon for Street Illumination

Antonio Joaquín Franco-Mariscal, José Manuel Hierrezuelo-Osorio, María José Cano-Iglesias, and Ángel Blanco-López

2.1 Introduction

Developing critical thinking (hereinafter, CT) skills in students is currently a priority objective of general education and science education (Osborne, 2014). CT is part of non-university and university curricula and teacher training and is promoted by various international organisations.

Despite its importance, the development of CT is a challenge for teachers in science education because, among other aspects, there has yet to be a widely accepted model on how to deal with it successfully in the classroom for several reasons. The first is that CT is a complex construct involving cognitive factors, skills and dispositions (attitude/motivation) (Vieira & Tenreiro, 2016). Another obstacle is that there needs to be more consensus in science education on the dimensions that comprise it. Furthermore, the nature of the context in which CT is intended to be developed plays an important role in promoting some dimensions or others (Franco-Mariscal et al.,

A. J. Franco-Mariscal (✉) · J. M. Hierrezuelo-Osorio · M. J. Cano-Iglesias · Á. Blanco-López
Science Education, Universidad de Málaga, Málaga, Spain
e-mail: anjoa@uma.es

J. M. Hierrezuelo-Osorio
e-mail: jose.hierrezuelo@uma.es

M. J. Cano-Iglesias
e-mail: mjcano@uma.es

Á. Blanco-López
e-mail: ablancol@uma.es

A. J. Franco-Mariscal (ed.), *Critical Thinking in Science Education and Teacher Training*, Contemporary Trends and Issues in Science Education 64,
https://doi.org/10.1007/978-3-031-78578-8_2

2025). However, despite all these limitations, there is agreement on the importance of promoting CT for action, both from the curricular framework and educational research (Puig & Jiménez-Aleixandre, 2022).

As highlighted in Chap. 1 (Franco-Mariscal et al., 2025), recent literature has shown different theoretical approaches for the development of CT in science education (Blanco-López et al., 2017; Solbes & Torres, 2012; Vázquez & Manassero, 2020) around the use of socio-scientific issues (Ariza et al., 2024; Cebrián-Robles et al., 2021a, 2021b; Evagorou et al., 2012). Although these developments are necessary, it is important to have these theoretical approaches and examples of experiences that show science teachers their designs and how to adapt these integrated approaches to CT to the reality of the classroom, as they are not yet widespread. Analysing these experiences will help teachers identify the dimensions of CT in the activities they propose, integrate them, and provide them with references for new designs.

The choice of the problem to be addressed is also important. Energy is a case in point as a key concept in science curricula (Park & Liu, 2016). Its teaching is still primarily conceptual, although it is increasingly necessary to address it as a socio-scientific issue given that it is a complex and global challenge that requires mechanisms of action, which must be carried out both globally and individually (DeBoer, 2011; Martín-Gámez & Erduran, 2018).

Therefore, this chapter proposes a design for integrating CT into socio-scientific issues and analysing its potential. This aspect needs to be explored in science education, as exemplified in a current case related to energy use.

2.2 Critical Thinking Development Through the Socio-Scientific Issue of Implementing an Artificial Moon to Illuminate a City

The use of energy in its different aspects (consumption, production, transport, use, alternative energy sources, etc.) constitutes a socio-scientific issue due, among other aspects, to greenhouse gas emissions as the gases responsible for climate change. Like any socio-scientific issue, it is a real and complex problem that affects all spheres (personal, local and global) because scientific-technological, economic, social, environmental, ethical, etc. aspects converge (Martín-Gámez & Erduran, 2018).

In turn, these issues are controversial and raise debate in society and the media (Díaz & Jiménez-Liso, 2016) about the convenience or not of their use or implementation for different reasons (safety, environmental impact, economic cost, etc.). The position adopted by each citizen will also depend on several factors: their scientific knowledge of the issue, the critical analysis they make of the information they receive or seek from different sources, the possibility they have to confront their ideas with other people and argue their own, etc., all of which are important dimensions in the development of CT in citizens (Blanco-López et al., 2017).

In the current context, the issues surrounding the use of nuclear energy have once again been placed high on the political agenda and in the media as it seeks to be labelled as green energy in the process of ecological transition (United Nations, 2015). Along these lines, the 2030 Agenda considers climate change as one of the most significant challenges facing the international community (United Nations, 2015) and includes a Sustainable Development Goal (SDG13, Climate Action) designed explicitly for this purpose. This goal calls for urgent action involving all citizens to combat climate change and its effects.

Some countries, like China, are already promoting measures such as implementing an artificial moon with a coating to reflect sunlight to illuminate cities at night (SpaceChina.com, 2018). The problem, which appeared in different media, consisted of replacing the current lighting system in a town with an artificial satellite (called an artificial moon) built with solar panels, which would allow, according to the company responsible for the project, to illuminate a city at night with considerable energy savings.

Next, we examine the contributions that the ENCIC-CT model (Chap. 1) can make to developing CT, delving into the dimensions that citizens need to activate to address this issue and make decisions about it.

To face this problem, the first dimension to develop is the critical analysis of information. In this case, data from environmental, social, economic, technological, energy, ethical, etc. perspectives on the impact of its construction and use would be necessary.

Among the environmental factors to be considered are whether it is an environmentally friendly alternative for using solar energy as renewable energy and what impact it will have on the ecosystems of the area, especially on the behaviour of the flora and fauna as it receives light 24 h a day, and light pollution, among others.

Personal factors to consider include how constant light at night affects people's sleep cycle or their rhythm of life and the daily activities organised around day and night. From a social point of view, one could analyse how it would affect work, trade, and cultural activities, as well as the possibility of the artificial moon becoming a new tourist attraction that would affect the economy.

As economic factors, it would have to be considered whether removing the lighting would save the city electricity or how long it would take to amortise the cost of construction, installation, and maintenance.

The technological/energy aspects to be assessed would be related to the feasibility of the project being carried out successfully, the energy needed to supply the moon with light, and the control that this light only illuminates the city. Finally, at the scientific/technological level, the possibility of closing down existing astronomical observatories in the area should be considered, among other consequences.

Considering all these factors together would also involve assessing whether it would be ethical to implement a moon that could bring about abrupt changes in how people live.

With all this information in hand, critically analysed, citizens would be better able to make more informed and responsible decisions. In reality, the analysis of such a volume of information is not usually common, as it would require a great deal of time and effort for the citizen and, in turn, because the information available in the media and social networks usually appears as one-off news items and from very few perspectives, which can sometimes even correspond to fake news (Puig et al., 2021). In this way, the decision may lead us to a position different from the one the citizens would have if they had a more complete view of the issue. It is, therefore, essential to promote this critical analysis of information at school (Blanco-López et al., 2015) to help students improve and develop their informal reasoning, which is used in the field of socio-scientific issues (Sadler & Zeidler, 2005).

On the other hand, citizens generally do not approach these issues and their analysis individually but interact with different people close to them in their family, personal sphere, etc., so that the decision-making above. However, it can be understood as personal and is very much conditioned by the social environment (Zeidler, 1997). Bearing that social interaction is very important, CT also requires the development of personal autonomy because they do not allow themselves to be uncritically dragged along by currents of thought or predominant positions on a given problem.

Another important dimension for tackling problems such as the one discussed here is scientific argumentation, which is fundamentally based on evidence (Erduran & Jiménez-Aleixandre, 2008). In this case, as the artificial moon has not yet been implemented, the necessary data are unavailable to assess the positive and negative aspects. Therefore, citizens will have to base their arguments on the scientific-technical reports available on the possible impacts or the arguments provided by scientists and people with a solid background in this field.

Nowadays, citizens' involvement in such issues also entails participation in various discussion forums, such as social networks, associations, and activist projects. CT demands that these citizens communicate their decisions using appropriate language according to the context and goals or intentions.

Likewise, the development of CT requires the application of scientific knowledge to the issues addressed. In this example, the knowledge involved is the concept and use of renewable energies, their use, the environmental impacts of their construction, use and maintenance, the technological aspects related to implementing the moon as a geostationary satellite, etc.

2.3 Research Questions

This chapter poses the following research questions (RQ):

- RQ1: Which CT dimensions are displayed by preservice science teachers in an experience addressing the socio-scientific issue of implementing an artificial moon to illuminate city streets?
- RQ2: In what ways can the development of these dimensions of CT be encouraged when presenting information about the problem through a headline or more detailed readings?

The hypothesis for RQ1 suggests that, although the experience focuses primarily on critical analysis of information, argumentation and decision-making about the socio-scientific issue addressed, it also facilitates the development of other dimensions of CT in preservice teachers. As for the hypothesis in RQ2, it is expected that the development of these dimensions will become more conducive as more information is provided to preservice teachers.

2.4 Method

2.4.1 Participants and Context

This experience is part of a broader programme focused on developing CT dimensions in Spanish preservice teachers (Hierrezuelo-Osorio et al., 2022) based on different socio-scientific dilemmas such as sugar consumption, the use of autonomous cars, the suitability of a vegan diet, the banning of single-use plastics or the implantation of an artificial moon.

This program was implemented with 80 preservice teachers, specifically in the Degree in Primary Education (N = 45) and two specialisations (Physics and Chemistry, and Biology and Geology) of the Master's Degree in Teaching of Secondary Education (N = 35) at the University of Malaga (Spain). For the preservice primary teachers, the average age was 21 years, with 88.8% female and 22.2% male, while for the preservice secondary teachers, the average age was 28 years, with 57.1% female and 42.9% male.

2.4.2 Experience Description

The experience associated with the issue of implanting an artificial moon was organised in the following sequence of tasks:

- *Instruction in argumentation.* The teacher conducted a training session explaining how different authors understand argumentation, emphasising its importance in everyday life, science education, and its contribution to CT. Toulmin's (1958) model of argumentation and Osborne et al.'s (2016) model were shown. The first step was to identify the essential elements of an argument (evidence, warrant and conclusion) through a simple example where they observe the tracks of an animal in the snow as evidence to give an argued answer to the question, *who is walking in the snow?* (Cebrián-Robles et al., 2021a). As a second step, it was proposed to argue whether a young redhead with white skin and freckles whose image was shown was likely to get sunburnt (Cebrián-Robles et al., 2021a). They aimed to recognise the importance of refuting arguments and learning about scientific preconceptions. Different dimensions of CT were also presented, focusing on the ENCIC-CT model (Franco-Mariscal et al., 2025).

- *Task 1.* Preservice teachers must make an initial reasoned decision about the issue by showing their agreement or disagreement with the headline: *A Chinese city is building its own "artificial moon" to illuminate its streets* without the possibility of accessing any information. In addition, this headline was provided online, and the preservice teachers had to formulate three arguments for and three against, including the corresponding warrants.
- *Task 2.* Preservice teachers are provided with the same headline accompanied by two adapted readings that present opposing arguments (one for and one against) (Revista Digital El Debate, 2018; Romero, 2018) (see Annex) so that they can identify them and, if they think it is appropriate, use them to develop new written arguments on the issue. The texts contain different evidence and warrants around four key aspects for understanding the issue regarding socio-economic, technological, energy-saving and environmental impacts. For its design, a search and analysis of information from different sources was carried out to ensure that the readings included arguments for and against the issue.
- *Task 3.* Preservice teachers must return to a final decision by reasoning based on a critical analysis of the information presented in the readings. Preservice teachers were required to identify arguments for and against the issue in the readings and use them to make a final reasoned decision.
- *Task 4.* An oral discussion is held in the classroom to find out about the different positions adopted.

2.4.3 ENCIC-CT Model Application

This experience employs the ENCIC-CT model (Franco-Mariscal et al., 2025) to examine the energy issue through the potential implementation of an artificial moon to illuminate a city (Layer 1). This socio-scientific issue is particularly relevant as it spans several domains, including technological, social, economic, energy, and environmental aspects, among others. The activity is designed to develop a critical analysis of information, argumentation and decision-making dimensions within the skills domain (Layer 2). Nevertheless, it also has the potential to advance other dimensions of the ENCIC-CT model across the knowledge, skills, and disposition domains, as explored in research question RQ1. The scientific practice of argumentation (Layer 3) addresses and critically analyses the issue. Finally, media sources (Layer 4) are integrated into the teaching strategy, specifically through two reading press articles, providing a contextual foundation that enriches understanding, enhances knowledge application, and supports well-informed decision-making.

2.4.4 Data Collection and Analysis

The data collection instrument was the preservice teachers' written productions, which were generated as responses to tasks 1 and 3 described in the experience description. These written responses were systematically collected using Google Forms, ensuring that all preservice teachers' responses were stored consistently, facilitating easier analysis and review.

The productions were subjected to a rigorous qualitative analysis to identify evidence aligned with the ENCIC-CT model's dimensions. To ensure a comprehensive analysis, a collaborative and iterative approach was employed. This involved several rounds of analysis, during which the authors worked together to refine and develop analysis units. These units served as the foundational elements for identifying and categorising the various dimensions of CT present in the participants' productions. Table 2.1 presents the analysis units used.

For each participant, the frequency of each unit of analysis in each task, and whether it referenced each domain of CT (knowledge, skills, and dispositions), was calculated. Additionally, the percentage of preservice teachers in each group (primary or secondary preservice teachers) that referenced each dimension was determined.

Two analyses were conducted for each group of preservice teachers between the initial and final tasks. First, the McNemar test was used to identify statistically significant differences in each dimension of CT, as it involved related samples and dichotomous nominal variables. Second, the Wilcoxon test was applied to determine significant differences in the number of CT domains referenced in the tasks, given that the study dealt with related samples and ordinal variables.

Two statistical tests were used to analyze possible differences between the two groups of preservice teachers. The chi-square test was employed to identify significant differences in each dimension of CT in both the initial and final tasks, given that the samples were independent and the variables were nominal. Additionally, the Mann–Whitney U test was used to assess possible differences in the number of

Table 2.1 Units of analysis related to the ENCIC-CT model

Domains	Dimensions	Analysis units
Knowledge	Knowledge application	KA
	Vision of science/epistemic knowledge about science	VS
Skills	Comprehensive analysis of the problem	CAP
	Critical analysis of information	CAI
	Argumentation	A
	Decision-making	DM
	Communicative skills	CS
Dispositions	Personal autonomy/metacognition/reflection	PA
	Social engagement/activism	SE
	Emotional engagement	EE

CT dimensions referenced and the number of domains, considering that the samples were independent and the variables were ordinal.

2.5 Results

2.5.1 Results for Research Question 1

This section presents evidence found in the productions of preservice teachers regarding the development of the dimensions of CT proposed by the ENCIC-CT model, illustrating them with several examples in the different domains:

- Knowledge Domain
  - *Scientific Knowledge Application and Knowledge About Science* (KA).

    Through the readings, the activity provides the scientific knowledge involved in the socio-scientific issue (renewable energies, environmental impact, satellite deployment, etc.) that the participants must understand and apply to their arguments.

    > I am against introducing the artificial moon because it can have a considerable environmental impact. On the one hand, it modifies the photoperiod of plant organisms, which can cause an imbalance in the ecosystem. It can also negatively impact animal organisms, as many have nocturnal cycles, and humans, whose circadian cycles can be influenced and affect the production of certain hormones, such as those that encourage sleep when there is no light (Preservice secondary teacher 20).

  - *Vision of Science/Epistemic Knowledge About Science* (VS).

    This dimension is present in the issue. First, the headline of the activity and, subsequently, the information provided by the two readings should make preservice teachers reflect on the role of scientific research in implementing the artificial moon and its direct impact on environmental and socio-economic aspects.

    > Based on the information provided, I would not pursue the project. I am not saying it could not be carried out at some point. Still, first, I would insist on continuing the research and see if the moon would negatively affect certain aspects, such as fauna and flora or intense light pollution (Preservice secondary teacher 14).

    This participant applies to the analysis of this issue the need to obtain solid evidence as a characteristic feature of scientific research before considering the construction of the artificial moon.

- Skills Domain
  - *Comprehensive Analysis of the Problem* (CAP).

    The readings provide the necessary information to understand the issue from all perspectives (scientific-technological, environmental, social, economic, etc.), which should help preservice teachers form their arguments. Information is provided from opposing positions to make them reflect on different solutions.

> My position favours the artificial moon, as it could significantly reduce light pollution and non-renewable energies to invest in other renewable energies. However, in defending this, I have not considered all the living beings that could be affected, remain awake, and carry out their evolutionary process and reproduction at night. I do not consider this a problem for human beings since darkness can be achieved in many ways without the need for darkness outside, but for living beings whose habits depend on the outside world and the cycle of the day, it is a problem. Therefore, it is necessary to read much more about it to know the advantages and disadvantages of this issue and to conclude with a personal point of view subject to valid arguments (Preservice primary teacher 42).

In its response, this preservice teacher reasonably comprehensively analyses the problem by including several perspectives (environmental, social, and scientific-technological) and recognising the need for more information to obtain a complete picture of the issue.

– *Critical Analysis of Information* (CAI).

The readings' content allows them to assess the credibility of the information provided, as they include data backed up by reliable sources, such as astrophysicist Dr. Sánchez of the University of Exeter (see Annex). Also, in situations of uncertainty, preservice teachers can turn to other sources to explore the topic further.

> Firstly, accepting the widely held view that implementing an artificial moon would not cause serious environmental damage is counterproductive. The ecological impacts are severe and palpable. Secondly, the nocturnal activity of some living beings would be altered. Animal species and ourselves would be affected, so the "positive" consequences alluded to in the texts are refuted, in my opinion.
>
> Before concluding, I would like to allude to the argument about the economic savings benefits mentioned in both news items. For me, the background could be called "social-political opportunism". The second text comments on the tourist attraction that the implantation of an artificial moon would bring about. It is somewhat ironic how they make us see that, on the one hand, it is suitable for the population (because the government saves capital that it invests in public lighting), but on the other hand, they present us with the counterpoint that it benefits tourism and, therefore, the country's economy. In short, my position is contradictory since, in general terms, introducing the artificial moon in China goes against the natural and biological process of developing the species. (Preservice primary teacher 10).

– *Argumentation* (A).

It is proposed as a key dimension integrated into all the tasks set for different purposes and in two formats (written and oral). Thus, in the initial and final decision-making tasks, preservice teachers must provide written arguments supported by evidence. In the first case, based on evidence or personal experiences related to the possible impacts that the implantation of the artificial moon could produce, and, in the final decision, based on possible arguments offered in the readings. Similarly, the readings themselves, where it is not made explicit whether one is for or against, should enable participants to identify different arguments and their elements (evidence, warrant and conclusion) and the quality they present within them. The aim is for them to be able to create solid arguments once they have a broad view of the problem. The final pooling allows them to develop oral arguments and counter-arguments and to discuss them in the classroom.

> The arguments favour using this artificial moon could be stronger than the significant disadvantages it produces. The lives of human beings and many animals and plants would be seriously affected, which are fundamental to any ecosystem. These facts are hidden in the second news item, which includes arguments such as an increase in the city's economy, tourism, or energy savings. We could agree on the latter measure of saving energy and reducing the pollution generated by coal production. However, these factors cannot be compared with the repercussions and consequences for nature and, above all, for living beings, as this is even more important than the economy or tourism of a city. It would even impact astronomical research, which could find a more viable solution to create another satellite that is not so harmful to the ecosystem (Preservice primary teacher 09).

This answer shows a high level of argumentation by recognising the arguments supporting both positions (for and against the construction of the artificial moon) and counter-arguing those who take a proposition.

- *Decision-Making* (DM).

  This dimension takes place at two points: at the beginning and the end of the experience. Only the individual's ideas intervene in the initial decision, while the arguments may influence the final decision in the activity. The aim is for the preservice teacher to rely on the arguments given or on their arguments to support their initial decision or to change their position on the problem after learning about other arguments.

  > I am against artificial moon implantation. However, despite the many benefits it may have, the negative impact is more significant. Moreover, other countries could take this initiative as a reference and implement these artificial systems, which would increase the negative impact (difficulty in visualising the stars, alteration of ecosystems, alteration of citizens' circadian cycles, etc.) (Preservice secondary teacher 23).

- *Communicative skills* (CS).

  Although the experience focuses mainly on written communication, it also provides opportunities for oral communication. Thus, the sharing aims to encourage dialogue in the classroom so that preservice teachers can present their decisions using appropriate scientific language and, later, be able to do so in other contexts, using scientifically based arguments.

• Dispositions Domain

- *Personal Autonomy/Metacognition/Reflection* (PA).

  The proposed tasks are intended to enable preservice teachers to develop independent opinions and reflect on them. It is also important for them to assess whether they already have enough information to decide or reflect on their degree of certainty.

  > This is a complicated answer. Considering its effects on the environment, light pollution and possible effects on humans, no. However, there are other aspects to consider, such as the research and innovation needed to carry it out and the information we can obtain from this experience. However, there are different aspects to consider, such as the research and innovation required to carry it out and the information we can obtain

from this experience, so I would favour implementation to advance technologically (Preservice secondary teacher 10).

Some preservice teachers do this autonomously by seeking additional information. However, to promote this dimension for all participants, they can be asked to seek further information on the problem that has had the most significant impact on them.

- *Social Engagement/Activism* (SE) *and Emotional Engagement* (EE).
  These two dimensions of CT were not found in the analysis.

2.5.2 Results for Research Question 2

The analysis of the development of these dimensions about the amount of information provided to the preservice teachers revealed more concise reasoning in cases where only the headline was provided (task 1). In most cases, these reasonings focused on the information provided, as can be seen in the responses of the same participant for both moments:

> Using an artificial moon would decrease the expenditure of street lamps worldwide, as the planet would always be illuminated. (Preservice primary teacher 16, production from information provided by the newspaper headline in task 1).

> I am in favour of implementation because, firstly, through this initiative, we would be more environmentally friendly and a large part of the energy expenditure would be saved, as the satellite would act as an artificial moon and provide light for the city. Secondly, the light from the satellite would be carefully controlled so that it would only illuminate from 10 to 80 kilometres above the ground. In this way, the satellite would not illuminate outside this city. Finally, its deployment would increase tourism by becoming an attraction that would bring people from all over the world. (Preservice primary teacher 16, production from information provided by the readings in task 3).

Table 2.2 shows the findings of the analysis of the preservice teachers' productions, focusing on the different units of analysis.

Regarding each group of preservice teachers, the McNemar and Wilcoxon tests did not find statistically significant differences in preservice secondary teachers, but they did in preservice primary teachers. The McNemar test revealed statistically significant differences in the critical analysis of information (CAI) dimension ($p = 0.016$), favoring the task where the preservice primary teachers group accessed information about the problem. Similarly, the Wilcoxon test identified differences in the skills domain ($Z = -3.138$; $p = 0.002$), favoring this same task and group of preservice teachers.

The chi-square test indicated that the two groups were homogeneous, as no statistically significant differences were found between them in the initial task. However, after accessing the information, this test identified significant differences in preservice primary teachers, particularly in the knowledge application (KA) ($\chi^2 = 8.889$; $p = 0.003$) and critical analysis of information (CAI) ($\chi^2 = 4.007$; $p = 0.045$) dimensions. Furthermore, following the activity, the Mann–Whitney U test showed differences in both the knowledge ($U = 634.00$, $p = 0.013$) domain and the total

Table 2.2 Percentages of the units of analysis for both groups of preservice teachers

	CT domains	Knowledge		Skills					Dispositions		
	CT dimensions	KA	VS	CAP	CAI	A	DM	CS	PA	SE	EE
Preservice primary teacher	Initial (Task 1)	8.9	4.4	22.2	84.4	53.3	77.8	0.0	2.3	0.0	0.0
	Final (Task 3)	22.2	4.4	33.3	100.0	73.3	88.9	0.0	0.0	0.0	0.0
Preservice secondary teacher	Initial (Task 1)	8.6	2.9	20.0	80.0	40.0	88.6	0.0	0.0	0.0	0.0
	Final (Task 3)	2.9	2.9	22.9	91.4	60.0	94.3	0.0	0.0	0.0	0.0

Note KA Knowledge Application, *VS* Vision of Science, *CAP* Comprehensive Analysis of Problem, *CAI* Critical Analysis of Information, *A* Argumentation, *DM* Decision-making, *CS* Communicative Skills, *PA* Personal Autonomy, *SE* Social Engagement, *EE* Emotional Engagement

domains ($U = 635.00$, $p = 0.013$), both favoring the preservice primary teacher's group.

2.6 Discussion and Educational Implications

This chapter has highlighted the importance of classroom experiences and practical examples that immerse preservice teachers in practice and foster CT development. Emphasis has also been placed on the approach to socio-scientific issues, as they can play different roles in developing various dimensions of CT and on the need to link them to a specific approach capable of integrating them, in this case, the ENCIC-CT model.

The qualitative analysis has shown that addressing the socio-scientific issue of implementing an artificial moon to illuminate city streets can indirectly develop the ENCIC-CT model's CT dimensions, particularly in the knowledge and skills domains (RQ1).

Regarding RQ2, the findings (Table 2.2) indicate an increase in the percentage of preservice primary and secondary teachers in the development of most CT dimensions when they were provided with readings, compared to when they only had access to the news headline. These findings suggest that the experience was effective in enhancing their CT.

Decision-making (DM), critical analysis of information (CAI), and argumentation (A) are the three predominant units of analysis in both the initial and final tasks for both groups. These dimensions align with the expected outcomes of the experience's design. Notably, a significant improvement is observed in critical analysis of information (CAI), wherein 100% of the preservice primary teachers are achieved in the final task. The observed improvements in addressing the issue comprehensively suggest that the readings offered a broader perspective than what was not available from the newspaper headline.

Although the two groups show improvement across most dimensions, preservice secondary teachers begin with higher percentages in decision-making (DM) compared to preservice primary teachers. This could indicate differences in prior experience or specialized training between the groups in this dimension.

Some dimensions, such as vision of science (VS), communicative skills (CS), personal autonomy (PA), social engagement (SE), and emotional engagement (EE), show little or no change in both groups. This indicates that these dimensions were not as relevant to the tasks performed and that, despite their presence, the preservice teachers could not effectively apply them.

It is important to highlight that incorporating additional information significantly enhances the CT dimensions under analysis. Access to more extensive data and knowledge enables a deeper understanding and more effective application of these dimensions. Moreover, this additional information not only strengthens existing skills but also promotes the development of new ones, thereby expanding the range of competences. In essence, providing more information results in a qualitative improvement in the development of CT dimensions.

However, these improvements are only statistically significant for preservice primary teachers, specifically in the critical analysis of information (CAI) dimension and the skills domain, favoring the final task. Additionally, group comparisons reveal that preservice primary teachers perform statistically better than preservice secondary teachers in the final task, particularly in the knowledge application (KA) and critical analysis of information (CAI) dimensions, as well as in the knowledge domain and across the total domains. Since the described activity was the first in the training program for both groups, the observed differences could be attributed to the greater prior familiarity of the preservice primary teacher participants with the analysis of written news, particularly in language-related subjects. We have no evidence that preservice secondary teachers had engaged in similar tasks during their science degree studies.

Experiences for this purpose should provide teachers with examples of developing CT, which will help them put them into practice and design new ones. Once the experiences have been designed and evaluated, they can be transferred to other teachers (Leach & Scott, 2002). It is essential to present the experience as fully as possible, explaining the ideas underlying their design and their contribution to developing CT (Muñoz-Campos et al., 2020). It is equally important for this analysis to form part of initial and in-service training programmes for science teachers.

One limitation of the experience lies in the selection of items used. Both assume that implementing the artificial moon is technologically possible, but this may be different as it would mean putting it into a geostationary orbit, requiring a considerable moon size to capture enough light and reflect it. To assess whether implementation is possible requires advanced physical and astronomical knowledge that was outside the reach of the preservice teachers, hence the decision to present readings that assumed the project was feasible to argue the pros and cons. However, this fact was not noted by the participants, and if it had happened, it would be further evidence of the degree of development of their CT.

Finally, after analysing and discussing the findings, the authors consider that this type of experience is valid for fostering CT among students at different educational stages, not only in initial teacher training. Moreover, it would be interesting for these future teachers to offer them more opportunities to confront their points of view with those of other people. This can be done by including specific tasks that demand this confrontation beyond identifying arguments for and against the issue. In summary, it would be a matter of providing them with global training, both as learners, through their development of CT using this type of experience, and as teacher trainees, providing them with educational strategies where they can transfer the knowledge acquired to their future educational practice.

Acknowledgements This work is part of the Spanish National Plan R+D+i Project entitled "Citizens with critical thinking: A challenge for teachers in science education", reference PID2019-105765GA-I00, funded by MCIN/AEI/10.13039/501100011033.

Annex

Article 1 Published 19 October 2018 in El debate digital magazine (https://www.debate.com.mx/mundo/luna-artificial-china-lazamiento-2020-consecuencias-20181019-0055.html).

The dire consequences of the artificial moon being sent by China

The satellite was designed to replace street lighting in cities or urban areas, saving an estimated 1.2 billion yuan (US$ 170 million) annually in electricity.

China plans to launch its own "artificial moon" by 2020 to replace urban lighting and reduce electricity costs in cities, state media reported on Friday, but the idea is worrying many dark-sky advocates as it could have dire consequences for the environment because of the enormous light pollution it could cause. By reflecting sunlight, the satellites could replace urban lighting in cities or metropolitan areas, saving an estimated 1.2 billion yuan ($170 million) annually in electricity in the town of Chengdu when the artificial moon illuminates an area of 50 km^2, the official said. However, experts believe that this could have severe consequences for the environment. According to Hypertextual, one of the first to be affected would be the astronomical observatories near the area. Astrophysicist Alejandro Sanchez, at the University of Exeter and leader of the Cities at Night project, says some observatories have been on the verge of closing for much less serious reasons. A good example is Siding Spring in Australia, which had serious problems with the illumination of a flame from a nearby oil station; he explained that the brightness of the flame was equivalent to that of the Moon, not eight times brighter. Furthermore, the light tended to be reddish. In contrast, the sunlight reflected by this artificial satellite would tend to be blue, making it much more problematic for astronomical observations and the ecosystems in the area. He also explained that this new "artificial moon" illumination source would not affect observatories only. "With our information, we can foresee that the illuminated area will radically change the ecosystems present," says Alejandro Sánchez. Both plants and animals could be affected. The former, for example, would have enough light to photosynthesise at night, which would change their growth cycles; they could also flower earlier. Consequently, less cold-adapted spices would freeze during periods of low temperatures. For animals, "darkness is a refuge", said Sánchez, who has found that many species are becoming increasingly nocturnal to escape light and human activity. Deprived of this darkness, they could find it challenging to rest and feed, endangering their survival. On the other hand, by remaining in permanent dusk, some species, such as mosquitoes, would be active for many more hours, with the danger that this poses even for humans. Humans would be forced to close their blinds to sleep, as the constant light could affect their circadian rhythms, causing, at the very least, sleep problems. It is understood that circadian cycles are periods that allow synchronisation with environmental rhythms that have a period value between 20 and 28 h, such as light and temperature cycles. These are just some of the problems experts have seen with launching an "artificial moon" as China intends. It is worth remembering that China is not the first country to try to capture sunlight. In the 1990s, Russian scientists reportedly used giant mirrors to reflect light from space in an experimental project called Znamya or Banner.

Article 2 Published in Revista Digital Muy Interesante (S. Romero. https://www.muyinteresante.es/ciencia/articulo/china-tendra-su-propia-luna-artificial-291539943211).

China will have its artificial moon

The aim is, above all, ecological: to replace street lamps with this one.

Lighting a large city at night can be pretty costly and leave a massive carbon footprint, so it is worth listening to alternative ideas on how to light the streets of a city in a more environmentally friendly way. This must be what the Chinese city of Chengdu, which represents one of the three most populous cities in western China, must have been thinking. The idea to solve this problem is unusual and original, to say the least: to place a second artificial moon in the sky.

Chengdu city officials have announced plans to build an artificial moon and launch it in 2020 to hover in the sky over Sichuan province. The city has already approved a plan a private aerospace company proposed to launch this sizeable reflective satellite into orbit that could reflect light rays from the Sun for the town uninterrupted. The satellite would act as an artificial moon that would flood the city with a "sunset glow" all night long, thus rendering all streetlights and street lighting useless at night. Since coal-fired power plants generate much electricity in Chengdu, the plan offers a carbon–neutral alternative. It is an idea regarding solar energy that has yet to be considered. At least less ambitious. This artificial moon, they explain, will reduce the costs of operating and maintaining streetlights. What would this artificial moon look like? The satellite would have solar panels with a super-reflective coating. Supposedly, the illumination generated by the reflective surface would be about 8 times what one would expect from the real moon, and the light would have a similar hue to that of darkness. In addition, the light beam would be carefully controlled only to illuminate an area 10 to 80 kms above the ground. In this way, it would not be visible to anyone outside the city limits. The researchers also claim that the light would not interfere with any nocturnal wildlife activity and would have minimal ecological impact, although this remains to be seen. It is hard to imagine how an unnatural night-time glow would not have some impact on the behaviour of animals and even plants. This is something that the artificial lighting we currently have also does, they hope not only to be more environmentally friendly but also to save money on street lighting and increase tourism. This has happened before such an initiative has been carried out. In 2013, the Norwegian town of Rjukan placed three large computer-controlled mirrors on the mountains surrounding the city to track the sun's movement and reflect its rays onto the town square. Rjukan never sees the sun in the winter due to the protection of the surrounding mountains. Of course, placing mirrors on mountaintops and placing them in orbit are considerably different engineering challenges, so it will be interesting to see if this plan will finally be realised in 2020. Russia also put forward a similar project in the 1990s, with the launch of a solar reflection system, a "space mirror", intended to produce light "equivalent to three to five full moons" covering an area of approximately 5 kms. Another Russian attempt to launch a space mirror in 1999 failed before liftoff. The company developing the idea also believes this fake moon could become a tourist attraction, bringing people worldwide to experience Chengdu's unusual nightlife.

References

Ariza, M. R., Quesada, A., & Estepa, A. (2024). Promoting critical thinking through mathematics and science teacher education: The case of argumentation and graphs interpretation about climate change. *European Journal of Teacher Education, 47*(1), 41–59. https://doi.org/10.1080/02619768.2021.1961736

Blanco-López, A., España-Ramos, E., & Franco-Mariscal, A.J. (2017). Estrategias didácticas para el desarrollo del pensamiento crítico en el aula de ciencias [Teaching strategies for the development of critical thinking in the teaching of science]. *Ápice, Revista de Educación Científica, 1*(1), 107–115. https://doi.org/10.17979/arec.2017.1.1.2004

Blanco-López, A., España-Ramos, E., González, F. J., & Franco-Mariscal, A. J. (2015). Key aspects of scientific competence for citizenship: A Delphi study of the expert community in Spain. *Journal of Research in Science Teaching, 52*(2), 164–198. https://doi.org/10.1002/tea.21188

Cebrián-Robles, D., Franco-Mariscal, A. J., & Blanco-López, A. (2021a). Secuencia de tareas para enseñar argumentación en ciencias a profesorado en formación inicial a través de CoRubric. Ejemplificación en una actividad sobre una central salina [Task sequence for teaching argumentation in science to pre-service teachers through CoRubric. Exemplification in an activity about a saline power plant]. *Didáctica de las Ciencias Experimentales y Sociales, 40*, 149–168. https://doi.org/10.7203/dces.40.18178

Cebrián-Robles, D., Franco-Mariscal, A. J., Lupión-Cobos, T., Acebal, M. C., & Blanco-López, A. (Coords.) (2021b). *Enseñanza de las ciencias y problemas relevantes de la ciudadanía. Transferencia al aula* [Science education and relevant citizenship issues. Transfer to the classroom]. Graó.

DeBoer, G. E. (2011). The globalization of science education. *Journal of Research in Science Teaching, 48*(6), 567–591. https://doi.org/10.1002/tea.20421

Díaz, N., & Jiménez-Liso, M. R. (2016). Searching for socioscientific issues in the press and how often they appear. In J. Lavonen, K. Juuti, J. Lampiselkä, A. Uitto, & K. Hahl (Eds.), *Electronic proceedings of the ESERA 2015 conference. Science education research: Engaging learners for a sustainable future* (pp. 1125–1130). ESERA.

Erduran, S., & Jiménez-Aleixandre, M. P. (2008) (Eds.). *Argumentation in science education: perspectives from classroom-based research.* Springer. https://doi.org/10.1007/978-1-4020-6670-2

Evagorou, M., Jiménez-Aleixandre, M. P., & Osborne, J. (2012). Should we kill the grey squirrels? A study exploring students' justifications and decision making. *International Journal of Science Education, 34*(3), 401–428. https://doi.org/10.1080/09500693.2011.619211

Franco-Mariscal, A. J., Cano-Iglesias, M. J., España-Ramos, E., & Blanco-López, A. (2025). The ENCIC-CT model for the development of critical thinking. In A. J. Franco-Mariscal (Ed.), *Critical thinking in science education and teacher training* (Chap. 1) (pp. 3–42). Springer. https://doi.org/10.1007/978-3-031-78578-8_1

Hierrezuelo-Osorio, J. M., Franco-Mariscal, A. J., & Blanco-López, Á. (2022). Uso de dilemas socio-científicos para el desarrollo de habilidades de pensamiento crítico en docentes en formación inicial. Percepciones del profesorado [Use of socio-scientific dilemmas for the development of critical thinking skills in pre-service teachers. Teachers' perceptions]. *Revista Interuniversitaria de Formación del Profesorado, 36*(1), 99–122. https://doi.org/10.47553/rifop.v97i36.1.92435

Leach, J., & Scott, P. (2002). Designing and evaluating science teaching sequences: An approach drawing upon the concept of learning demand and a social constructivist perspective on learning. *Studies in Science Education, 38*(1), 115–142. https://doi.org/10.1080/03057260208560189

Martín-Gámez, C., & Erduran, S. (2018). Understanding argumentation about socio-scientific issues on energy: A quantitative study with primary pre-service teachers in Spain. *Research in Science & Technological Education, 36*(4), 463–483. https://doi.org/10.1080/02635143.2018.1427568

Muñoz-Campos, V., Franco-Mariscal, A. J., & Blanco-López, A. (2020). Integration of scientific practices into daily living contexts: A framework for the design of teaching-learning sequences. *International Journal of Science Education, 42*(15), 2574–2600. https://doi.org/10.1080/09500693.2020.1821932

Osborne, J. (2014). Teaching critical thinking. New directions in science education? *School Science Review, 352*, 53–62.

Osborne, J., Henderson, J. B., MacPherson, A., Szu, E., Wild, A., & Yao, S. (2016). The development and validation of a learning progression for argumentation in science. *Journal of Research in Science Teaching, 53*(6), 821–846. https://doi.org/10.1002/tea.21316

Park, M., & Liu, X. (2016). Assessing understanding of the energy concept in different science disciplines. *Science Education, 100*(3), 483–516. https://doi.org/10.1002/sce.21211

Puig, B., Blanco-Anaya, P., & Pérez-Maceira, J. J. (2021). "Fake news" or real science? Critical thinking to assess information on COVID-19. *Frontiers in Education, 6*, 646909.

Puig, B., & Jiménez-Aleixandre, M. P. (2022). *Critical thinking in biology and environmental education. Facing challenges in a post-truth world.* Springer Cham. https://doi.org/10.1007/978-3-030-92006-7

Revista Digital El Debate (2018). Las terribles consecuencias de la luna artificial que enviará China [The terrible consequences of the artificial moon that China will launch]. www.debate.com.mx/mundo/luna-artificial-china-lazamiento-2020-consecuencias-20181019-0055.html

Romero, S. (2018). China tendrá su propia luna artificial [China will have its own artificial moon]. *Revista Digital Muy Interesante.* www.muyinteresante.es/ciencia/articulo/china-tendra-su-propia-luna-artificial-291539943211

Sadler, T. D., & Zeidler, D. (2005). Patterns of informal reasoning in the context of socioscientific decision making. *Journal of Research in Science Teaching, 42*(1), 112–138. https://doi.org/10.1002/tea.20042

Solbes, J., & Torres, N. (2012). Análisis de las competencias de pensamiento crítico desde el aborde de las cuestiones sociocientíficas: un estudio en el ámbito universitario [Analysis of critical thinking competencies through the approach of socioscientific issues: A study in the university context]. *Didáctica de las Ciencias Experimentales y Sociales, 26*, 247–269. https://doi.org/10.7203/dces.26.1928

SpaceChina.com (2018). *Artificial moon to illuminate urban cities in China.* https://www.springwise.com/artificial-moon-to-illuminate-urban-cities-in-china/

Toulmin, S. E. (1958). *The uses of argument* (2003, 3rd ed.). Cambridge University Press.

United Nations (2015). *Transforming our world: The 2030 agenda for sustainable development.* A/RES/70/1. https://documents-dds-ny.un.org/doc/UNDOC/GEN/N15/291/89/PDF/N1529189.pdf?OpenElement

Vázquez, A., & Manassero, M.A. (2020). Pensamiento científico y pensamiento crítico: Competencias transversales para aprender [Scientific Thinking and Critical Thinking: Cross-Curricular Competencies for Learning]. In A. Vilches (Coord.), *Veinte años de avances y nuevos desafíos en la Educación CTS para el logro de Objetivos de Desarrollo Sostenible. VII Seminario Iberoamericano CTS* [Twenty Years of Progress and New Challenges in STS Education for Achieving Sustainable Development Goals. VII Ibero-American STS Seminar], (pp.519–522). CTS.

Vieira, R. M., & Tenreiro, C. (2016). Fostering scientific literacy and critical thinking in elementary science education. *International Journal of Science and Mathematics Education, 14*, 659–680. https://doi.org/10.1007/s10763-014-9605-2

Zeidler, D. L. (1997). The central role of fallacious thinking in science education. *Science Education, 81*(4), 483–496. https://doi.org/10.1002/(SICI)1098-237X(199707)81:4%3c483::AID-SCE7%3e3.0.CO;2-8

Chapter 3
Critical Thinking to Face Uncertain and Complex Health Issues: COVID-19 Vaccination

Blanca Puig and Noela Rodríguez Losada

> *Critical thinking does not simply place demands on students, it also requires teachers to show by example that learning in action means that not all of us can be right all the time, and that the shape of knowledge is constantly changing. (Bell Hooks, 2010, p. 10)*

3.1 Introduction

Uncertainty is part of science and represents an intrinsic characteristic of scientific activity since it is through it and the process of searching for answers to new questions that science can advance (Kampourakis & McCain, 2020). The COVID-19 pandemic made real-time science available to a large proportion of the population for the first time (Puig & Evagorou, 2023). The emergence of the novel coronavirus pandemic and the search for an effective vaccine are clear examples of situations in which science does not generate certainties or offer definitive answers but rather represents the best knowledge available at a given time (Villapol, 2023). However, citizens are not used to viewing science as a dynamic activity that includes various successes, mistakes, progresses, delays, obstacles, etc. (Kampourakis, 2018).

The development of critical thinking (hereinafter, CT) in the context of teachers' training in science education plays a crucial role in helping teachers preparing critical citizens who can recognize and addressing uncertainty, an embedded characteristic

B. Puig (✉)
Universidade de Santiago de Compostela (USC), Santiago de Compostela, Spain
e-mail: blanca.puig@usc.es

N. Rodríguez Losada
Science Education, Universidad de Málaga, Málaga, Spain
e-mail: noela@uma.es

A. J. Franco-Mariscal (ed.), *Critical Thinking in Science Education and Teacher Training*, Contemporary Trends and Issues in Science Education 64,
https://doi.org/10.1007/978-3-031-78578-8_3

of scientific activity and socio-scientific issues (hereinafter, SSI) that affect human life. This scenario highlights the need to investigate how preservice teachers tackle uncertainty in the context of addressing complex health issues that require decision-making, such as COVID-19 vaccination.

SSI comprise established scientific knowledge and other factors (such as social, ethical, and economic issues), but they also frequently entail complex and uncertain situations in which science is under construction (Latour, 2021). According to López-Fernández et al. (2022), SSI do not provide definitive answers to problems that affect us, what is difficult to manage for individuals and makes the instruction of CT essential in SSI contexts.

CT is a key competence for addressing uncertainty and confronting various conspiracy theories and types of misinformation that affect decision-making on vaccination (Kampourakis & McCain, 2020). Like any type of competence, the development of CT requires continuous practice, and it can be enhanced through argumentation (Kuhn, 2019). This chapter presents a case study in which a group of preservice preschool teachers focus on a health education-related subject that engaged them in the practice of CT with the goal of addressing complex health issues such as COVID-19 vaccination.

The work was conducted at a time of great uncertainty regarding the pandemic, namely, before the vaccines had become available to the entire population of Spain. The practice of CT has previously been explored in the context of vaccination and the COVID-19 pandemic (e.g., Duncan et al., 2022; Puig & Ageitos, 2022; Puig et al., 2021); however, as far as we know, previous research did not adopt the perspective of this study, which focuses not only on CT skills but also on attitudes.

This study seeks to expand our knowledge about the practice of CT in health contexts and to improve our understanding of how scientific and social knowledge and CT are enacted in the context of decision-making about vaccination.

3.2 Critical Thinking Aimed at Complex Health Issues

CT is identified as one of the twenty-first century competencies and understood as a set of skills and dispositions that students need to develop to make decisions and actively participate in postmodern societies (McComas, 2014). A notion of CT that is coherent with this view comes from Russell's philosophy, which anticipated many of the insights to be found in the recent CT literature, and involves not only skills and dispositions, but also attitudes. Among the skills, he included the abilities to form an opinion for oneself, to find an impartial solution, and to identify and question assumptions (Hare, 2001). According to this view, CT denotes a component of evaluation according to which knowledge, skills and attitudes are articulated, as Halpern and Danna (2023) point out.

CT is a competence that can be taught and learned in diverse disciplines and contexts. Biology education is considered a "privileged" context that encompasses a wide variety of SSI that facilitate the development of CT (Jiménez-Aleixandre & Puig, 2022).

Concurring to Halpern (1996), CT is a type of thinking that is immersed in reasoning processes that require problem solving, decision-making, establishing inferences, and recognizing the uncertainty that is inherent in science. These processes are inherent to scientific practices. However, the development of CT in the context of scientific training for students and teachers cannot be limited to scientific activity (García-Carmona, 2023). CT must be articulated in the identification of fake news and application of appropriate criteria to evaluate information. Additionally, it involves acting responsibly to face health problems that affect us, as the context of vaccination, in which CT requires recognizing social inequalities that deprive people from the most disadvantaged countries of access to vaccines. The pandemic highlighted the importance of valuing vaccines as a fundamental public right (Villapol, 2023), which is related to CT-related dispositions (Facione et al., 1995), such as having an open mind and social empathy. This view of CT, which attends to the ability to criticize social inequalities, is aligned with the characterization proposed by Jiménez-Aleixandre and Puig (2012, 2022), according to which CT includes a set of components that are related to reasoned judgment (including skills and dispositions) and another set of components that are linked to citizen action. Therefore, CT can be developed by argumentation processes (Kuhn, 2019), which, in some cases, such as in the context of this study, engage students in decision-making.

Furthermore, critical thinkers must be able to evaluate the results of their own thinking processes (Halpern, 1996), identify their own limitations (Osborne & Pimentel, 2023), and evaluate the thinking processes of others. For instance, CT requires determining how good a decision you have made has been (Halpern & Danna, 2023), being aware of how you processed the information or evidence available at the moment in question. Also, the extent to which you have recognized uncertainty as an intrinsic characteristic of science. All these aspects are particularly relevant in the context of this study, which refers to a situation in which citizens wanted certainties regarding COVID-19 vaccines, a demand that was not in line with the uncertain scientific reality. Figure 3.1 shows the knowledge, skills, attitudes (dispositions) and values that integrate the development of CT in uncertain and complex health contexts, such as the context of vaccination addressed in this chapter. Despite these dimensions are presented separated in Fig. 3.1, they can overlap.

According to Anderman et al. (2012), CT requires specific knowledge of the topic under discussion; for example, vaccination involves scientific knowledge of health-related notions such as vaccines and the immune system (theoretical knowledge), procedural knowledge concerning vaccine production, and epistemic knowledge, which refers to knowledge about the characteristics of the science underlying vaccines, such as uncertainty. However, young people are largely dependent on other people's testimony to obtain knowledge about the world (Sandoval et al., 2016), thus highlighting the social nature of children's knowledge acquisition processes (Clément, 2016). Accordingly, in coherence with epistemic cognition (Chinn et al., 2011), CT in the context of SSI also involves the articulation of social knowledge that can be retrieved from personal experiences, social media, etc., or other social

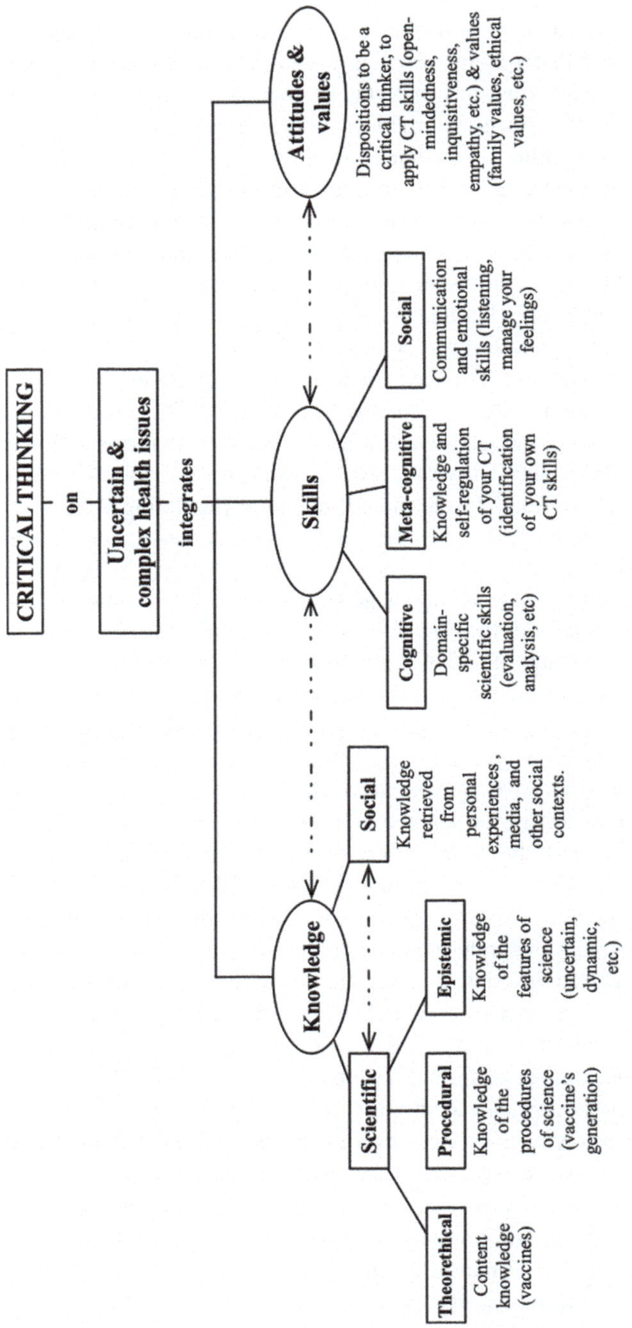

Fig. 3.1 CT in uncertain and complex health contexts as COVID-19 vaccination

sources (see Fig. 3.1). As Greene and Yu (2016) indicate, personal theories about the process of knowing and knowledge can activate or deactivate the cognitive skills that are necessary for the practice of CT, which is why CT is dependent on knowledge of the context at hand (Pithers & Soden, 2000). A previous study on CT in initial teacher education about the anti-vaccine movement (Puig & Ageitos, 2022) showed that a higher level of knowledge of vaccines did not necessarily imply greater CT skills among preservice teachers.

Complex problems featuring a high degree of uncertainty, such as the COVID-19 pandemic, require a more sophisticated CT that allows students to evaluate different perspectives (Zeidler et al., 2013). This is necessary for the stimulation of cognitive and meta-cognitive skills (Puig et al., 2021), since CT requires reasoned and reflective thinking. These skills also depend on CT-related dispositions (Facione, 1990) (see Fig. 3.1).

3.2.1 Being a Critical Thinker to Address Uncertainty Regarding Health Issues

Being a critical thinker involves being capable of reflecting on one's own beliefs and those of others and thus being able to act responsibly (Bencze et al., 2020). According to Ennis (1985), "CT is a reasonable reflective thinking that is focused on deciding what to believe or do" (p.45); therefore, being a critical thinker involves acting in accordance with this reality. It also requires individuals to be predisposed to think critically; thus, CT teaching and learning should consider the attitudes that stimulate the practice thereof (Halpern & Danna, 2023).

Despite the beneficial effects of vaccines, some people tend to deny that they work or attribute adverse side effects to them (Villapol, 2023) due to a view of science that is far from reality, understood, in terms of the domain of "right" and "wrong." (Kampourakis & McCain, 2020). The dynamic nature of scientific knowledge is widely recognized within the scientific community (Feinsten & Waddington, 2020), however, citizens do not perceive science in the same way. As Kampourakis and McCain (2020) noted, "we need to accept epistemic uncertainty and learn to live with it" (p. 28). Citizens in general tend to oppose this claim, since it is not easy to assume, such as in the case of COVID-19, that science does not know everything and that scientists may have different points of view regarding this health problem (Jiménez-Aleixandre & Puig, 2022).

To face this problem, it is important to consider that humans tend to believe what they want to believe is true as well as people's preference for anecdotes over scientific evidence (Halpern & Danna, 2023). Additionally, as Xiao and Sandoval (2017) noted, students' conceptualizations of the nature of science influence the ways in which they evaluate relevant information. For instance, if students consider SSI to be a multidimensional issue that can be interpreted in more than one way, in which context knowledge is usually subordinated to other sources of thinking, such as ethical judgments, they may exhibit a better understanding of its dynamic and uncertain nature. The problem in this context is that the science taught in school still focuses on teaching "what is true" rather on "thinking about why it is true" or "knowing and thinking what is best to do". Attitudes and perceived values in the context of SSI influence how we confront such issues and make corresponding decisions. For instance, the ways in which students perceive the science underlying vaccines may affect their decisions concerning vaccination. Teacher training should address this issue and shift toward new approaches, tools, and learning environments that provide learners with opportunities to practice CT (Vincent-Lacrin et al., 2019) in contexts that can help them understand that uncertainty is an inherent characteristic of science and serves as the engine for its continued advancement.

Bearing in mind the uncertainty that is inherent to a complex health SSI such as COVID-19 vaccination could allow people to develop a better understanding of what we should trust and what we should not trust. People who appreciate the uncertainty associated with receiving a vaccine against a disease, which has a very high probability of preventing possibly serious effects and a very low probability of leading to harmful side effects, are better prepared to make informed decisions (Kampourakis & McCain, 2020). The problem, as these authors have emphasized, is that scientific education has not focused on addressing uncertainty. In SSI instruction, this task requires, among other aspects, future teachers to be equipped with the types of knowledge, skills and attitudes that can allow them to identify uncertainty and make decisions based on the available evidence, which is one aspect of CT in this context (see Fig. 3.1).

3.3 Research Questions

The research questions (RQ) are:

- RQ1. What type of knowledge regarding vaccines is mobilized by a group of preservice preschool teachers in an uncertain context during the COVID-19 pandemic?
- RQ2. What CT skills and dispositions were displayed by participants in their arguments?

3.4 Method

This work is qualitative and uses discourse analysis to investigate the practice of CT. We are interested in exploring CT performance in teacher training in an uncertain context that required decision-making about COVID-19 vaccination in a time the vaccines had not yet been provided to the entire population, including the participants.

3.4.1 Participants and Context

The study was conducted in the second semester of the 2020–2021 academic year, i.e., during the most severe period of the COVID-19 pandemic. Participants included a group of preservice preschool teachers (N = 81) that involved mostly women (98% women, 2% men) aged between 19 and 20 years. The study was integrated into the compulsory course "Health, Hygiene and Child Nutrition", which was part of the second year of the preschool degree program. In this course, information related to health and diseases, hygiene and health habits was introduced, with a particular emphasis on the role of the teachers, as future health agents in the promotion of healthy habits among preschoolers; thus, CT was crucial for such action.

The scenario in which the study was implemented featured great uncertainty and was characterized by the pace of vaccination in Spain. Due to the situation of the pandemic at that time, teaching was developed in a dual mode; thus, in-person work sessions in small groups were combined with large group remote teaching. The distribution of the groups was determined in accordance with health measures aimed at reducing the transmission of SARS-CoV-2. A total of 20 groups consisting of five people each participated in the research while following relevant safety protocols such as wearing masks and maintaining minimum interpersonal distances. The virtual theoretical sessions were synchronous; thus, half of the preservice teachers (N = 40) attended the classroom in person, whereas the other half (N = 41) did so virtually from their homes.

3.4.2 Experience Description

This chapter describes the initial activity (Do you agree with COVID-19 vaccination? Would you do it?) about COVID-19 vaccination of a teaching sequence that was designed with the goal of engaging preservice teachers in CT through argumentation and modeling-based activities. The activity presented preservice teachers with a piece of information that had been published in the local press[1] concerning the first elderly person to be vaccinated against COVID-19 on December 27, 2020, which occurred in the same location as the study. The survey included four questions (see below Table 3.1) that asked participants to express their knowledge about the COVID-19 vaccine and their position on vaccination as well as to decide whether they would be vaccinated at that moment, simultaneously providing arguments to support their decision (Table 3.1). The news was read individually, and the participants responded to each of the questions individually. Due to the need to protect against COVID-19, the preservice teachers in the classroom wore masks and maintained wide safety distances; thus, interactions among them were limited.

[1] https://www.malagahoy.es/malaga/vacunas-Covid-llegan-residencia-Palo_0_1532546986.html

Table 3.1 Activity 1. Do you agree with COVID-19 vaccination? Would you do it?

Q1. After reading the news about the first woman to be vaccinated in our region, who was 88 years old, what do you think about the COVID-19 vaccine?
Q2. What do you know about the COVID-19 vaccine?
Q3. Do you agree with vaccination (in general, not just the COVID vaccine)? Yes/No. Provide an argument for your position
Q4. If your health center called you today to receive a vaccination, would you do it? Explain and justify your response

3.4.3 ENCIC-CT Model Application

The methodological framework is coherent with the ENCIC-CT model for CT development, discussed in Chap. 1 (Franco-Mariscal et al., 2025). This research explicitly addresses three out of the four layers that integrate this model. SSIs are essential for promoting CT and constitute the core of the ENCIC-CT model. In this study, the SSI is on the controversy of COVID-19 vaccination (Layer 1), a complex health issue that requires the articulation of knowledge, CT skills, and dispositions—three domains of CT embedded in the second layer of the model. The analysis of the two research questions focuses on this, specifically on the application of scientific and social knowledge on vaccination and health-related issues (see Fig. 3.1) and the practice of CT by preservice preschool teachers. CT skills are enacted in argumentation and decision-making processes, described in the ENCIC-CT model, and comprises a range of cognitive, meta-cognitive, and social skills, as well as dispositions necessary for the practice CT, as detailed in the model. Finally, argumentation as a scientific practice (Layer 3) is essential to this process, as it involves the formulation and evaluation of evidence-based claims and counterclaims. This practice facilitates a deeper understanding of complex issues and fosters the ability to reason critically about them.

3.4.4 Data Collection and Analysis

The data collected for this research included individual written reports of activity 1. The analysis of RQ1, i.e., *What type of knowledge regarding vaccines is mobilized by a group of preservice preschool teachers in an uncertain context during the COVID-19 pandemic?* involves the examination of individual responses to Q2 (see Table 3.1). First, preservice teachers' responses were divided into two main categories: (A) scientific knowledge and (B) social knowledge (Fig. 3.1) in an iterative process that involved an analysis of the content of individual written responses. Second, different subcategories emerged in interaction with data. For category A, the process was inductive, since we draw from previous categories defined in PISA 2015 science framework (OECD, 2018). Particularly, we based upon the view of scientific knowledge as "consisting of three distinguishable but related elements: content knowledge, procedural knowledge and epistemic knowledge" (OECD, 2018, p. 74). For category B, the process of analysis was deductive, categories were not pre-established, thus, they emerged in interaction with data.

Category A comprises these three subcategories: (A.1) *theoretical,* which corresponds to preservice teachers' scientific knowledge about vaccines and/or immunization; (A.2) *procedural,* which refers to knowledge of the procedures followed to produce vaccines; and (A.3) *epistemic*, which pertains to knowledge of the features of science. We investigated whether the preservice teachers' responses included some characteristics of the science underlying the production of vaccines. Specifically, if these responses referred to the provisional nature of scientific knowledge of vaccines at that time, the changing nature of such knowledge, its dependence on the evidence available at that time, the existence uncertainty, or similar issues. It was viewed as necessary to consider the historical moment in which the study was conducted. Responses that referred to knowledge regarding the speed at which the vaccines were developed were also included in this subcategory.

Regarding subcategory (A.1) *Theoretical knowledge*, two levels were distinguished: *immature* knowledge vs. *mature* knowledge. *Immature* knowledge "of" science includes responses that merely mentioned scientific notions without explaining them, in contrast to *mature* knowledge, which developed these notions more fully. For example, biological processes related to cellular infection, dissemination, or resolution of a viral infection were considered in this level.

Category B, Social Knowledge, includes responses that used scientific terms when the source of such knowledge was preservice teachers' previous experiences, others' experiences or information disseminated by the media (including internet, social networks, and newspapers).

The analysis of RQ2, i.e., *What CT skills and dispositions were displayed by participants in their arguments concerning COVID-19 vaccination?* focused on the examination of individual responses to Q3 (see Table 3.2). Content analysis (Hsieh & Shannon, 2005) was used to identify CT skills and dispositions based on Facione's framework (1990) of CT skills and dispositions. The categories were then adapted considering a previous study conducted by Puig and Ageitos (2022).

The data were coded and analyzed independently by the two authors with respect to both research questions. Categories were also identified in each response to calculate the final frequencies shown in Tables 3.2 and 3.3 in the Results section.

3.5 Results

3.5.1 Knowledge of Vaccines and COVID-19 Vaccination

The results regarding RQ1 are summarized in Table 3.2. Notably, the same response could be coded into several categories and subcategories.

Table 3.2 Results of type of knowledge mobilized by participants

Categories	Subcategories	Levels or aspects	Frequencies (%) (N = 81)
A. Scientific knowledge	A.1 *Theoretical*	Mature knowledge	12
		Immature knowledge	24
	A.2 *Procedural*	Production of vaccines	7
	A.3 *Epistemic*	Uncertainty of vaccines	24
B. Social knowledge	B.1 *Retrieved from personal experiences*		11
	B.2 *Based on media information*		51

Table 3.3 CT skills and dispositions identified in preservice teachers' answers to Q3

CT skills	Frequencies (%)	CT-related dispositions	Frequencies (%)
Interpretation	54	Self-confidence	60
Evaluation	50	Analyticity	30
Explanation	44	Open-mindedness	20
Inference	26	Inquisitiveness	10
Self-regulation	8		

A. Scientific Knowledge

The findings show that participants used theoretical, procedural, and epistemic knowledge, with epistemic knowledge and theoretical knowledge being the most frequent and procedural knowledge being the least frequent.

A.1 *Theoretical Knowledge*

Most of the participants (24%) used immature knowledge, i.e., they mentioned scientific notions but did not develop them, whereas a minority (12%) of them were coded as using "mature" knowledge. Examples are discussed below.

Immature knowledge:

> What the vaccine contains is not the virus itself but rather a modification created in a laboratory (Preservice teacher 09).

This preservice teacher noted that the COVID-19 vaccine is not of viral origin "but a modification created in the laboratory." Although she included scientific terms such as "virus" in her response, she did not develop her response. This quotation indicated that "the vaccine is not the virus itself" but did not explain its composition or the notion of an "artificial" origin, nor did it provide further specification.

Mature knowledge:

> A vaccine consists of inoculating your body with a part of the virus or bacteria that causes the infection with little or no infectious capacity. After injection, the cells in charge of our immune defense respond by generating and producing antibodies and memory T cells. These memory cells are responsible for recognizing viruses and knowing how to act if you become infected, so normally you can pass the disease with very mild symptoms. If these cells are not produced, the disease can reinfect you, leading to all kinds of symptoms [...] (Preservice teacher 01).

This example reflects mature knowledge regarding COVID-19 vaccine and immunization. The preservice teacher explained how the immune system responds to vaccination. She employed scientific terminology, such as viruses, antibodies, and memory T cells. She used these notions appropriately to explain the process of immunization.

A.2 ***Procedural Knowledge***

The survey included responses that referred to procedures, particularly the tests and safety measures that are needed to incorporate any vaccine into a vaccination program, as the example that follows.

> Personally, I believe that if this vaccine goes onto the market, it would be after passing the appropriate clinical tests successfully (Preservice teacher 46).

A.3 ***Epistemic Knowledge***

It corresponds to responses that mentioned features of science beyond the level of vaccines, particularly those pertaining to uncertainty. Table 3.1 shows that these aspects were considered by 24% of participants. Specifically, uncertainty, which was linked to potential side effects and their effectiveness regarding preventing the contagion and transmission of viruses among people, as these examples show.

> These vaccines could have side effects, but for now… not many have come to light, or those that have come to light are not true (Preservice teacher 66).

> This vaccine, after being tested on certain people, has been successful, although it might cause some side effects similar to those that COVID-19 can give you… (Preservice teacher 51).

B. **Social Knowledge**

This category contains responses that mobilized knowledge drawn from social sources, such as media, personal experiences or a combination of both. The participants explicitly mentioned the source of information, as shown in the example below.

> The information I have about vaccines comes from the news. It is a source that the whole world has, and today, we give a lot of importance to everything they report about COVID and the information I have about vaccines indicates that they have a high percentage of validity for people […] (Preservice teacher 15).

3.5.2 Critical Thinking in Practice

Table 3.3 displays the CT skills and dispositions identified in the preservice teachers' responses to Q3 (see Table 3.1). Several categories of skills and dispositions may be associated with the same response.

3.5.2.1 CT Skills

Participants were able to perform six type of CT skills; *interpretation and evaluation* were the most frequently appearing skills, followed by *explanation,* as Table 3.3 shows. In contrast, *self-regulation* was the least frequently appearing skill. These skills are discussed alongside examples in the following paragraphs.

Interpretation: To express the meaning or significance of COVID-19 vaccines during the pandemic.

> Yes, I support vaccination because with the vaccines, we can solve the pandemic or... at least reduce its effects. The high number of deaths should be enough to prevent people from questioning vaccination (Preservice teacher 13).

This preservice teacher presents her position regarding COVID-19 vaccination and expresses her understanding of the effects of the vaccine. She considers them to be a solution to the pandemic or a way of mitigating its effects. She also provides an example based on the number of deaths, which supports her position and shows how this preservice teacher related the lack of vaccines to deaths caused by COVID-19.

Evaluation: To evaluate the role of vaccines or vaccination, particularly regarding their potential benefits or side effects.

> I agree with vaccination because it is the best method to become immunized against a disease. Thanks to vaccinations, most people worldwide do not catch viruses or diseases that are present around the world, and, above all, they help people have a better life (Preservice teacher 17).

This preservice teacher supported vaccination and justified her position since *people do not catch viruses or diseases that are present around the world*; she viewed vaccination as beneficial for people's health.

Explanation: To justify your position regarding vaccination and vaccines by using previous experiences or other sources of data as evidence.

> In general, yes, I agree (with COVID-19 vaccination), since in my case I have been vaccinated at all stages of my life, and I did not have any problems, for instance, with the flu or with chickenpox. I was in contact with relatives who were ill when I was young, and because I was vaccinated, nothing happened to me. In my opinion, vaccines help, at least, to avoid aggravating certain diseases (Preservice teacher 12).

This preservice teacher agreed with the COVID-19 vaccination and justified her position based on her previous experience with vaccines for other infectious diseases. She used as evidence of the benefits of vaccines the fact that she had not been infected with diseases when she encountered relatives who had been infected by them.

Inference: To develop conjectures or hypotheses regarding COVID-19 vaccination and to identify and consider relevant information to draw reasonable conclusions.

> I totally agree; I believe that it is of great importance (being vaccinated) because it helps prevent diseases and avoid suffering their consequences. In addition, it is also an act of solidarity because it helps other people not be affected by these diseases. For example, if parents vaccinate their children when they go to school, they will be protected, and they will also protect younger children who have not yet received those doses, who will be safe from contracting the disease because everyone around them will be immunized (Preservice teacher 09).

This preservice teacher expressed his position on vaccination. He considered it to be important to *prevent diseases and avoid suffering their consequences* and to protect others. Considering this information and using vaccination in schools as an example, he concluded that vaccination could generate social immunity.

Self-regulation: To express explicitly your own knowledge on vaccination and vaccines and to question your own judgments and position on the COVID-19 vaccine based on your own cognitive limitations and ability to understand and/or critically evaluate information spread by media.

> I believe that vaccines are fundamental and that people who decide not to vaccinate themselves or their children can probably survive thanks to the vaccination of the rest of the population. People who deny this scientific evidence (I am referring to the need for vaccination) do so because of their ignorance, since it is a way of training our body, which is like teaching a person self-defense. However, I cannot lie and say that I have clear thoughts and knowledge regarding the COVID-19 vaccine since I feel some concern about the bombardment of information from social networks, which adopts different points of view about this virus in contrast with the very selective information that is provided to us through the media (Preservice teacher 20).

The preservice preschool teacher considered vaccines to be fundamental and questioned people who would deny the "scientific evidence" and benefits related to such vaccines. She used "teaching a person self-defense" as an analogy for vaccination. However, despite her clear position on vaccines, she expressed her own doubts regarding the COVID-19 vaccine due to the amount and different types of information disseminated by social media in contrast to other media.

3.5.2.2 CT Attitudes (Dispositions)

Participants displayed several attitudes linked to CT development (Fig. 3.1), four in total (Table 3.2). *Self-confidence* was the most frequent disposition (60%), while *inquisitiveness* was least frequent (10%).

Self-confidence: To trust one's own position and judgments concerning vaccines or vaccination.

> Clearly this question (about COVID-19) vaccination for me is a resounding yes! First, among many other reasons, because vaccines are one of the tools, we must fight diseases and save lives (Preservice teacher 01).

In his answer, this preservice teacher exhibited self-confidence regarding COVID-19 vaccination and considered vaccines to be tools that could be used "to fight diseases and save lives".

Analyticity: The disposition to use reasons and evidence concerning vaccination to generate expectations regarding its potential benefits among the population.

> I believe that the vaccine will be a method to ensure that a smaller population becomes infected with the disease, and it would be difficult for there to be as many infections as there are now. In addition, we would prevent older people from continuing to die and have less impact on young people's health since it has been noted that this disease has serious consequences (Preservice teacher 15).

This preservice teacher analyzed the effects of COVID-19 vaccination and generated expectations regarding its potential benefits on older and younger people's health.

Open-mindedness: To exhibit tolerance and empathy regarding divergent views, beliefs and even others' decisions concerning vaccination.

> Yes, I agree with vaccination since it can make the difference between a disease being fatal or not. However, I respect people who do not want to be vaccinated and who do not vaccinate their children until they are old enough to decide (Preservice teacher 42).

This preservice teacher agreed with vaccination and expressed tolerance for others' decisions, particularly people who did not want to vaccinate their children.

Inquisitiveness: To exhibit intellectual curiosity or the desire to learn more about vaccines. To be open to the generation of new knowledge.

> I believe that it (the COVID-19 vaccine) should be studied and improved, but if it does not have negative consequences and is somewhat effective, then even if it is not definitive, it is progress (Preservice teacher 34).

This preservice teacher noted that the development of COVID-19 vaccines was in progress and that new knowledge was being generated. The response reflects an open attitude towards the development of knowledge on COVID-19 vaccines.

3.6 Discussion and Educational Implications

In this chapter, we presented an experience of CT that was used to identify the knowledge, skills and attitudes that are integrated into the practice of CT in the context of uncertain and complex health issues, particularly the COVID-19 vaccination. The purpose of this study was to explore how these dimensions of CT were articulated by a group of preservice preschool teachers and to explore how they managed uncertainty. That is, this chapter aimed to explore whether participants accepted uncertainty as an intrinsic feature of the science underlying vaccines or not.

Participants were able to articulate diverse types of knowledge (RQ1), although most of the theoretical knowledge they expressed was *immature*, which is coherent with the context in which this study was conducted; namely, the development of COVID-19 vaccines was ongoing at the time of this research.

Concerning the epistemic knowledge, uncertainty was the only feature that was acknowledged by participants in their responses; they recognized the benefits of vaccination about protecting the population against the coronavirus, slowing contagion, and controlling the pandemic, despite the potential for side effects (which characterize every vaccine). We concur with Kampourakis and McCain's (2020) claim that science education should address the uncertainty associated with SSIs since CT development can be affected by this issue, among other factors. Further research is needed to examine how preservice teachers' acceptance of uncertainty in the context of SSI knowledge affects their CT, which can be assessed by reference to their CT skills and dispositions, as suggested by previous studies developed in our research group on CT in the context of health issues (Puig & Ageitos, 2022; Puig et al., 2021).

Notably, when this study implemented, a great deal of disinformation regarding COVID-19 was spreading through social media; thus, the identification of credible information concerning vaccines was difficult. Preservice teachers' articulation of social knowledge allowed us to determine that most of their knowledge was drawn from media rather than their personal experiences. Epistemic cognition emphasizes the social sources of knowledge and critical source evaluation (Chinn et al., 2011) since selective trust in the sources of knowledge facilitates preservice teachers' understanding of multiple conflicting perspectives on controversial SSIs. Although this topic was not the focus of this analysis, most preservice teachers did not seem to question the source of their knowledge; therefore, a deep examination of the ways in which future teachers come to trust or mistrust sources of knowledge (Sandoval et al., 2016) is also needed to enrich our understanding of CT in this context.

The analysis of CT allowed us to confirm that participants were able to exhibit diverse CT skills, as the results of a previous study on the anti-vaccine movement (Puig & Ageitos, 2022). However, in contrast to this study, in which *explanation* was the most frequent skill, *interpretation* and *evaluation* were more common here. This difference can be explained by the contexts and questions posed, which drove preservice teachers to apply these skills. Concerning the analysis of attitudes toward CT attitudes, it is notable that despite the uncertain situation related to vaccines, preservice teachers showed *self-confidence* in their responses to and position on the notion of receiving the vaccine. In fact, they did not express any need to consult experts to improve their knowledge or make decisions concerning vaccination, something that appeared in the previous study. *Inquisitiveness* was not frequent, which might be related to this result.

At the time of this study, individuals were required to make decisions on COVID-19 vaccination and to put the benefit of society above their individual benefit (Iordanou, 2022). In this work, we asked participants what they would do, and all of them were open-minded regarding receiving a vaccination against COVID-19 and embracing their social responsibility.

CT involves reflective thinking that can help individuals decide what to believe or to do (Ennis, 1985). The main question associated with our study and under investigation by the authors is: how we can prepare future teachers to be aware of their own CT following a developmental progression, as suggested by Kuhn (2022). Being a critical thinker involves various forms of meta-level thinking (Kuhn, 2022), in which context epistemological thinking related to students' view of their own knowledge and the process of knowing are crucial to their development of CT.

Acknowledgements Work supported by the Spanish National Plan R+D+ i Projects: "Critical thinking for action to face emergent real-world challenges in science education", reference PID2022-138166NB-C21, funded by MCIN/AEI/10.13039/50110001103/FEDER, developed at Universidade de Santiago de Compostela (USC) in RODA (Reasoning, Discourse and Argumentation) research group, GI-1667 and "Citizens with critical thinking: A challenge for teachers in science education", reference PID2019-105765GA-I00, funded by MCIN/AEI/10.13039/501100011033 supported by University of Málaga, UMA-Tech, group PAIDI-CTS156.

References

Anderman, E. M., Sinatra, G. M., & Gray, D. L. (2012). The challenges of teaching and learning about science in the twenty-first century: Exploring the abilities and constraints of adolescent learners. *Studies in Science Education, 48*(1), 89–117. https://doi.org/10.1080/03057267.2012.655038

Bencze, J. L., El Halwany, S., & Zouda, M. (2020). Critical and active public engagement in addressing socioscientific problems through science teacher education. In M. Evagorou, J. A. Nielsen, & J. Dillon (Eds.), *Science teacher education for responsible citizenship. Contemporary trends and issues in science education* (Vol. 52, pp. 63–83). Springer. https://doi.org/10.1007/978-3-030-40229-7_5

Chinn, C. A., Buckland, L. A., & Samarapungavan, A. (2011). Expanding the dimensions of epistemic cognition: Arguments from philosophy and psychology. *Educational Psychologist, 46*(3), 141–167. https://doi.org/10.1080/00461520.2011.587722

Clément, F. (2016). *Social cognition.* In J. A. Greene, W. A. Sandoval & I. Bråten (Eds.), *Handbook of epistemic cognition* (pp. 86–99). New York, NY: Routledge.

Duncan, R. G., Caver, V. L., & Chinn, C. A. (2022). The role of evidence evaluation in critical thinking. In B. Puig, & M. P. Jiménez-Aleixandre (Eds.), *Critical thinking in biology and environmental education. Facing challenges in a post-truth world* (pp. 75–92). Springer. https://doi.org/10.1007/978-3-030-92006-7

Ennis, R. (1985). A logical basis for measuring critical thinking skills. *Educational Leadership, 43*(2), 44–48.

Facione, P. A. (1990). *Critical thinking: A statement of expert consensus for purposes of educational assessment and instruction.* California State University.

Facione, P. A., Sánchez, C. A., Facione, N. C., & Gainen, J. (1995). The disposition toward critical thinking. *Journal of General Education, 44*(1), 1–25. https://www.jstor.org/stable/27797240

Feinstein, N. W., & Waddington, D. I. (2020). Individual truth judgments or purposeful, collective sensemaking? Rethinking science education's response to the post-truth era. *Educational Psychologist, 55*(3), 155–166. https://doi.org/10.1080/00461520.2020.1780130

Franco-Mariscal, A. J., Cano-Iglesias, M. J., España-Ramos, E., & Blanco-López, A. (2025). The ENCIC-CT model for the development of critical thinking. In A. J. Franco-Mariscal (Ed.), *Critical thinking in science education and teacher training* (Chap. 1) (pp. 3–42). Springer. https://doi.org/10.1007/978-3-031-78578-8_1

García-Carmona, A. (2023). Scientific thinking and critical thinking in science education. *Science & Education.* https://doi.org/10.1007/s11191-023-00460-5

Greene, J. A., & Yu, S. B. (2016). Educating critical thinkers: The role of epistemic cognition. *Policy Insights from the Behavioral and Brain Sciences, 3*(1), 45–53. https://doi.org/10.1177/2372732215622223

Halpern, D. F. (1996). *Thought and knowledge: An introduction to critical thinking* (3rd ed.). Lawrence Erlbaum Associates.

Halpern, D. F., & Danna, D. S. (2023). *Though and knowledge. An introduction to critical thinking* (Vol. 2, 6th ed.). Routledge.

Hare, W. (2001). Bertrand Russell on critical thinking. *Journal of Thought, 36*(1), 7–16. http://www.jstor.org/stable/42589643

Hsieh, H. F., & Shannon, S. (2005). Three approaches to qualitative content analysis. *Qualitative Health Research, 15*(9), 1277–1288. https://doi.org/10.1177/104973230527668.

Iordanou, K. (2022). Supporting critical thinking through engagement in dialogic argumentation: Taking multiple considerations into account when reasoning about genetically modified food. In B. Puig, & M. P. Jiménez-Aleixandre (Eds.), *Critical thinking in biology and environmental education. Facing challenges in a post-truth world* (pp. 93–111). Springer Cham. https://doi.org/10.1007/978-3-030-92006-7_6

Jiménez-Aleixandre, M. P., & Puig, B. (2012). Argumentation, evidence evaluation and critical thinking. In B. Fraser, K. Tobin, & C. McRobbie (Eds.), *Second international handbook for science education* (Vol. 2, pp. 1001–1017). Springer.

Jiménez-Aleixandre, M. P., & Puig, B. (2022). Educating critical citizens to face post-truth: the time is now. In B. Puig, & M. P. Jiménez-Aleixandre (Eds.), *Critical thinking in biology and environmental education. Facing challenges in a post-truth world* (pp. 3–19). Springer. https://doi.org/10.1007/978-3-030-92006-7

Kampourakis, K. (2018). Science and uncertainty. *Science & Education, 27*(9), 829–830. https://doi.org/10.1007/s11191-018-0019-3

Kampourakis, K., & McCain, K. (2020). *Uncertainty*. Oxford University Press.

Kuhn, D. (2019). Critical thinking as discourse. *Human Development, 62*(3), 146–164. https://doi.org/10.1159/000500171

Kuhn, D. (2022). Metacognition matters in many ways. *Educational Psychologist, 57*(2), 73–86. https://doi.org/10.1080/00461520.2021.1988603

Latour, B. (2021). *After lockdown: A metamorphosis* (J. Rose, Trans.). Polity Press.

López-Fernández, M. M., González-García, F., & Franco-Mariscal, A. J. (2022). How can socioscientific issues help develop critical thinking in chemistry education? A reflection on the problem of plastics. *Journal of Chemical Education, 99*(10), 3435–3442. https://doi.org/10.1021/acs.jchemed.2c00223

McComas, W. F. (2014). Critical thinking. In W. F. McComas (Ed.), *The language of science education*. Sense Publishers. https://doi.org/10.1007/978-94-6209-497-0_24

OECD. (2018). PISA for development science framework. In *PISA for development assessment and analytical framework: Reading, mathematics and science*. OECD Publishing. https://doi.org/10.1787/9789264305274-en

Osborne, J., & Pimentel, D. (2023). Science education in an age of misinformation. *Science Education, 107*(3), 553–571. https://doi.org/10.1002/sce.21790

Pithers, R. T., & Soden, R. (2000). Critical thinking in education: A review. *Educational Research, 42*(3), 237–249. https://doi.org/10.1080/001318800440579

Puig, B., & Ageitos, N. (2022). Engaging in argumentation as critical evaluation of the anti-vaccination movement. In K. Korfiatis, & M. Grace (Eds.), *Current research in biology education. Contributions from biology education research* (pp. 19–29). Springer. https://doi.org/10.1007/978-3-030-89480-1_2

Puig, B., & Evagorou, M. (2023). COVID-19: A context to promote critical thinking and argumentation in secondary and university students. In N. Rezaei (Ed.), *Brain, decision making and mental health. integrated science* (Vol. 12, pp. 219–236). Springer. https://doi.org/10.1007/978-3-031-15959-6

Puig, B., Blanco-Anaya, P., & Pérez-Maceira, J. J. (2021). "Fake news" or real science? Critical thinking to assess information on COVID-19. *Frontiers in Education, 6*, 646909. https://doi.org/10.3389/feduc.2021.646909

Sandoval, W. A., Greene, J. A., & Bråten, I. (2016). Understanding and promoting thinking about knowledge: Origins, issues, and future directions of research on epistemic cognition. *Review of Research in Education, 40*(1), 457–496. https://doi.org/10.3102/0091732X16669319

Villapol, S. (2023). *Consciencia. A ciencia que nos salvou da pandemia.* Xerais.

Vincent-Lacrin, S., González-Sancho, C., Bouckaert, M., de Luca, F., Fernández-Barrera, M., Jacotin, G., Urgel, J., & Vidal, Q. (2019). *Fostering students' creativity and critical thinking: What it means in school.* OECD Publishing. Educational Research and Innovation. https://doi.org/10.1787/62212c37-en

Xiao, S., & Sandoval, W. A. (2017). Associations between attitudes towards science and children's evaluation of information about socioscientific issues. *Science & Education, 26*(3–4), 247–269. https://doi.org/10.1007/s11191-017-9888-0

Zeidler, D. L., Berkowitz, M. W., & Bennett, K. (2013). Thinking (scientifically) responsibly: The cultivation of character in a global science education community. In M. P. Mueller, D. J. Tippins, & A. J. Stewart (Eds.), *Assessing schools for generation R (responsibility) a guide for legislation and school policy in science education* (pp. 83–99). Springer. https://doi.org/10.1007/978-94-007-2748-9_7

Chapter 4
Personal Autonomy and Decision-Making in Preschool Through the Naturalization of Spaces with the Think-Pair-Share Technique

Alba Ramos-Solano and Daniel Cebrián-Robles

4.1 Introduction

Developing scientific competences and critical thinking (hereinafter, CT) is a fundamental pillar for sustaining full and responsible citizenship (Blanco-López et al., 2015). From an early age, the unquestionable need for all people to acquire the ability to approach with rigour and discernment the multiple socio-scientific controversies present in their local and global context (Levinson, 2006) is particularly evident when considering some of the main environmental challenges we face. In this sense, as Lipman (2003) points out, CT is a fundamental skill in forming a citizenry capable of approaching different social situations rationally and democratically. It constitutes a central axis of scientific practice (Osborne, 2014).

Defining CT is a complex task, and its conception has evolved and adopted different perspectives (Solbes & Torres, 2012). It is worth highlighting the paradigm shift between McPeck (1981), who defines it as the individual's willingness and ability to engage in an activity with reflective scepticism, passing through the complexity of the set of dispositions and skills encompassed in CT, according to Facione (1990), and Kuhn's (2019) conceptualisation as a social and dialogical activity that is developed in continuous interaction with others, this being the notion that currently prevails in the context of science education (Puig et al., 2023). Furthermore, Bargiela et al. (2022) identify that the need for context-specific knowledge is a common element in many of the characterisations of CT made by various authors,

A. Ramos-Solano (✉) · D. Cebrián-Robles
Science Education, Universidad de Málaga, Málaga, Spain
e-mail: albaramsol@uma.es

D. Cebrián-Robles
e-mail: dcebrian@uma.es

A. J. Franco-Mariscal (ed.), *Critical Thinking in Science Education and Teacher Training*, Contemporary Trends and Issues in Science Education 64,
https://doi.org/10.1007/978-3-031-78578-8_4

pointing out the importance of jointly considering factors from the affective and social domains. In line with this, some of the most recent characterisations of CT believe various facets relating to cognitive skills, dispositions, and cultural and social aspects, including democratic participation and social justice (Puig & Jiménez-Aleixandre, 2022).

Therefore, when considering a methodology for developing CT in science education, as suggested by Vila et al. (2023), it is essential to create a classroom environment that fosters communication, collaboration and dialogue between peers, along with authentic problem-solving, such as issues related to and affecting society. Although this can be particularly challenging in preschool and research on the development of CT at this stage is relatively scarce (Bargiela et al., 2022), some work has identified how early learners can develop different abstract thinking and argumentation skills (Lipman et al., 2010) and how teachers can incorporate different teaching strategies for the development of CT in preschool classrooms (León, 2015).

In order to promote CT in preschoolers, this paper proposes the initiative of involving them in discussion and decision-making about the organisation of their classroom (Manassakis, 2019) and, specifically, about the creation of a corner (Bottini & Grossman, 2005) dedicated explicitly to nature (Alabay, 2009). This initiative is underpinned by the need for place-based learning (Smith & Sobel, 2010), enabling students to connect with their natural and cultural environment. It is further supported by the importance of creating a scientific environment in preschool classrooms that fosters curiosity-based science experiences (Tu, 2006). The potential of naturalised spaces for developing scientific competences in the classroom is also highlighted (Sanz et al., 2021). Incorporating a nature space in the classroom would allow them to explore and experiment directly with elements of nature, which, according to Hodson (2011), could contribute significantly to developing their environmental awareness and re-evaluating their environmental values. In previous work, Ramos-Solano (2022), analysing good practices aimed at naturalising educational spaces, pointed to incorporating a nature corner as one of the most straightforward and direct strategies to encourage children's direct contact with nature and thus strengthen their educational experience.

To achieve the goal of involving preschoolers in the discussion and decision-making about the incorporation of a nature corner in their classroom, it is interesting to consider cooperative learning methods, such as the *Think-Pair-Share* (TPS) technique (Sharma & Saarsar, 2018). This technique is organised in three phases. Students must begin by individually reflecting (*thinking*) on an issue and then *pair up* with a partner and agree on a common position they must finally *share* with the rest of the class. This technique has been used with very positive results to promote the CT development at the university level (Kaddoura, 2013) and in primary education (Silva et al., 2022). To our knowledge, previous studies have yet to implement it in preschool for this purpose. Therefore, the present study is an exploratory but pioneering inquiry into the feasibility and potential of using the TPS technique to develop CT in preschool. Furthermore, to improve the adaptation of this technique to the ages of the children for whom the experience is intended, modeling elements were introduced (drawings, plasticine, models, etc.) whose potential at this stage has recently been highlighted, among others, by Cruz-Guzmán and Martínez-Maqueda (2022).

4.2 Personal Autonomy and Decision-Making as Critical Thinking Dimensions to Be Developed with the Think-Pair-Share Technique

Given the breadth and complexity of the construct of CT, it is necessary to break it down into more specific subcategories, whether skills and dispositions (Ennis, 2015; Facione, 2023), competences (Paul & Elder, 2007) or components (Jiménez-Aleixandre & Puig, 2022), which allow, among others, to guide and orient in a more concrete way the training proposals that pursue its development. Complementing the above, the ENCIC-CT model for science education (Franco-Mariscal et al., 2025) (Chapter 1) includes personal autonomy as a disposition and decision-making as a skill, among other dimensions.

On the one hand, personal autonomy involves developing an independent opinion and acquiring the power to reflect on and participate in society and, on the other hand, decision-making consists of making rational choices and informed judgements as elements of the decisions they use to solve problems (Blanco-López et al., 2017, p. 109). Both dimensions are closely interrelated, as fostering personal autonomy promotes decision-making, and decision-making should become more informed and responsible as children grow older. In turn, the ability to make decisions and think independently is closely linked to autonomy (Kamii, 1991), as it implies the ability to process information, evaluate options and make appropriate decisions to solve problems or cope with everyday situations. Delval (2006) emphasises that the relationship between autonomy and decision-making is manifested through responsibility, understood as the ability to decide freely and to assume the consequences of such decisions. On the other hand, these two dimensions align with one of the components that Jiménez-Aleixandre and Puig (2022) relate to citizenship education, which consists of forming independent opinions and questioning one's group or pre-established ideas.

In terms of educational strategies to promote autonomy in childhood, Piaget (1973) highlighted the importance of creating a learning environment that fosters decision-making and independence, guaranteeing the active participation and collaboration of students throughout the life of the school. On the other hand, decision-making is, according to Rodríguez-Melero et al. (2023), a CT skill that is especially challenging to develop in preschool, as it involves, in turn, other high-level cognitive skills such as argumentation. In this regard, and despite the lack of corroborating studies, the TPS method can play a valuable role, as it teaches children to reason, analyse, and reflectively make decisions, involving them in issues that affect them in their daily lives.

4.3 Research Questions

This chapter aims to study the TPS technique as an educational strategy to develop CT corresponding to personal autonomy and decision-making in preschoolers through democratic participation and involvement in the naturalisation of their classroom. To achieve this objective, the following research questions (RQ) are specifically addressed:

- RQ1: To what extent does the TPS technique develop the decision-making autonomy of preschoolers in dealing with the naturalisation of their classroom?
- RQ2: What natural elements are proposed and incorporated by most preschoolers when they are involved in the process of reflection and decision-making about the spatial organisation of their classroom?

4.4 Method

This qualitative research is a case study with an exploratory design (Yin, 2013). By the characteristics of this research methodology (Lederman et al., 2023), an in-depth study of the potential of the TPS technique is carried out in the real context of a preschool classroom, paying attention to the specific characteristics of this educational context to investigate broader aspects of the CT development.

4.4.1 Participants and Context

The instructional module of this study was carried out during the 2021–2022 school year with a group of 24 children aged 5 to 6 years from the Huertas Viejas preschool in Coín (Málaga), a medium-sized rural municipality in southern Spain. In this instructional module, the aim is to develop the children's autonomy and decision-making. In addition, the instructional module served as a preliminary phase to a subsequent training programme focused on the collective activism approach based on inquiry (Cebrián-Robles et al., 2021), in which the students, together with their families and the educational community, faced a socio-scientific issue in their immediate environment and culminated, after a process of inquiry, consensual planning and decision-making, with a step towards real action.

The design of the instructional module in this study consists of two different activities in which the TPS technique was adapted in contrasting ways (Fig. 4.1).

The first activity asks what the children want to incorporate into their classroom, simplifying the TPS technique. In this adaptation, individual reflection and pairing are replaced by reflection in small groups and the production of consensual drawings, followed by a reasoned presentation of their proposal to the rest of their classmates. In

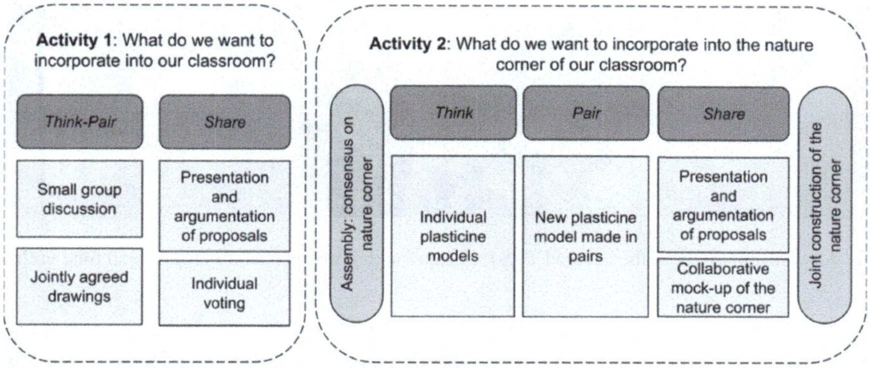

Fig. 4.1 Outline of the design of the instructional module

addition, after the presentations, an individual vote is held to decide which proposals have the most support.

The second activity involves a preliminary debate on the suitability of incorporating a nature corner in the classroom and, based on this already agreed idea, carrying out the TPS technique to decide on the elements to include in the classroom nature corner. In addition, we propose modeling the aspects the children wish to incorporate in plasticine, both in individual reflection and pairing and presentation to the rest of the classmates. The different plasticine models are combined into a collaborative model of the nature corner. Finally, considering the ideas and models produced, the construction of the nature corner begins with the elements that the children themselves and their families contribute.

4.4.2 Experience Description

The first activity was carried out in the classroom the week before the Christmas holidays, and following the proposed design, the preschoolers were presented with the open question of what they would like their classroom to have, and they were asked to discuss it in small groups and to draw a picture of what each group decided. Afterwards, each group presented their drawings and explained what they thought was missing in the classroom and why. In order to see which proposals had the most support, they voted using interlocking pieces, one per student, with which they constructed a bar chart of each of the drawings associated with the different proposals (Fig. 4.2).

Great difficulties were observed in agreeing and acting together, which were accentuated by a competitive and individualistic attitude. Moreover, the behaviour was associated with a general lack of familiarity of the children with the principles and ways of proceeding in democratic societies (dialogue, consensus building, participation, decision-making, etc.).

Fig. 4.2 Development of the adapted TPS technique with drawings, small groups, and final voting

Fig. 4.3 Development of the adapted TPS technique by incorporating modeling with plasticine

In the second activity, a special effort was made to encourage all the children to contribute their ideas to the design of this corner. It was very positive to observe how the children, working in pairs, talked to each other more than in the small groups of the first activity and even actively tried to reach agreements, although this was still very difficult for them. The pupils were a bit fed up with the drawings, so it was a good idea to adapt this technique by replacing the drawings as a means of expression with plasticine models (Fig. 4.3). This made the activity more manipulative and attractive for the children. However, it also increased the difficulty of pooling the pairs' proposals as they had to combine their initial individual three-dimensional models into a single one and reach agreements.

After getting an idea of what we wanted to include in our nature corner, the children wrote a message to their families asking them to provide natural elements for the corner. This got the preschoolers fully involved in the construction of the nature corner, and, in this way, many of them transmitted orally to their families the need to contribute natural elements more insistently and effectively than a letter written by the teacher alone would have been.

4.4.3 ENCIC-CT Model Application

The instructional module adopts the ENCIC-CT model (Franco-Mariscal et al., 2025), which provides a structured and comprehensive framework for developing CT in educational settings. The module applies this model's four-layered approach, beginning with the core level (Layer 1), where preschoolers engage with a daily life problem—in this case, incorporating nature into their classroom environment. In the second layer, the module focuses on two key domains: the skill domain and the disposition domain. Within the skill domain, the decision-making dimension is highlighted, guiding students to critically evaluate various options and make informed choices about incorporating natural elements into their learning spaces. Simultaneously, the disposition domain is addressed through the personal autonomy dimension, encouraging students to take ownership of their learning process, develop self-directed learning habits, and exercise autonomy in their decision-making. The third layer of the ENCIC-CT model is emphasized through the implementation of scientific practices, particularly argumentation and modeling. These practices are seamlessly integrated into the instructional activities, enabling students to articulate their reasoning, and develop models that reflect their understanding of the natural world. The TPS strategy, associated with Layer 4, is employed to support these practices and enhance cooperative learning in an environment conducive to the development of CT.

4.4.4 Data Collection and Analysis

A triangulation of different sources of information was carried out to provide further rigour and a more holistic perspective to our research. Firstly, the participant observation carried out by the researcher herself (first author) was collected in a field diary with the development of the activities. On the other hand, different student productions were collected: group drawings, which were categorised according to their content, and the results of the voting in the first activity; graphic evidence of the plasticine models, individual and in pairs, which were compared to analyse the evolution in decision-making, together with the collaborative model made during the different phases of the TPS technique in the second activity. In addition, recordings were made of the pairs' presentations during the *sharing* phase of the second activity to facilitate their subsequent analysis, and recorded interviews were conducted in small groups, both before the start of the second activity and after the end of the nature corner. Finally, graphic evidence was also collected on the evolution of the construction of the corner, which allowed us to corroborate the involvement and personal autonomy of the children in the process.

The data collected from these data sources were subjected to a content analysis (Lincoln, 1995) through an inductive process from which emerged narrative descriptions of the main patterns identified in the two activities about the dimensions of CT studied and the natural elements proposed by the children for incorporation into the classroom.

4.5 Results

In the first activity, the children made six group proposals, which they presented with the support of drawings composed collaboratively by all the members of each group. Table 4.1 shows the proposals with a breakdown of their content, which could be divided into two main categories, depending on whether they had cultural or natural environment content, their individual or collective nature and the support received regarding the number of votes.

Among the proposals on what to incorporate into the classroom, cultural elements related to popular festivals, specifically Christmas, and natural elements, primarily trees, often combined in the same drawing, dominated. Likewise, most of the drawings showed a diversity of elements distributed equally, demonstrating high collaboration and participation by all group members. However, when arguing their proposals in small groups, the exhibition was mainly taken up by the child who had initially proposed it. This indicates that, in most cases, there was not a fair debate; instead, one of the initial proposals was imposed on each group, and the others accepted it and were relegated to a secondary role.

Both during the presentation and decision-making, the children generally did not argue the reason for their proposal or vote. If any did, they justified it simply because it was what they liked best. On the other hand, they showed a certain degree of personal autonomy and reflection when making their votes since, unlike what would be expected in the absence of voting, many opted for other groups' proposals rather than their own. Interestingly, the most voted options were those related to the Christmas holidays, which were very close at the time of the activity. Although this denotes their strong interest in cultural elements, it also shows the significant influence that the hegemonic social and cultural context has on decision-making from an early age.

Table 4.1 Results of the first activity, categorisation of its content and voting results

Proposal	Christmas tableau	Tree Christmas	Tree Christmas	Flute	Box Nature	Flowers
Content cultural	Artistic and popular festivals	Popular festivals	Popular festivals	Musical	Artistic	–
Natural content	Tree	Trees and animal	Tree and atmospheric elements	Ecosystem	Ecosystem	Ecosystem
Character	Collective: various paintings	Collective: several trees	Individual: single central tree	Collective: various instruments and natural elements	Collective: various elements of the ecosystem	Collective: two natural proposals
Votes	9	4	4	3	2	2

In the interviews recorded before the second activity, the children answered unanimously in the affirmative when asked if they would like to have more nature in the school and the classroom. Then, when asked how this could be achieved, they were able to identify a variety of natural elements (leaves, flowers, trees, etc.) that could be incorporated, either in the classroom or in the school playground, and also associated them with increased play opportunities and pro-environmental actions, such as planting plants and recycling. During the assembly in which the incorporation of the nature corner was debated, the children showed remarkable coherence with their answers, unanimously positioning themselves in favour of it and autonomously identifying numerous benefits and utilities of such a corner: research, learning about nature, relaxing, breathing, making new friends, enjoying the outdoors, not polluting, etc.

In the individual reflection and modeling phase, the children produced a wide range of elements, all perfectly coherent and plausible for incorporation into the nature corner. These elements are listed in Table 4.2, both in the individual and paired models during the TPS technique, and the main arguments they put forward when presenting the models to the rest of the pupils.

As can be seen (Table 4.2.), flora and fauna elements predominate, although a house (anthropic element) and a magnifying glass (investigative function of the corner) are also present. Moving on to the pairing phase, a closer and more equal interaction and discussion could be observed in the pairs than in the previous activity's small groups. In this regard, there was a greater manifestation of personal autonomy and a greater depth of decision-making. This was reflected in the fact that, although the proposals put forward by the pairs coincided in many cases with some of the individual proposals of their members, these were enriched with individual contributions. In many instances, new proposals emerged, the result of combining the individual ones. A clear example was that a well was created from a worm and a house to obtain water for the animals to drink. In addition, during the presentation phase, there was more balanced participation, and most pairs included more complex arguments, even concerning the joint decision-making process itself.

It is worth noting that the subsequent construction of the nature corner, taking into account the children's contributions, helped, to a large extent, to foster their autonomy and their self-perception of their real capacity to make decisions. This was evident in the high level of involvement of the children in the construction of the nature corner (Fig. 4.4) and in the enthusiasm with which they responded in the interview after the incorporation of the nature corner when asked about what they thought of it. In these same interviews, they made numerous references to the elements they provided for this corner and their role in deciding how it was constructed. This is further evidence of how this final phase, in which what they had devised through the TPS technique was implemented, helped foster their autonomy in decision-making.

Table 4.2 Proposals presented by the children during the TPS technique.

Individual models of the members of each couple (*Think*)	Models in pairs (*Pair*)	Arguments of the exhibitions (*Share*)
Cake/rabbit with snail	Fossil and dog	*To be in nature and because we like animals* *For very sleepy animals* *A dog sleeping in the fossil*
Flower/tarantula	Spider	*To keep the bugs in the wild*
Snail/corner	Snail	*To be free*
Ladybird, snake and beetle/snail	Butterfly	*Because we agreed and it seemed easy to us. We did it because there was much plasticine to make a butterfly, and we would like to have it in the corner*
Worm/house	Well	*They gave us plasticine, and we made a well with water for the animals. We did not make a river out of rubbish because we needed to know how and more plasticine*
Flower/spider	Stones stuck in the ground	*We thought of making a stone because it came in handy with the pink and blue modeling clay. Also, for the spiders to climb on*
Snail/flower	Snakes in a tree	*We did that because it occurred to us, and we needed more time*
Snake/snail	Snake	*Because she and I have decided. We want a snake to play with*
Snake/worm and spider	Snake	*To make jokes*
Worm and magnifying glass/ snail	Snail with a snake	*We would like to have it in the corner*
A cake with flowers/ flower	Snail on a cake	*Because we want to* *Because the snail wants to eat the cake* *Because if they do not eat, they die, and if they do not have water, they dry up and die, but if they have water from the well, they die*

Fig. 4.4 Construction of a nature corner as a final phase after the TPS technique

4.6 Discussion and Educational Implications

The enormous potential of the TPS technique and its feasibility for this stage is proven, as is the extent to which the TPS technique develops the decision-making autonomy of preschoolers when dealing with the naturalisation of their classroom. Our results align with those of Silva et al. (2022), who showed that the TPS and other cooperative learning methods were much more effective than traditional ones in developing CT in primary school students.

Moreover, it was observed that the simplification of the TPS technique used in the first activity, working in small groups, meant that the participation of all students was different. It should also be noted that in this first activity, there were significant shortcomings in the students' argumentation skills, coinciding with the observation of Rodríguez-Melero et al. (2023), who found difficulties in five-year-old students in deciding, in an argued manner, on the objects to be saved during the eruption of a volcano.

On the other hand, in the second activity, in which the three phases of the TPS technique were carried out, greater involvement of the students and a better argumentation of their decision-making during the presentation phase were observed. The greater participation can be interpreted as progress in their autonomy, reflected in their more remarkable ability to put forward their points of view and not simply be carried along by what was presented by a classmate, as was observed in the first activity. In addition, students gained confidence in their ability to make decisions about their immediate environment, thus promoting interesting synergies between the two dimensions of CT in this study. As Bargiela et al. (2022) suggested, the level of guidance required by preschoolers during activities aimed at developing CT should progress from more to less structured. Therefore, the TPS technique is a good starting point to begin to foster the dimensions considered, as it offers a structure through which to involve them from an early age in making decisions in an autonomous and participatory way. In this way, it could have been more beneficial to carry out the second activity first, as it maintains the complete structure of the TPS technique, and then to carry out the first one, in which simplifying the phases of the TPS makes the structure that guides the children's decision-making process more flexible.

As for the RQ2, it was observed that the elements they wished to incorporate into the spatial organisation of their classroom were all associated with cultural elements, whether related to artistic or musical expression, popular festivals, or natural elements. Moreover, the children were unanimously in favour of adding a nature corner to their classroom and could autonomously identify the benefits and utilities the changes could bring them. Interestingly, many of the elements suggested by the children, following the autonomous decision-making process, such as the presence of natural plants and animals, as well as the use of magnifying glasses to investigate them, coincide with several of the materials that Sanz et al. (2021) identified as suitable for integration into a naturalised school playground.

Therefore, the nature corner presents itself as a splendid candidate to promote not only the development of CT but also to guarantee the presence in the classroom of a space that supports the exercise of children's innate passion for discovery, which, according to Tu (2006), is essential to promote scientific thinking. Moreover, as this study shows, it is feasible to involve students in the design and construction of this corner in the classroom, overcoming any limitations they may have due to their young age. This provides an excellent opportunity to foster their autonomy and decision-making. This approach takes on particular importance when considering Manassakis's (2019) assertion that children have the right to be heard and felt in all matters that concern them, always remembering their real capacity and competence to interpret their environment.

In short, as Osborne (2014) points out, the opportunities to engage in critique, argumentation and questioning, which can be provided to early childhood learners by adapting the TPS technique to a learning approach, not only help them to approach and understand science but also contributes to developing their ability to reason critically and scientifically.

Acknowledgements This research was funded by the Ministry of Science, Innovation and Universities through the University Teacher Training Programme (FPU22/01428) and by the Spanish National Plan R+D+i Project entitled "Citizens with critical thinking: A challenge for teachers in science education", reference PID2019-105765GA-I00, funded by MCIN/AEI/10.13039/501100011033.

References

Alabay, E. (2009). Analysis of science and nature corners in preschool institutions (Example of Konya province). *Procedia Social and Behavioral Sciences, 1*, 857–861. https://doi.org/10.1016/j.sbspro.2009.01.154

Bargiela, I. M., Blanco, P., & Puig, B. (2022). Las preguntas para la indagación y activación de pensamiento crítico en educación infantil [Questions for Inquiry and the Activation of Critical Thinking in Early Childhood Education]. *Enseñanza de las Ciencias, 40*(3), 11–28. https://doi.org/10.5565/rev/ensciencias.5470

Blanco-López, Á., España-Ramos, E., & Franco-Mariscal, A. J. (2017). *Estrategias didácticas para el desarrollo del pensamiento crítico en el aula de ciencias* [Teaching strategies for the development of critical thinking in the teaching of science]. *Ápice, Revista de Educación Científica, 1*(1), 107–115. https://doi.org/10.17979/arec.2017.1.1.2004

Blanco-López, Á., España-Ramos, E., González-García, F. J., & Franco-Mariscal, A. J. (2015). Key aspects of scientific competence for citizenship: A Delphi study of the expert community in Spain. *Journal of Research in Science Teaching, 52*(2), 164–198. https://doi.org/10.1002/tea.21188

Bottini, M., & Grossman, S. (2005). Center-based teaching and children's learning: The effects of learning centers on young children's growth and development. *Childhood Education, 81*(5), 274–277. https://doi.org/10.1080/00094056.2005.10521305

Cebrián-Robles, D., España-Ramos, E., & Reis, P. (2021). Programa formativo sobre el activismo colectivo basado en la indagación para la formación inicial del profesorado [Training program on collective activism based on inquiry for initial teacher education]. In D. Cebrián-Robles, A. J. Franco-Mariscal, T. Lupión-Cobos, M. C. Acebal-Expósito, & A. Blanco-López (Eds.), *Enseñanza de las ciencias y problemas relevantes de la ciudadanía. Transferencia al aula*

[Science education and relevant issues of citizenship. Transfer to the classroom.] (pp. 175–186). Graó.

Cruz-Guzmán, M., & Martínez-Maqueda, E. (2022). Iniciación a las prácticas científicas en educación infantil: Aprendiendo sobre el sistema digestivo por indagación basada en modelos [Scientific practical initiation in Kindergarten: the digestive system through inquiry and modeling based learning]. *Revista Eureka sobre Enseñanza y Divulgación de las Ciencias, 19*(1), 1–21. https://doi.org/10.25267/rev_eureka_ensen_divulg_cienc.2022.v19.i1.1202

Delval, J. (2006). *Hacia una escuela ciudadana* [Towards a citizenship school]. Ediciones Morata.

Ennis, R. H. (2015). Critical thinking: A streamlined conception. In M. Davies & R. Barnett (Eds.), *The Palgrave handbook of critical thinking in higher education* (pp. 31–47). Springer.

Facione, P. A. (1990). *Critical thinking, a statement of expert concensus for purposes of educational assessments and instruction. Executive summary. The Delphi Report*. California Academic Press.

Facione, P. A. (2023). Critical thinking: What it is and why it counts. *Insight Assessment,* 1–31.

Franco-Mariscal, A. J., Cano-Iglesias, M. J., España-Ramos, E., & Blanco-López, A. (2025). The ENCIC-CT model for the development of critical thinking. In A. J. Franco-Mariscal (Ed.), *Critical thinking in science education and teacher training* (Chap. 1) (pp. 3–42), Springer. https://doi.org/10.1007/978-3-031-78578-8_1

Hodson, D. (2011). *Looking to the future*. Sense Publishers.

Jiménez-Aleixandre, M. P., & Puig, B. (2022). educating critical citizens to face post-truth: The time is now. In B. Puig, & M. P. Jiménez-Aleixandre (Eds.), *Critical thinking in biology and environmental education. Facing challenges in a post-truth world* (pp. 3–19). Springer. https://doi.org/10.1007/978-3-030-92006-7

Kaddoura, M. A. (2013). Think Pair Share: A teaching learning strategy to enhance students' critical thinking. *Educational Research Quarterly, 36*(4), 3–24.

Kamii, C. (1991). Toward autonomy: The importance of critical thinking and choice making. *School Psychology Review, 20*(3), 382–388. https://doi.org/10.1080/02796015.1991.12085561

Kuhn, D. (2019). Critical thinking as discourse. *Human Development, 62*(3), 146–164. https://doi.org/10.1159/000500171

Lederman, N. G., Zeidler, D. L., & Lederman, J. S. (2023). *Handbook of research on science education*. Routledge. https://doi.org/10.4324/9780367855758

León, J. M. (2015). A baseline study of strategies to promote critical thinking in the preschool classroom. *Gist Education and Learning Research Journal, 10*, 113–127.

Levinson, R. (2006). Towards a theoretical framework for teaching controversial socio-scientific Issues. *International Journal of Science Education, 28*(10), 1201–1224. https://doi.org/10.1080/09500690600560753

Lincoln, Y. S. (1995). Emerging criteria for quality in qualitative and interpretive research. *Qualitative Inquiry, 1*(3), 275–289. https://doi.org/10.1177/107780049500100301

Lipman, M. (2003). *Thinking in education*. Cambridge University Press. https://doi.org/10.1017/CBO9780511840272

Lipman, M., Sharp, A. M., & Oscanyan, F. S. (2010). *Philosophy in the classroom*. Temple University Press.

Manassakis, E. (2019). Children's participation in the organisation of a kindergarten classroom. *Journal of Early Childhood Research, 18*(1), 18–28. https://doi.org/10.1177/1476718x19882714

McPeck, J. E. (1981). *Critical thinking and education*. Routledge.

Osborne, J. (2014). Teaching critical thinking? New directions in science education. *School Science Review, 352*, 53–62.

Paul, R., & Elder, L. (2007). *A guide for educators to critical thinking competency standards: Standards, principles, performance indicators, and outcomes with a critical thinking master rubric*. Rowman & Littlefield Publishers/The Foundation for Critical Thinking.

Piaget, J. (1973). *To understand is to invent*. Viking [Original work published 1948].

Puig, B., Blanco-Anaya, P., & Bargiela, I. M. (2023). Integrar el pensamiento crítico en la educación científica en la era de la post-verdad [Integrating Critical thinking in Science Education at the Post-truth Era]. *Revista Eureka sobre Enseñanza y Divulgación de las Ciencias, 20*(3), 330101–330117. https://doi.org/10.25267/rev_eureka_ensen_divulg_cienc.2023.v20.i3.3301

Puig, B., & Jiménez-Aleixandre, M. P. (2022). The integration of critical thinking in biology and environmental education. Contributions and further directions. In B. Puig, & M. P. Jiménez-Aleixandre (Eds.), *Critical thinking in biology and environmental education. Facing challenges in a post-truth world* (pp. 269–276). Springer.

Ramos-Solano, A. (2022). La naturaleza en la organización espacial de las escuelas infantiles [Nature in the spatial organisation of nursery schools]. In D. Madrid, & R. Pascual (Coords.), *Buenas Prácticas en Educación Infantil* (pp. 606–616). Dykinson.

Rodríguez-Melero, A. M., Algar, N., Bujedo, C., Cabello, N., & Franco-Mariscal, A. J. (2023). Desarrollo de pensamiento crítico en niños y niñas de educación infantil a través de la toma de decisiones. El caso de ¿qué salvamos ante la erupción de un volcán? [Development of critical thinking in children in early childhood education through decision-making. The case of 'What do we save during a volcano eruption?']. In P. Membiela, & M. I. Cebreiros (Eds.), *Expectativas y prácticas docentes en la enseñanza de las ciencias* [Teacher Expectations and Practices in Science Education] (pp. 421–432). Educación Editora.

Sanz, J. S., Zuazagoitia, D., Lizaso, E., & Pérez, M. (2021). ¿Promueven los patios naturalizados el desarrollo de la competencia científica? Un estudio de caso en la educación infantil [Do Naturalized Playgrounds Promote the Development of Scientific Competence? A Case Study in Early Childhood Education]. *Revista Eureka sobre Enseñanza y Divulgación de las Ciencias, 18*(2), 1–19. https://doi.org/10.25267/rev_eureka_ensen_divulg_cienc.2021.v18.i2.2203

Sharma, H. L., & Saarsar, P. (2018). TPS Think-Pair–Share: An effective cooperative learning strategy for unleashing discussion in classroom interaction. *International Journal of Research in Social Sciences, 8*(5.1), 91–100.

Silva, H., Lopes, J., Domínguez, C., & Morais, E. (2022). Think-pair-share and roundtable: Two cooperative learning structures to enhance critical thinking skills of 4th graders. *International Electronic Journal of Elementary Education, 15*(1), 11–21. https://doi.org/10.26822/iejee.2022.274

Smith, G. A., & Sobel, D. (2010). *Place- and community-based education in schools.* Routledge.

Solbes, J., & Torres, N. (2012). Análisis de las competencias de pensamiento crítico desde el aborde de las cuestiones sociocientíficas: Un estudio en el ámbito universitario [Analysis of Critical Thinking Competencies from the Approach of Socio-Scientific Issues: A Study in the University Context]. *Didáctica De Las Ciencias Experimentales y Sociales, 26*, 247–269. https://doi.org/10.7203/dces.26.1928

Tu, T. (2006). Preschool science environment: What is available in a preschool classroom? *Early Childhood Education Journal, 33*(4), 245–251. https://doi.org/10.1007/s10643-005-0049-8

Vila, L., Márquez, C., & Oliveras, B. (2023). Una propuesta para el diseño de actividades que desarrollen el pensamiento crítico en el aula de ciencias [A didactic proposal for the development of critical thinking in the high school science classroom]. *Revista Eureka sobre Enseñanza y Divulgación de las Ciencias 20*(1), 1302. https://doi.org/10.25267/Rev_Eureka_ensen_divulg_cienc.2023.v20.i1.1302

Yin, R. K. (2013). *Case study research: Design and methods* (5th ed.). Sage Publications.

Chapter 5
The Comprehensive Analysis of the Space Care Issue Through Controversy Mapping in Architecture Students

Paloma España-Naveira, Eva Morales-Soler, Ángel Blanco-López, and Daniel Cebrián-Robles

5.1 Introduction

This study considers that critical thinking (hereinafter, CT) develops as individuals progress, particularly in their ability to approach problems holistically, taking into account their complexity and the various aspects—scientific, technical, ethical, cultural, philosophical, social, environmental, economic, etc.—and the relationships between them.

We consider that controversy mapping has the potential to enhance the comprehensive analysis of issues, such as those related to care spaces in the city. This is because it allows us to approach the complexity of the issue, thereby facilitating the development of projects that integrate these spaces. Controversy mapping makes the network of actants involved in these controversies visible. The development of digital tools further enables the collaborative mapping of these actants and their influences (Hervé, 2014). Thus, mapping becomes an opportunity to make visible not only the

P. España-Naveira (✉) · E. Morales-Soler
Art and Architecture Department, Universidad de Málaga, Málaga, Spain
e-mail: palomanaveira@uma.es

E. Morales-Soler
e-mail: eva.morales@uma.es

Á. Blanco-López · D. Cebrián-Robles
Science Education, Universidad de Málaga, Málaga, Spain
e-mail: ablancol@uma.es

D. Cebrián-Robles
e-mail: dcebrian@uma.es

A. J. Franco-Mariscal (ed.), *Critical Thinking in Science Education and Teacher Training*, Contemporary Trends and Issues in Science Education 64,
https://doi.org/10.1007/978-3-031-78578-8_5

actants involved but also the relationships and conflicts between a dominant commodified model associated with productive activities and a model that prioritizes the care of life by valuing reproductive activities.

The ability to deal with complex issues can be assessed by examining the levels of complexity outlined in the relations of the controversy maps. The study of Cabello-Garrido et al. (2023) presents a scheme for analysing these levels of complexity, which can be understood as a framework for the progression of the CT dimension related to the comprehensive analysis of issues.

5.2 Controversy Mapping to Develop Critical Thinking About Socio-scientific Issues

In our society, socio-scientific issues (hereinafter, SSIs) emerge that interest citizens and generate controversies, which are present in various contexts, such as science and society, and notably in the media and social networks (Bencze et al., 2020; Sadler, 2011). Moreover, these SSIs are open and complex issues related to values and knowledge from different fields, such as scientific-technical, political, economic, ethical, etc. (Bencze et al., 2020). This variety of fields and their diverse interests make SSIs particularly complex to study. It is essential to first recognize this complexity in order to approach them effectively, avoiding simplistic views and being open to different perspectives (Morin, 2007), which often translate into controversial issues. According to Levinson (2006), a controversial issue is one that lacks a single, universally accepted perspective and typically polarizes society, with significant groups offering conflicting explanations and solutions.

It is interesting to address SSIs in education to develop citizens capable of making responsible decisions about controversies related to science and technology (Levinson, 2006; Reis, 2014). However, the overabundance of information in digital media and social networks, including disinformation and fake news, presents challenges. In addition, the recent impact of artificial intelligence raises questions about the veracity of the information it generates (Cooper, 2023; Zhai, 2023).

An example of SSI is that of spaces for care in the city, understood as an issue that starts from self-care to common care from an environmental, social and feminist perspective of the city (Chinchilla, 2020; Gutiérrez-Valdivia, 2021; Muxi, 2021), especially in the face of the challenges of the twenty-first century, such as pandemic situations like COVID-19 (Martínez & Short, 2021), the environmental challenges that cities must face (Klein, 2014; Kukina, 2019), the social challenges aggravated by various crises, etc. In short, conflicts arise in cities where various controversies affect care spaces in urban and living spaces. Care spaces value reproductive and life-sustaining activities, assuming that we are interdependent beings and, therefore, it is important to build community and mutual support.

From sociology, the Actor-Network theory states that all those who participate in a controversy, calling them *actants* (human and non-human), influence it and must therefore be considered (Callon et al., 1986; Latour, 2005). This theory has been adapted as an educational tool under the name of Cartography of Controversy (Latour, 2005), which, in addition to addressing an issue and studying its complexity, makes it possible to follow a process of inquiry into SSIs. From an educational point of view, the aim is for students to conduct an inquiry to understand a controversy and represent its complexity by visualising how those involved act (Christodoulou et al., 2021).

As mentioned above, SSIs are complex because of the variety of interconnected fields (scientific, technological, political, economic, etc.) and the difficulty in conducting media research. The overabundance of information, some of it rigorous and some of it false, currently present in the media, places citizens in infodemia (Eysenbach, 2020). Much of this information is hoaxes or rumours, known as fake news (Cebrián-Robles, 2020; Puig et al., 2021), which are based on misunderstood or purposely misinterpreted scientific topics and make it difficult for people to find reliable sources and guidance when they need it. All these aspects, which make it difficult to conduct media inquiries (Valverde-Crespo et al., 2022), represent enormous challenges for this educational work.

Controversy mapping as part of the Cartography of Controversy approach (España-Naveira et al., 2023) represents and makes visible the complexity of controversies through groupings that include the actants involved in it and the relationships that represent the interactions that take place between the actants, constituting the whole as a network. We prefer to speak of controversy mapping rather than controversy map to emphasise that "it is neither a product nor a process, but a movement or, more precisely, a series of movements" (Venturini et al., 2015, pp. 83–84). For its elaboration, this chapter proposes some platforms such as MIRO[1] or MURAL[2] to facilitate cooperation, analysis and visualisation of the controversy (Cabello-Garrido et al., 2023; España-Naveira et al., 2023) since, as Venturini (2010) points out, it is important to take advantage of the potential of digital technologies.

This chapter presents controversy mapping as a tool to develop the capacity of Spanish university students, specifically undergraduate architecture students, to deal with complex issues, considered an important dimension for developing CT on the SSI of spaces for care in the city.

5.3 Research Questions

This chapter studies the usefulness of controversy mapping for developing the ability to deal with complex issues with students of the Degree in Foundations of Architecture. For this purpose, the SSI controversy maps of care spaces in the city mapped by the students are analysed. The following research questions (RQ) are suggested:

[1] MIRO (https://miro.com/) is a visual collaboration platform for teamwork.

[2] MURAL (https://www.mural.co/) is a platform for visual information work in teams.

- RQ1: What levels of complexity are students able to establish using the SSI controversy mapping of care spaces in the city?
- RQ2: To what extent do students perceive that controversy mapping makes it easier to deal with complex issues in architectural design, in this case, in the analysis of care spaces in the city?

5.4 Method

5.4.1 Participants and Context

The controversy mapping experience was implemented with 17 students (53% women and 47% men, aged between 19 and 20) of the subject Architectural Projects 2 of the second year of the Fundamentals of Architecture Degree at the University of Malaga (Spain) during the academic year 2021–2022.

This experience was part of a more extensive training program in which students, after this experience, carried out a broader inquiry into the SSI of care spaces in the city of Málaga, following an extended version of the cartography of controversies approach (España-Naveira et al., 2023). The training program was part of the Domestic Spaces for Care in Everyday Life and Environment course project.

5.4.2 Experience Description

As a preliminary step to the experience, the professor worked with the students to choose and explore the current and local contextualised controversy. This step is crucial, as it is necessary to have a previous approach to the controversy to gather and understand the perspectives of all the actants involved, thus allowing the creation of their map with a more complete vision.

The experience began by asking the students to answer individually: Are there adequate spaces for care in the city? They had to justify their answer. Subsequently, groups of 4–5 students worked on mapping controversies using the MURAL platform as a didactic resource. This platform facilitates collaboration through collaborative sketching and drawing. Teams can create canvases that function as digital whiteboards, adding actants represented by squares with descriptive text and establishing relationships using arrows to link the various actants together. At the same time, groupings are created, which we will call poles, to bring together actants with common interests in the SSI. Technologies make it possible, if computers and Internet connection are available, to collaboratively map disputes online, recording all interactions with the map for later analysis. If it is not possible to use technologies for this task, it can be done using posters placed on a blackboard and linked with arrows and circles to make groupings.

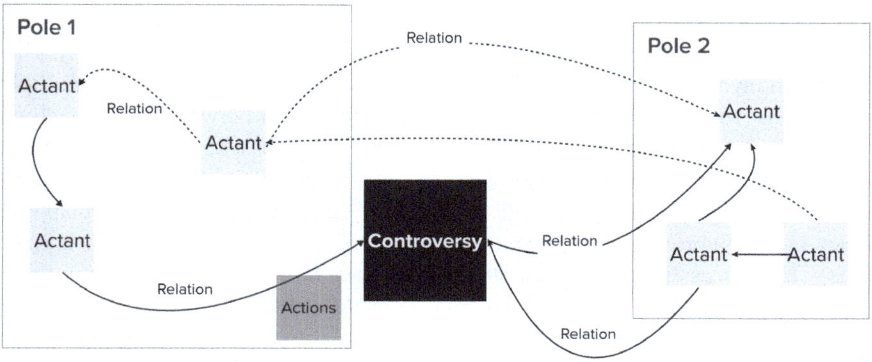

Fig. 5.1 Example of an illustrative template of elements of the controversy map provided to students

Figure 5.1 illustrates the template that students are taught, for example, about the elements of the controversy map (actants, relations and poles) and how to represent them in the MURAL platform.

In the controversy mapping experience, students followed these phases:

- *Phase 1. Identification of actants and their grouping*. In this phase, each participant individually identifies at least 2–3 relevant actants. Subsequently, they are grouped according to justifications agreed upon within the group.
- *Phase 2. Creation and naming of poles*. It consists of naming the groupings of actants into poles according to criteria determined and agreed upon by the students. During this phase, it is flexible to add new actants as needed and rearrange their grouping if perceived as appropriate.
- *Phase 3. Relationship between actants*. It involves the creation of potential relationships, where the nature of the connection is explained, indicating whether it is favourable or unfavourable. The positive or negative influence one actants can exert on another is considered.
- *Phase 4. Possible actions and measures*. This phase is a further step towards decision-making once the complexity of the issue has been made visible through the network of actants. It consists of proposing actions to help resolve the conflicts between the actants identified in the previous phases of the mapping.

5.4.3 ENCIC-CT Model Application

This study explores the analysis of controversies within the framework of the ENCIC-CT model (Franco-Mariscal et al., 2025). In this context, the SSI of spaces for care in the city forms the first layer of the model, representing the core topic around which the controversy unfolds. These kinds of issues, which encompass both social and scientific aspects, demand an interdisciplinary approach and critical reflection from students.

The focus of the experience is on the CT dimension known as comprehensive analysis of problems, which belongs to the skills domain (layer 2 of the model). This dimension involves not only identifying the issue but also tackling it from multiple perspectives, recognizing its complexity and the interactions between various factors.

The development of this dimension requires modeling as a key scientific practice (layer 3), enabling students to map out the problem's actants and their interrelationships. Modeling is applied through the teaching strategy of controversy mapping (layer 4), which provides a clear visualization of the actants involved and their positions. This process is essential for students to grasp how decisions regarding these care spaces affect both the social and environmental contexts, encouraging CT that integrates the scientific, ethical, social, and political dimensions of the controversy.

5.4.4 Data Collection and Analysis

The controversy maps made by the students constitute the data collection instrument of this study. These data will be analysed in this chapter to address RQ1 and to find out what levels of complexity they can establish when using controversy mapping concerning whether there are adequate spaces for care in the city.

For instance, Fig. 5.2 shows the result of the controversy mapping of a group of students. It shows how students have grouped the actants into five poles (infrastructure, culture, citizenship, environment and communication) and the existing relationships.

For the analysis, and following Christodoulou et al. (2021), we consider the connections between actants as indicators of students' understanding of the complexity of the controversy. These controversy maps are analysed according to the levels of complexity of the relationships established in them, using the analysis scheme of Cabello-Garrido et al. (2023) (Fig. 5.3).

As illustrated in Fig. 5.2, relationships at levels 2, 3, 4, and 5 have been identified. There are six within-pole relationships (Level 2), such as the connection between *rufett places* and *shade*, both within the infrastructure pole. Fifteen between-pole relationships (Level 3) are present, for instance, the relationship between the actant *recycling points* from the infrastructure pole and *clean and hygienic* from the recycling pole. Three multiple between-pole relationships (Level 4) have been identified, such as the actant *bike lanes* from the communication pole, which connects with two actants from different poles: *bicycle* from the environment pole and *youth* from the citizenship pole. Finally, there are four combinations of within-pole and between-pole relationships (Level 5), such as the actant *shade*, which connects with *roofed places* within the infrastructure pole and with the actants *walks* and *pavements* from the communication pole.

For RQ2, an open-ended question was designed and posed to the students at the end of the experience: "How has the actant network map [controversy map] helped you to develop your mapping of care in your daily life" (activity to be developed later). This question was categorised by the authors of this chapter through inductive

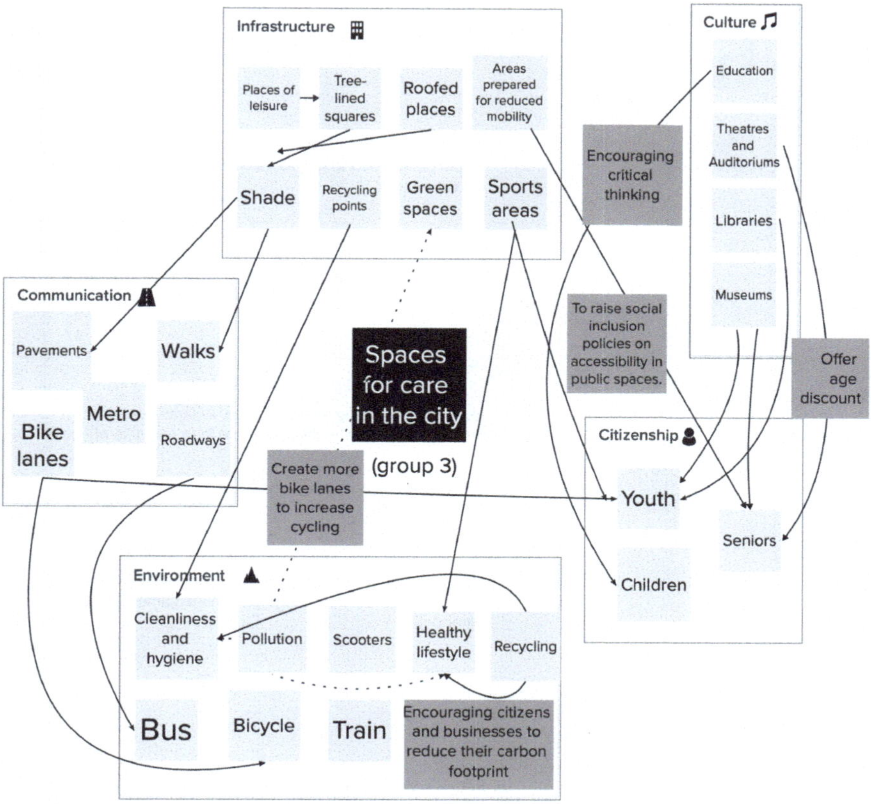

Fig. 5.2 Example of controversy mapping for a student group

analysis of the responses, allowing categories to emerge from the data rather than from a pre-established category system. A thorough reading of all responses was done to obtain a general understanding of the content. It was followed by open coding, where each response was examined to identify emerging issues and ideas. These aspects and ideas were grouped into broader categories to reflect the key aspects of the impact of the actor-network map on the development of individual care mapping. For the analysis, the frequency of students in each category was determined and illustrated with examples.

5.5 Results

The findings for each RQ are shown below. The first section presents the complexity levels the students achieved in the different controversy maps, and the second section presents the students' assessment after carrying out the mapping activity.

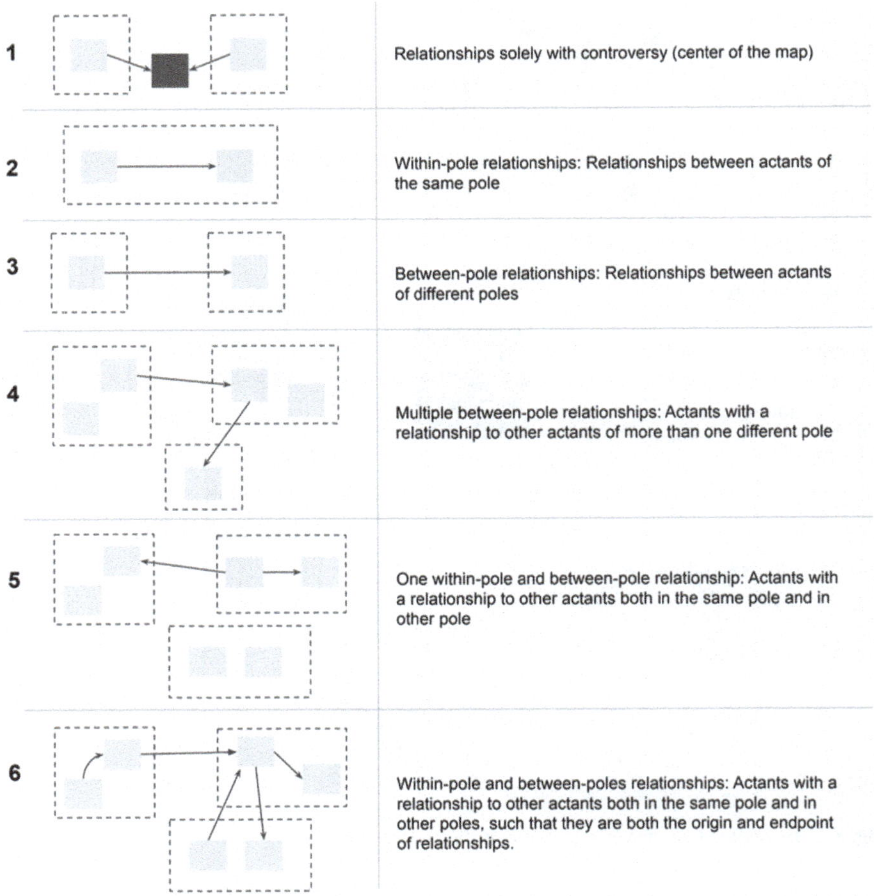

Fig. 5.3 Levels of complexity in the relationships established when mapping the controversy (1: lowest level; 6: highest level) (adapted from Cabello-Garrido et al., 2023)

5.5.1 Levels of Complexity Represented by Students on the Issue

Table 5.1 shows the levels of complexity represented in the four controversy maps produced by the students, analysed following an adaptation of the scheme proposed by Cabello-Garrido et al. (2023).

The results (Table 5.1) reveal that the students made the controversy maps at an intermediate level of complexity (levels 3 and 4). Practically, there is an absence of connection of the actants with the central issue (level 1) since it was directly related to the actants in this case. We believe that, as in the work of Cabello-Garrido et al. (2023), establishing relations between actants (above level 1) is a somewhat more complicated exercise than directly linking it to the main controversy since the most

Table 5.1 Results of the levels of complexity of the controversy maps produced by the students

Level of complexity		Number of relationships between actants				
		Group 1	Group 2	Group 3	Group 4	Total
1	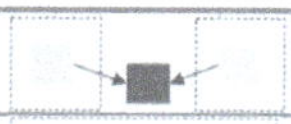	1	0	0	0	1
2	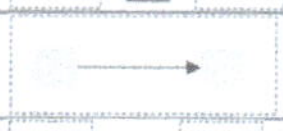	7	1	6	2	16
3	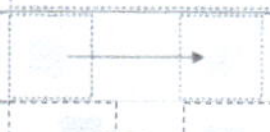	13	27	15	11	66
4		4	14	3	7	28
5	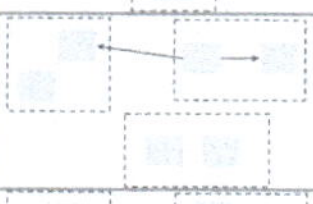	8	4	4	2	18
6	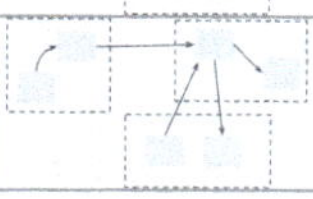	3	4	0	2	9
Total		36	50	28	24	138

difficult thing is to see to what extent there are competing interests between actants. Moreover, it can be assumed that, as the chosen dispute was unique, linking the actants to the disagreement was unnecessary since, to some extent, all the actors could be considered connected to it.

As indicated above, poles are groupings of actants with common interests, so one would expect that it is easier to find relationships between actants of the same pole (intrapole, level 2). However, it was found that the connection between actants from different poles (inter-pole, level 3 onwards) was more common than those from the same pole. Perhaps we can think that the students mainly looked for relations between actants with opposing ideas instead of establishing the connections by the affinities between actants. Furthermore, we consider it highly complex to connect the same actant with intra-pole and inter-pole relations (levels 5 and 6), which is confirmed by the minority of intra-pole and inter-pole relations established.

5.5.2 *Student Assessment After the Controversy Mapping Process*

Table 5.2 shows the students' assessment of whether the controversy mapping helped them to investigate the SSI of care spaces in the city.

As can be seen (Table 5.2), the predominant perception among the students was that controversy mapping proved to be a tool that helped them to develop a more critical perspective on the issue of care in the city, as expressed by five of the participants (e.g., *It helped me a lot to go out in the street and see the things that need*

Table 5.2 Student feedback on the controversy mapping activity

Categories	Students frequency
Taking a more critical look at city care	5
Ordering and relating ideas and reflecting	4
Helping to represent information	3
Better understanding of care at home and in the city	3
No response or general/vague response	2

more attention and care and things that in our house we have in a very comfortable and accessible way in the streets there are not). It supports the idea that controversy mapping helps to develop CT, specifically in the comprehensive analysis of complex issues, as shown in the response: *By graphically representing relationships that previously did not seem important, they acquire a new perspective.*

In addition, there was a perceived development of skills of ordering and relating ideas and reflection (four students), with expressions such as *to order and understand many concepts that they knew but were very disordered, as* well as their help in representing information (three students), (*to represent in a personal way how one would like those spaces of care to be*); both activities being necessary skills of the process to deal with the complexity of the issues.

Finally, it also served for the perception of an understanding of the SSI in terms of care spaces at home and the city (three students) (*although at that time, I did not yet know what the activity was about and what it was about, it was quite useful for me to ask myself what was missing at home that I missed in the city*).

5.6 Discussion and Educational Implications

This chapter shows that the analysis of the controversy maps with an adaptation of the Cabello-Garrido et al. (2023) scheme can be helpful to approximate students' understanding of the complexity of the controversy when the issue of whether there are adequate spaces for care in the city is posed to them.

From the results obtained, it can be concluded that controversy mapping is a tool that makes it possible to highlight the different levels of complexity with which students approach a controversy about an SSI, specifically in the comprehensive analysis of the problems. From it, the professor can discuss in class or provide information on hidden voices not represented (Bazzul & Siry, 2019), unestablished relationships, identify misinformation (Cebrián-Robles, 2019), etc., in the controversy maps developed to emphasise the need to recognise the complexity of the issue in order to be able to address it. It is essential to initiate the creative process of appropriate architectural design adapted to a city's complex and changing reality. However, one of the difficulties in evaluating the students' controversy maps is that it was difficult to count the relationships when the number of arrows was high.

On the other hand, only the number and level of relationships established were taken into account, and we did not go into the characteristics of the actants, the poles that grouped these actants or the solutions proposed. Therefore, we believe that the assessment instruments for the controversy maps should be improved and expanded in order to obtain a complete view of how the students show their understanding of the complexity of the SSIs, which will help to the comprehensive analysis of the problems.

The students' assessment after the implementation supports the idea that controversy mapping develops CT, in terms of critical capacity about the issue, and develops specific skills necessary for the comprehensive analysis of complex issues such as the ordering, relationship and reflection of ideas, as well as the visual representation of the issue. As future lines of research, it is necessary to extend this work to evaluate how controversy mapping contributes to developing skills for urban and architectural design in a complex and integral way.

Acknowledgements This work has been supported by the Spanish Ministry of Universities through the University Teacher Training Programme (FPU19/04507) and by the Spanish National Plan R+D+i Project entitled "Citizens with critical thinking: A challenge for teachers in science education", reference PID2019-105765GA-I00, funded by MCIN/AEI/10.13039/501100011033.

References

Bazzul, J., & Siry, C. (Eds.) (2019). *Critical voices in science education research: Narratives of hope and struggle* (Vol. 17). Springer. https://doi.org/10.1007/978-3-319-99990-6

Bencze, L., Pouliot, C., Pedretti, E., Simonneaux, L., Simonneaux, J., & Zeidler, D. (2020). SAQ, SSI and STSE education: Defending and extending "science-in-context." *Cultural Studies of Science Education, 15*(3), 825–851. https://doi.org/10.1007/s11422-019-09962-7

Cabello-Garrido, A., Cebrián-Robles, D., España-Ramos, E., González-García, F. J., Cruz-Lorite, I. M., España-Naveira, P., & Blanco-López, Á. (2023). The dominant model of meat production and consumption as a socially acute question for activist education. *Cultural Studies of Science Education, 18*(3), 911–935. https://doi.org/10.1007/s11422-023-10188-x

Callon, M., Rip, A., & Law, J. (1986). *Mapping the dynamics of science and technology: Sociology of science in the real world*. Palgrave Macmillan.

Cebrián-Robles, D. (2019). Identificación de noticias falsas sobre ciencia y tecnología por estudiantes del grado de Primaria [Identification of false news about science and technology by pre-service elementary science teachers]. *Píxel-Bit, Revista de Medios y Educación, 55*, 23–36. https://doi.org/10.12795/pixelbit.2019.i55.02

Cebrián-Robles, D. (2020). Desarrollo del pensamiento crítico con el análisis de las noticias falsas de contenido científico y tecnológico [Development of critical thinking through the analysis of fake news with scientific and technological content]. *Comunicación y Pedagogía, 319–320*, 75–78.

Chinchilla, I. (2020). *La Ciudad de los Cuidados. Salud, Economía y Medio Ambiente* [The City of Care. Health, Economy, and Environment]. Los Libros de la Catarata.

Christodoulou, A., Levinson, R., Davies, P., Grace, M., Nicholl, J., & Rietdijk, W. (2021). The use of cartography of controversy within socioscientific issues-based education: Students' mapping of the badger-cattle controversy in England. *International Journal of Science Education, 43*(15), 2479–2500. https://doi.org/10.1080/09500693.2021.1970852

Cooper, G. (2023). Examining science education in ChatGPT: An exploratory study of generative artificial intelligence. *Journal of Science Education and Technology, 32*, 444–452. https://doi.org/10.1007/s10956-023-10039-y

España-Naveira, P., Cruz-Lorite, I. M., Cebrián-Robles, D., Cabello-Garrido, A., España-Ramos, E., González-García, F. J., & Blanco-López, Á. (2023). Enfoque de cartografía de controversias para abordar cuestiones socialmente vivas desde la enseñanza de la ciencia y la tecnología [Cartography of controversy approach to address socially acute questions from science and technology education]. *Revista Eureka sobre Enseñanza y Divulgación de las Ciencias, 20*(3), 3101. https://doi.org/10.25267/Rev_Eureka_ensen_divulg_cienc.2023.v20.i3.3101

Eysenbach, G. (2020). How to fight an infodemic: The four pillars of infodemic management. *Journal of Medical Internet Research, 22*(6), e21820. https://doi.org/10.2196/21820

Franco-Mariscal, A. J., Cano-Iglesias, M. J., España-Ramos, E., & Blanco-López, A. (2025). The ENCIC-CT model for the development of critical thinking. In A. J. Franco-Mariscal (Ed.), *Critical thinking in science education and teacher training* (Chap. 1) (pp. 3–42). Springer. https://doi.org/10.1007/978-3-031-78578-8_1

Gutiérrez-Valdivia, B. (2021). *La ciudad cuidadora: Calidad de vida urbana desde una perspectiva feminista* [The Caring City: Urban Quality of Life from a Feminist Perspective] [Doctoral Thesis, Universitat Politècnica de Catalunya, Spain]. https://www.tesisenred.net/handle/10803/671506

Hervé, N. (2014). Cartographier des controverses pour apprendre la complexité des technosciences: l'étude des gaz de schiste en lycée agricole. *Revue Francophone Du Développement Durable, 4*, 155–170.

Klein, N. (2014). *This changes everything: Capitalism vs. the climate.* Simon & Schuster.

Kukina, I. (2019). Research on the multidimensionality of the renovation of the architectural and urban environment in the "post-carbon" city. In *2019 International conference on architecture: Heritage, traditions and innovations (AHTI 2019)* (pp. 469–473).

Latour, B. (2005). *Reassembling the social: An introduction to actor-network-theory.* Oxford University Press.

Levinson, R. (2006). Towards a theoretical framework for teaching controversial socio-scientific issues. *International Journal of Science Education, 28*(10), 1201–1224. https://doi.org/10.1080/09500690600560753

Martínez, L., & Short, J. R. (2021). The pandemic city: Urban issues in the time of COVID-19. *Sustainability, 13*(6), 3295. https://doi.org/10.3390/su13063295

Morin, E. (2007). *Introduction to complex thinking.* Sulina.

Muxi, Z. (2021). *Beyond the threshold. Women, houses, and cities.* DPR-Barcelona.

Puig, B., Blanco, P., & Pérez, J. J. (2021). "Fake news" or real science? Critical thinking to assess information on COVID-19. *Frontiers in Education, 6*, 1–10. https://doi.org/10.3389/feduc.2021.646909

Reis, P. (2014). Promoting students' collective socio-scientific activism: Teachers' perspectives. In J. Bencze, & S. Alsop (Eds.), *Activist science and technology education. Cultural studies of science education* (Vol. 9, pp. 547–574). Springer. https://doi.org/10.1007/978-94-007-4360-1_31

Sadler, T. D. (2011). *Socio-scientific issues in the classroom: Teaching, learning and research.* Springer Science & Business Media. https://doi.org/10.1007/978-94-007-1159-4

Valverde-Crespo, D., De-Pro-Bueno, A., & González-Sánchez, J. (2022). La fiabilidad de la información sobre ciencia de Internet y criterios utilizados para justificarla por parte de estudiantes de educación secundaria [Reliability of Internet science information for high school students and robustness of the criteria used to justify it]. *Revista Eureka sobre Enseñanza y Divulgación de las Ciencias, 19*(3), 3103–3103. https://doi.org/10.25267/Rev_Eureka_ensen_divulg_cienc.2022.v19.i3.3103

Venturini, T. (2010). Diving in magma: How to explore controversies with actor-network theory. *Public Understanding of Science, 19*(3), 258–273. https://doi.org/10.1177/0963662509102694

Venturini, T., Ricci, D., Mauri, M., Kimbell, L., & Meunier, A. (2015). Designing controversies and their publics. *Design Issues, 31*(3), 74–87. https://doi.org/10.1162/DESI_a_00340

Zhai, X. (2023). ChatGPT for next generation science learning. *XRDS: Crossroads, The ACM Magazine for Students, 29*(3), 42–46. https://doi.org/10.1145/3589649

Chapter 6
Critical Analysis of Information on Pasteurisation by High School Biology Students

José Antonio García-Pérez, Daniel Cebrián-Robles, and Ángel Blanco-López

6.1 Introduction

In today's secondary education, media literacy, such as the ability to search for information online, has become essential in the age of misinformation (Osborne et al., 2022). Social and scientific factors play central roles, especially in scientific learning through everyday life problems (Muñoz-Campos et al., 2020) or socio-scientific issues (Wu & Tsai, 2007). Moreover, with the growing importance of social networks and technology in adolescents' lives, it is crucial to explore how these tools can be used for these educational purposes.

In turn, some studies in the context of secondary education have focused on the use of information on the Internet in settings outside the school context to foster debate on socio-scientific issues, concluding that the Internet offers a huge opportunity for our students to develop argumentative competence and improve their scientific literacy (Greenhow et al., 2015). It has also been observed that the Internet can be of interest within secondary education classrooms too, in addition to engaging young people in debates on socio-scientific issues (García-Bermúdez et al., 2022), identifying fake news (Arnold, 2019), improving the search for information for the justification of scientific knowledge on the Internet (Cheng et al., 2021; Wu & Tsai, 2007), etc.

In order to use the Internet in the classroom, it is necessary to have connected devices, with mobile phones being particularly relevant due to their versatility and

J. A. García-Pérez (✉) · D. Cebrián-Robles · Á. Blanco-López
Science Education, Universidad de Málaga, Málaga, Spain
e-mail: jagarciaperez@uma.es

D. Cebrián-Robles
e-mail: dcebrian@uma.es

Á. Blanco-López
e-mail: ablancol@uma.es

A. J. Franco-Mariscal (ed.), *Critical Thinking in Science Education and Teacher Training*, Contemporary Trends and Issues in Science Education 64,
https://doi.org/10.1007/978-3-031-78578-8_6

ease of use. Preliminary results of some studies have determined that 73% of teachers working at this educational stage allowed their students to use smartphones in their classrooms to search for information on the Internet (Black-Fuller et al., 2016), demonstrating the need for effective and sustained development of the use of this type of mobile devices in secondary schools.

Mobile devices, particularly smartphones, are relevant in secondary and science education today, especially for media literacy and information seeking (Ubben et al., 2023). This is one of the significant changes taking place in schools today. Education is a system of cultural transmission, and the educator's task is to integrate and mediate between the different agents of the school reality (Campos, 2016). This is where the use of the smartphone in the secondary classroom comes in. The problem is no longer the quantity of information available on the Internet but the quality of it.

This chapter presents and analyses the impact of an Internet-based information search activity using smartphones to find evidence supporting a statement on everyday life issues, conducted with high school students in a Biology class.

6.2 Critical Analysis of Information as a Critical Thinking Skill in the Use of Evidence

Nowadays, argumentation as a scientific practice is fundamental in science education (Erduran & Jiménez-Aleixandre, 2008; Fischer et al., 2018). According to Rietz et al. (2021), the possession of scientific knowledge by secondary school students does not necessarily guarantee the development of their argumentation skills. The use of evidence and the ability to evaluate such evidence are crucial skills of argumentation competence (Bravo-Torija & Jiménez-Aleixandre, 2018; Osborne et al., 2016), and it is one of the dimensions of critical thinking (hereinafter, CT) (Franco-Mariscal et al., 2025).

Searching for information online using smartphone devices can facilitate the search for evidence needed to justify our conclusions. Incorporating complex and ambiguous evidence-seeking and evaluation activities on the Internet teaches students to develop criteria for judging knowledge claims from diverse sources of information while providing them with skills useful for lifelong learning (Bell & Linn, 2000). In addition, among the skills required for such evaluation are the critique and refutation of ideas as part of developing CT about scientific issues (Driver et al., 2000).

The evaluation of the quality and reliability of the knowledge that can be found on the Internet and social networks is fundamental. According to Fornás (2003), although it is not possible to establish strict rules for the reliability of Internet content, any Internet user should have a series of guidelines that allow them to evaluate the information to determine its quality in such a way that a series of criteria can be established to assess the quality and reliability of online content (Fig. 6.1).

In line with Fornás (2003) (Fig. 6.1), the criteria for assessing the quality and reliability of Internet content are authority, which aims to identify who is responsible for

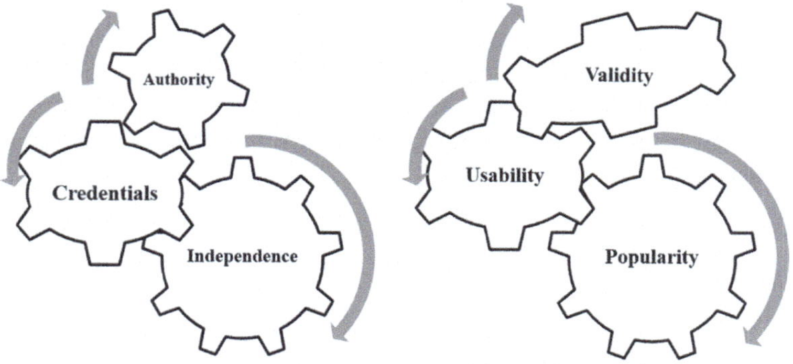

Fig. 6.1 Criteria for assessing the quality and reliability of Internet content (Fornás, 2003)

the contents of the website; credentials, which allow us to know the qualifications of the authors responsible for the contents; independence, which refers to avoiding any hint of confusion between the contents and the website's own commercial promotions or advertising messages; currency, which refers to the date of creation or the date of its last update or revision; usability, defined as ease of use; and finally, popularity, as there is no infallible correlation between the popularity of a website and the reliability of its contents.

Although applying these criteria does not guarantee that the contents accessed are reliable, they limit the possibility of using erroneous or outright false evidence and data. In any case, it is recommended not to use sources of information that cannot be checked or verified by other sources (Fornás, 2003).

Finally, previous research focused on promoting the understanding of evidence needed by secondary school students when they relate science to some fact in their daily lives, demonstrates the need for classroom practices that promote the search for, analysis and use of evidence in science teaching at this educational stage (Duncan et al., 2018). This fact, with the use of informal reasoning in the presentation of evidence (Bernal-Herrera et al., 2023; Ozturk & Yilmaz-Tuzun, 2017), as well as the implementation of activities of this type, encourage the use of medium–high quality evidence by students (Salcedo-Armijo et al., 2023). These findings reinforce the need to incorporate these practices in the educational environment.

6.3 Research Questions

This chapter looks at the development of argumentation as a dimension of CT through the critical analysis of information to support and justify claims from everyday life that students can find on the Internet. This suggests the following research questions (RQ):

- RQ1: What information sources do high school students use in Biology to support an everyday life statement about pasteurisation?
- RQ2: To what extent can a training proposal based on searching for information on the Internet using a smartphone help develop critical judgement to analyse and use information sources with high reliability and relevant evidence in their arguments?

6.4 Method

6.4.1 Participants and Context

This study involved 21 students aged 15–16 (71.4% female and 28.6% male) who took Biology in the first year of baccalaureate in a private educational centre in Málaga (Spain) during the 2018–2019 academic year.

This group can be considered a convenience sample due to the research team's ease of access to the student members (Jager et al., 2017). The teacher of this group (first author of this chapter) has a degree in biological sciences and more than 20 years of high school teaching experience in biology and geology. In addition, he received training in scientific argumentation before starting this research.

The sample of students was receiving a training programme based on scientific argumentation together with the contents of the subject. The programme was for two years and aimed to develop the student's competence in argumentation. All participants had yet to receive training on criteria for assessing the veracity of media information sources. This study presents a learning activity included in the first year of the training programme.

6.4.2 Experience Description

The learning activity analysed in this paper was carried out in two training programme sessions one hour each. In the first session, students were presented with a statement: *Pasteurisation is a key process in improving human food*, so each student, individually using their mobile devices, searched for data and evidence on the Internet to help them justify this statement. In addition to justifying, they were required to cite the sources of information from which they obtained the data and evidence.

After this first session, the teacher collected evidence from each student to analyse it with the research team and provide feedback to the students. After verifying the low reliability of the sources used, the research team decided to design a second session to improve the reliability of the students.

At the beginning of the second session, students were presented with the criteria for assessing the quality and reliability of Internet content proposed by Fornás (2003), together with an adaptation based on the reliability of the sources most used by secondary school students by Franco-Mariscal et al. (2014). Together with these criteria, different examples of each of the levels of reliability selected by the research team and other websites where they could be accessed were shown. For instance, they could use the Google Scholar website to access scientific articles practically unknown to most of them. After these first 20 min, they did the same practice again, collecting new arguments to justify the same assertion made in the first session and the new references used for this purpose.

6.4.3 ENCIC-CT Model Application

The learning activity is structured within the framework of the ENCIC-CT model (Franco-Mariscal et al., 2025). In the first layer of this model, an everyday life statement related to pasteurization, a key process in improving food safety, is addressed. In the second layer, the focus is primarily on the development of skills by critically analyzing information obtained from the Internet, as well as fostering personal autonomy as a key disposition in the learners. The third layer integrates the practice of scientific argumentation, where students must justify a previously provided statement based on evidence. Finally, in the fourth layer, mobile phones are used as a teaching strategy for searching data and evidence on the Internet. This approach encourages students to actively engage in independent research, enhances their ability to navigate and evaluate online sources, and fosters CT and digital literacy.

6.4.4 Data Collection and Analysis

Students' written productions showing the references used in their complete arguments during sessions 1 and 2 were used as a data collection instrument.

The research team analysed and categorised the level of reliability of each reference using reliability levels adapted from those proposed by Franco-Mariscal et al., (2014, p. 429) also for secondary school students, which are shown below in order from the lowest to the highest level of reliability of the referenced source:

- Dairy sector forums: low level of reliability.
- Blogs: medium–low level of reliability.
- Electronic encyclopaedia: medium level of reliability.
- Press: medium level of reliability.
- Specialised associations: medium–high level of reliability.
- Dairy professionals and universities: high level of reliability.
- Scientific research articles: very high level of reliability.

For example, we show how the references submitted by the same student in both sessions were categorised.

Session 1: *Mr Pasteur realised that the milk of some animals had certain bacteria that were difficult for humans to digest and found a way to kill them* (Quora.com) (Student 05).

After analysing this reference, it can be seen that it is a public place to obtain and share knowledge. It is a free platform where anyone can ask a question, and anyone else can answer it. This social network does not take responsibility for the possible questions and answers that may be given, nor does it consider the academic or work profile of the people who interact. However, it currently has more than 200 million users worldwide. This response was categorised as a public forum on a specific topic, in this case, the concept of pasteurisation, and therefore needs a higher level of reliability.

Session 2: *Pasteurisation, understood as a sterilisation process, allows milk to remain in perfect condition for months, even at room temperature, as long as the container is closed. Moreover, during this process, the nutritional properties of the product are not altered* (Spanish Association of Food Science and Technology) (Student 05).

In the second session, the student used data and evidence from an article authored by an association specialising in food processing. Therefore, it is a reference with a medium–high level of reliability.

The data analysis was carried out in two stages. First, the distribution of the references cited by the students was analysed in terms of their level of reliability, as shown in the examples above. Secondly, the number and relevance of the evidence used by the students to justify their claims were analysed, and their frequency was visualised in a bar chart, considering the comparison between the two sessions.

6.5 Results

The findings obtained after the analysis and categorisation of all the references used by the students in both sessions, together with the number of pieces of evidence used to justify the proposed statement and how many of them are relevant, are shown in Table 6.1.

It can be seen that, during the first session, of the total of 69 references cited by the students, the two most frequently used sources of information were electronic encyclopaedias (medium level of reliability) with 27.5% and forums on the subject dealt with (low level of reliability) with 24.6%.

This was followed by different blogs in the dairy sector (medium–low level of reliability) with 18.8%, as well as the press (medium level of reliability). Only 10.1%, as a result of the sum of information sources with medium–high, high and very high levels of reliability, were used by students to obtain data and evidence to present a complete argument to justify the statement.

Table 6.1 Results of the level of reliability of the relevant sources of information and evidence used by the students in each session

Source of information	Level of reliability	Session 1 (69 references) (%)	Number of total/relevant evidence	Session 2 (58 references) (%)	Number of total/relevant evidence
Scientific research articles	Very high	0.0	0/0	22.4	13/13
Dairy professionals/ Universities	High	2.9	2/2	60.3	34/33
Specialised associations	Medium–high	7.2	5/4	8.6	5/3
Press	Media	18.8	13/11	8.6	5/2
Electronic encyclopaedias	Media	27.5	20/11	0.0	0/0
Dairy or health blogs	Medium–low	18.8	10/2	0.0	0/0
Forums on the subject	Low	24.6	17/9	0.0	0/0
Total		100	67/39	100	57/51

However, the results obtained were very different during the second session, practically the reverse of the previous session. All sources of low, medium–low and medium reliability only reached 8.6%, while the sum of medium–high, high and very high-reliability levels was 91.3%. Among the latter, the use of data from industry professionals and universities (high level of reliability) stands out with 60.3%. This inversion of the results can be seen in Fig. 6.2, which compares the percentage of references according to the type of source used in both sessions.

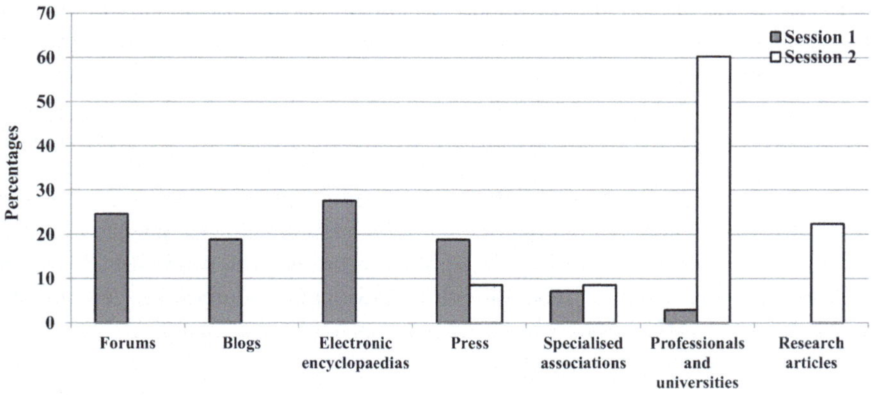

Fig. 6.2 Percentage of the type of reference used in each session

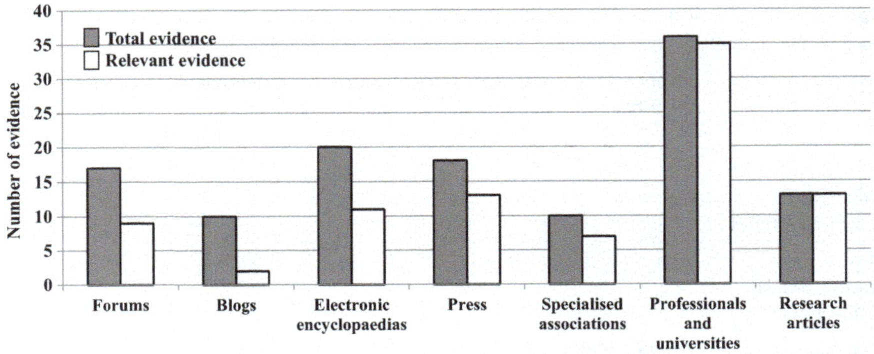

Fig. 6.3 Ratio of total/relevant evidence for each source of information in both sessions

Regarding the number of evidence used by the students to justify the statement provided, it can be seen that during the first session, they used a total of 67 evidence, which is more than in the second session, where they provided 57 evidence among all the justifications presented.

In addition, the number of relevant evidence was analysed for each case. There is a direct relationship between the level of reliability of the information source used by the students and the relevance of the evidence selected, with the more relevant evidence being used, the higher the level of reliability. Figure 6.3 shows this relationship considering the type of information source for both sessions.

6.6 Discussion and Educational Implications

The findings of this work, although limited by the sample of students with whom we worked, led us to believe that the number of sources selected by the students before they were aware of the criteria for their reliability was more significant in the first session than in the second, due to a narrowing down to improve the reliability of these sources of information.

If the evidence is more reliable, they help our students improve their justification of the problem. However, the scientific terminology in the high-reliability content was more complicated for them to understand than the terms used by the sources in the first session, based on the majority of comments collected by the teacher from his students. This fact reinforces the idea of strengthening the need to improve our students' scientific vocabulary. In addition, it has been shown that high school students do not question the reliability of the information sources they select daily (Singh & Lubbe, 2011). Similarly, students do not have a critical attitude about how to search for and select information from the Internet (Valverde & González, 2016) even though they have and handle smartphones daily. However, with some brief indications on criteria for selecting such sources with a high level of reliability and a series of chosen previous examples, students could discard some very commonly used sources of information for others who would not have realised their level of reliability until that moment.

Finally, and sticking to the research questions, we can conclude that:

> Regarding RQ1, without the appropriate criteria, students use references as sources of information that lack a medium–high level of reliability. Students of this age are not concerned about whether the data and evidence in these sources are accurate, nor do they check the veracity of the information shown, using forums, personal blogs, electronic encyclopaedias and press reports as their main sources.
> In relation to RQ2, devoting a session to introducing students to criteria and examples of searching for information in medium–high, high and very-high reliability sources allowed them to access and learn how to use these sources to support the proposed claim. This enabled them to exercise CT in the face of the wealth of information about the proposed claim available on the Internet. According to Bravo-Torija and Jiménez-Aleixandre (2012), the classroom is conceived as a learning community, i.e., classes in which students discuss and pose problems and try to solve them.

The analysis of the results concludes that using information sources with a high level of reliability allows students to use more relevant evidence in their justifications. Limiting the sources of information to high levels of reliability has likely had a positive influence on the quality and, thus, the relevance of the evidence selected. This should stimulate students to consider more critically the sources of information they regularly use to justify their claims.

Acknowledgements This work is part of the Spanish National Plan R+D+i Project entitled "Citizens with critical thinking: A challenge for teachers in science education", reference PID2019-105765GA-I00, funded by MCIN/AEI/10.13039/501100011033.

References

Arnold, R. (2019). *Fake news in science and education*. Rowman & Littlefield.

Bell, P., & Linn, M. C. (2000). Scientific arguments as learning artifacts: Designing for learning from the web with KIE. *International Journal of Science Education, 22*(8), 797–817. https://doi.org/10.1080/095006900412284

Bernal-Herrera, P., Cano-Iglesias, M. J., Franco-Mariscal, A. J., & Blanco-López, Á. (2023). Impacto de un debate sociocientífico en las habilidades argumentativas y en la toma de decisiones del profesorado de secundaria en formación inicial [Impact of a Socio-Scientific Debate on the Argumentative and Decision-Making Skills of Preservice Secondary Teachers]. *Enseñanza de las Ciencias, 41*(3), 113–132. https://doi.org/10.5565/rev/ensciencias.5861

Black-Fuller, L., Taube, S., Koptelov, A., & Sullivan, S. (2016). Smartphones and pedagogy: Digital divide between high school teachers and secondary students. *US-China Education Review, 6*(2), 124–131. https://doi.org/10.17265/2161-623X/2016.02.005

Bravo-Torija, B., & Jiménez-Aleixandre, M. P. (2012). Progression in complexity: Contextualizing sustainable marine resources management in a 10th grade classroom. *Research in Science Education, 42*, 5–23. https://doi.org/10.1007/s11165-011-9254-1

Bravo-Torija, B., & Jiménez-Aleixandre, M. P. (2018). Developing an initial learning progression for the use of evidence in decision-making contexts. *International Journal of Science and Mathematics Education, 16*(4), 619–638. https://doi.org/10.1007/s10763-017-9803-9

Campos, J. A. (2016). *El uso de las TIC, dispositivos móviles y redes sociales en un aula de la educación secundaria obligatoria* [The use of ICT, mobile devices, and social media in a compulsory secondary education classroom] [Doctoral Thesis, University of Granada, Spain]. http://hdl.handle.net/10481/42209

Cheng, C.-H., Bråten, I., Yang, F.-Y., & Brandmo, C. (2021). Investigating structural relationships among upper-secondary school students' beliefs about knowledge, justification for knowing, and Internet-specific justification in the domain of science. *Journal of Research in Science Teaching, 58*(7), 980–1009. https://doi.org/10.1002/tea.21689

Driver, R., Newton, P., & Osborne, J. (2000). Establishing the norms of scientific argumentation in classrooms. *Science Education, 84*(3), 287–312. https://doi.org/10.1002/(SICI)1098-237X(200005)84:3%3c287::AID-SCE1%3e3.0.CO;2-A

Duncan, R. G., Chinn, C. A., & Barzilai, S. (2018). Grasp of evidence: Problematizing and expanding the next generation science standards' conceptualization of evidence. *Journal of Research in Science Teaching, 55*(7), 907–937. https://doi.org/10.1002/tea.21468

Erduran, S., & Jiménez-Aleixandre, M. P. (2008). *Argumentation in science education. Perspectives from classroom-based research* (Vol. 35). Springer. https://doi.org/10.1007/978-1-4020-6670-2_1

Fischer, F., Chinn, C. A., Engelmann, K., & Osborne, J. (2018). *Scientific reasoning and argumentation: The roles of domain-specific and domain-general knowledge.* Routledge.

Fornás, R. (2003). Criterios para evaluar la calidad y la fiabilidad de los conocimientos en Internet [Criteria for assessing the quality and reliability of knowledge on the Internet]. *Revista Española de Documentación Científica, 26*(1), 75–80. https://doi.org/10.3989/redc.2003.v26.i1.226

Franco-Mariscal, A. J., Cano-Iglesias, M. J., España-Ramos, E., & Blanco-López, A. (2025). The ENCIC-CT model for the development of critical thinking. In A. J. Franco-Mariscal (Ed.), *Critical thinking in science education and teacher training* (Chap. 1) (pp. 3–42). Springer. https://doi.org/10.1007/978-3-031-78578-8_1

Franco-Mariscal, A.J., España-Ramos, E., & Blanco-López, A. (2014). Uso de Internet para analizar las relaciones entre drogas y salud bucodental. Una experiencia en 4º de E.S.O [Use of the Internet to analyze the relationships between drugs and oral health. An experience in 4th year of Secondary Education]. In E. Soriano-Ávila, A. J. González-Jiménez, & V. Cala (Eds.), *Retos actuales de educación y salud transcultural* [Current challenges in transcultural education and health] (pp. 1–8). University of Almería.

García-Bermúdez, S., Reis, P., & Vázquez-Bernal, B. (2022). Facebook como herramienta para promover el activismo ambiental en las clases de ciencias [Facebook as a Tool to Promote Environmental Activism in Science Classes]. *Enseñanza de las Ciencias, 40*(3), 51–70. https://doi.org/10.5565/rev/ensciencias.2935

Greenhow, C., Gibbins, T., & Menzer, M. M. (2015). Re-thinking scientific literacy out-of-school: Arguing science issues in a niche Facebook application. *Computers in Human Behavior, 53*(Supplement C), 593–604. https://doi.org/10.1016/j.chb.2015.06.031

Jager, J., Putnick, D. L., & Bornstein, M. H. (2017). More than just convenient: The scientific merits of homogeneous convenience samples. *Monographs of the Society for Research in Child Development, 82*(2), 13–30. https://doi.org/10.1111/mono.12296

Muñoz-Campos, V., Franco-Mariscal, A. J., & Blanco-López, Á. (2020). Integration of scientific practices into daily living contexts: A framework for the design of teaching-learning sequences. *International Journal of Science Education, 42*(15), 2574–2600. https://doi.org/10.1080/09500693.2020.1821932

Osborne, J., Henderson, J. B., MacPherson, A., Szu, E., Wild, A., & Yao, S. (2016). The development and validation of a learning progression for argumentation in science. *Journal of Research in Science Teaching, 53*(6), 821–846. https://doi.org/10.1002/tea.21316

Osborne, J., Pimentel, D., Alberts, B., Allchin, D., Barzilai, S., Bergstrom, C., Coffey, J., Donovan, B., Dorph, R., Kivinen, K., Kozyreva, A., Perkins, K., Perlmutter, S., & Wineburg, S. (2022). *Science education in an age of misinformation.* Stanford University. https://t.ly/BaAZ

Ozturk, N., & Yilmaz-Tuzun, O. (2017). Preservice science teachers' epistemological beliefs and informal reasoning regarding socio-scientific issues. *Research in Science Education, 47*, 1275–1304. https://doi.org/10.1007/s11165-016-9548-4

Rietz, L., Jönsson, A., & Lundström, M. (2021). Students' use of justifications in socioscientific argumentation. *Nordic Studies in Science Education, 17*(3), 247–264. https://doi.org/10.5617/nordina.8203

Salcedo-Armijo, E., Cano-Iglesias, M. J., Blanco-López, Á., & Franco-Mariscal, A. J. (2023). Uso de pruebas y decisiones de profesores en formación inicial en un debate sobre prohibición de plásticos [Use of evidence and decisions of pre-service teachers in a debate on plastics bans]. *Educación Química, 34*(4), 104–118. https://doi.org/10.22201/fq.18708404e.2023.4.84663

Singh, A., & Lubbe, S. (2011). E-lumination-evaluating the quality of free online information for decision-making. *Alternation, 18*(1), 252–275.

Ubben, M. S., Kremer, F. E., Heinicke, S., Marohn, A., & Heusler, S. (2023). Smartphone usage in science education: A systematic literature review. *Education Sciences, 13*(4), 345. https://doi.org/10.3390/educsci13040345

Valverde, D., & González, J. (2016). Búsqueda y selección de información en recursos digitales: Percepción de alumnos de Física y Química de Educación Secundaria Obligatoria y Bachillerato sobre Wikipedia [Search and selection of information in digital resources: Perception of secondary and high school Physics and Chemistry students about Wikipedia]. *Revista Eureka sobre Enseñanza y Divulgación de las Ciencias, 13*(1), 67–83. https://doi.org/10.25267/Rev_Eureka_ensen_divulg_cienc.2016.v13.i1.06

Wu, Y., & Tsai, C. (2007). High school students' informal reasoning on a socioscientific issue: Qualitative and quantitative analyses. *International Journal of Science Education, 29*(9), 1163–1187. https://doi.org/10.1080/09500690601083375

Part III
Argumentation and Critical Thinking

Chapter 7
Secondary School Students' Argumentation on Energy Production and Consumption

Joaquín Cañero-Arias, Ana Belén Ferrer-Baldomero, María José Cano-Iglesias, and Antonio Joaquín Franco-Mariscal

7.1 Introduction

Today's society needs a critical citizenry, literate in science and technology, with the skills to engage and participate in decision-making on the wide range of socio-scientific issues and controversies facing humanity (Colucci-Gray et al., 2006; Fensham, 2011; Hodson, 2003). One of these is the energy problem. Electricity consumption and mandatory generation are crucial to our society's economic and technological development, serving as key indicators of progress and well-being (Martín et al., 2013). The growing global demand for electricity poses significant environmental and social challenges. These challenges are linked to the growing need for electricity, the controversy surrounding various energy sources and their environmental impact (Alderman & Pronczuk, 2022), and increasingly pronounced global and local climate change. In addition, there are concerns about "energy poverty" (Sy & Mokaddem, 2022), which affects many families, especially during the winter months.

J. Cañero-Arias (✉) · M. J. Cano-Iglesias · A. J. Franco-Mariscal
Science Education, Universidad de Málaga, Málaga, Spain
e-mail: jqncarias@uma.es

M. J. Cano-Iglesias
e-mail: mjcano@uma.es

A. J. Franco-Mariscal
e-mail: anjoa@uma.es

J. Cañero-Arias · A. B. Ferrer-Baldomero
Colegio La Presentación, Málaga, Spain
e-mail: anabelenbaldomero@gmail.com

A. J. Franco-Mariscal (ed.), *Critical Thinking in Science Education and Teacher Training*, Contemporary Trends and Issues in Science Education 64,
https://doi.org/10.1007/978-3-031-78578-8_7

The environmental, economic, and social problems associated with the production and consumption of electricity can be understood by analysing the various energy sources available (Martín et al., 2013). Thus, fossil fuel sources such as coal, oil and natural gas have historically prevailed as the main sources of electricity and have undoubtedly contributed to technological development and improved quality of life. However, their combustion generates greenhouse gases that contribute significantly to climate change. On the other hand, the extraction of fossil resources damages ecosystems and nearby communities, as in the case of natural gas through *fracking*. Other energy sources, such as biomass and biogas, derived from organic waste, may be considered more sustainable than fossil fuels, but have the associated dilemma of food production and deforestation or fuel generation.

Electricity from nuclear power plants currently poses social acceptance problems due to misgivings about safety and waste treatment as green energy (Alderman & Pronczuk, 2022). Nevertheless, it is an energy source with very low carbon dioxide emissions, and research is currently being carried out into new fuel substitutes for uranium, such as thorium (International Atomic Energy Agency, 2022), and the reuse of waste.

In addition, solar, wind or hydroelectric power should be highlighted as renewable energy sources, whose environmental impact on emissions is much lower than those collected from fossil sources, but which present their environmental challenges, such as the extensive use of land for solar and wind installations (Spellman, 2014), and the potential disruption of ecosystems.

All these issues can be addressed in secondary education to promote critical thinking (hereinafter, CT) and evidence-based decision-making on relevant issues such as energy, a core and unifying concept across all science disciplines and grade levels (Park & Liu, 2016). Science education should raise awareness among new generations of the social and environmental problems associated with energy generation and consumption, encouraging the search for innovative and sustainable solutions to ensure a better future for our planet (Feinstein & Kirchgasler, 2015). This idea is included in the Spanish curriculum by introducing the development of scientific thinking and CT that enables students to confront the social problems surrounding them (Zeidler et al., 2005) and gain a greater understanding of the world.

The socio-scientific issues framework has significantly impacted the development of functional scientific literacy in students over the last few decades (Zeidler et al., 2019). Addressing these issues at the secondary school level is essential for informing citizens about current and future energy and environmental challenges. In this scenario, science teachers play an important role, as they must facilitate the acquisition of a solid scientific basis for their students, helping them to understand the knowledge associated with the different energy sources, their advantages and disadvantages, and their impacts. Always from an objective point of view and based on up-to-date scientific evidence, using real data, concrete examples and close to the students to illustrate not only the curricular content but also to address how decision-making about energy affects people and the planet.

In this context, this chapter presents the results obtained from Spanish secondary school students who approach the problem of energy production and consumption from the perspective of the development of CT.

7.2 Knowledge Application, Critical Analysis of Information, and Metacognition as Important Dimensions of Critical Thinking in the Energy Problem

The study by Martín et al. (2013) revealed that the treatment of the energy problem in secondary school textbooks lacks the information necessary to convey the controversial nature and the enormous relevance of the issue for our lifestyle and future. This is because textbooks do not clearly show the advantages and disadvantages of energy production activities and the social, environmental, and economic implications of energy consumption.

Although proposals can be found in the literature for addressing the socio-scientific issue of energy (Cruz-Lorite et al., 2022; Lee, 2016), many of these studies have focused on describing the attitudes (Kilinc et al., 2013), values, feelings, preferred sources of information or processes of informal reasoning, and decision making (Ercan et al., 2015), with scant attention being paid to the development of dimensions of CT such as scientific knowledge application, critical analysis of information, and metacognition. An interesting work in line with student thinking about the energy problem is the study by Lee (2016), who examined German students' knowledge about the contribution of different energy sources to their country. This research identified students' mistaken beliefs about the role of different energy sources, which could lead to biases in decision-making when supporting or rejecting energy proposals. In this context, the appropriate use of the involved scientific knowledge is essential for understanding the issue.

For this reason, strategies are needed to address the problem holistically by considering the different factors involved, thus promoting CT. Strategies such as brainstorming, classroom discussion or role-playing (Cruz-Lorite et al., 2022) help students to express their opinions, listen to others, and debate and argue from different perspectives of the problem (Jiménez-Aleixandre, 2010), which not only encourages CT but also raises awareness of the need to build a sustainable energy future.

It is, therefore, about providing learners with opportunities to engage in argumentation. In the context of science education, argumentation is essential for developing both CT and scientific understanding (Erduran & Jiménez-Aleixandre, 2007). This process involves not only constructing well-supported arguments that include evidence, warrants, conclusions, background information, and theoretical frameworks (Toulmin, 2003) but also critically evaluating and challenging the arguments presented by others.

Essentially, argumentation requires students to analyze, evaluate, and synthesize information to make informed decisions and solve complex problems (Bailin, 2002), in this case, related to energy. In this sense, the European Commission (2018) recently promoted the plan to combat misinformation to raise awareness in society, and therefore among students, of the need to confront the enormous amount of information we receive daily.

Achieving this challenge requires using real databases close to the context of the students, digital tools, and simulation as a tool for familiarisation with this type of information (Batanero & Díaz, 2011). In relation to critical information analysis, the strategy should provide up-to-date energy production and consumption data. An alternative could be to provide students with various reports from international organisations, such as the International Energy Agency (IEA, https://www.iea.org), which detail the state of energy at the global, European or national level. However, this data tends to become outdated quickly. A more efficient option is to use applications that provide real-time data on energy production and consumption, such as *Electricity Maps* (https://www.electricitymaps.com), a freely available application designed by Electricity Maps ApS that is available in several languages, including Spanish, which has been used in this study.

Concerning metacognition, this dimension encompasses personal autonomy and facilitates decision-making (Eggert et al., 2013; Franco-Mariscal et al., 2025). In science education, metacognition can enhance learning and develop CT, such as in scientific problem-solving or evaluating scientific evidence and detecting cognitive biases, i.e., misinterpretations of a reality (Vila et al., 2023).

The role of the teacher is also essential in student reflection, as the teacher must promote constructive feedback to help them reflect on their thinking. In this way, critical minds will be fostered. Furthermore, metacognition is interconnected with decision-making by enabling students to examine their thinking process to make informed decisions based on their knowledge (Eggert et al., 2013). All of this will contribute to the learner developing better arguments in their decision-making. Integrating these skills into the everyday classroom will help better prepare our students for current and future challenges (Sadler, 2009).

7.3 Research Question

This chapter addresses the research question: What evidence do Spanish secondary school students in eighth-grade use in their arguments when comparing electricity production and consumption between different countries before and after critically analysing energy information?

7.4 Method

7.4.1 Participants and Context

The study sample consisted of 58 eighth-grade Spanish secondary school students, with an average age of 13 years, enrolled in Physics and Chemistry in a publicly funded private school in Malaga (Spain) during the academic year 2022–2023. These students belonged to two class groups, 29 in group A (14 girls and 15 boys) and 29 in group B (12 girls and 17 boys). The student's academic performance was medium–high; they showed interest and dedication to science, with deficient levels of conflict and absenteeism.

The implementation was carried out by two teachers (the first and second authors of the chapter), who always coordinated its development in the classroom. The first author, with a solid background in science education and nine years of teaching experience, designed the experience. With two years' expertise, the second author contributed contributions that improved the sequencing of activities and served to outline the implementation in the classroom.

7.4.2 Electricity Maps Application

The experience addresses the issue of electricity production and consumption in Europe using the *Electricity Maps* application (Fig. 7.1). This application is a simulation that allows users to visualize the origin of the electricity consumed, the environmental impact of its generation, and even the interconnection between electricity systems. Additionally, it offers real-time, historical, and statistical data on the evolution of electricity demand and supply based on the different sources of production available in each country and/or region.

Students can check and compare each country's electricity production by energy source (nuclear, geothermal, biomass, coal, wind, solar, hydro, natural gas or oil), type of storage (hydro, battery), emissions associated with these sources, and other data such as the percentage of each source compared to the total installed power, or the contribution to carbon dioxide emissions of the country concerned.

In addition, they have access to other relevant data, such as carbon intensity, which refers to the amount of carbon dioxide emitted per kilowatt hour. They can also find out the percentage of all energy sources low in emissions (referring to carbon dioxide) and renewable sources. The application also provides information in different time slots. Figure 7.2 shows the electricity consumption for the same country during the night (left image) and during the day (right image). As can be seen, there is a variation in electricity consumption and carbon intensity with the addition of solar energy (150 g CO_2 eq/Kwh at night and 88 g CO_2 eq/Kwh in the presence of the sun).

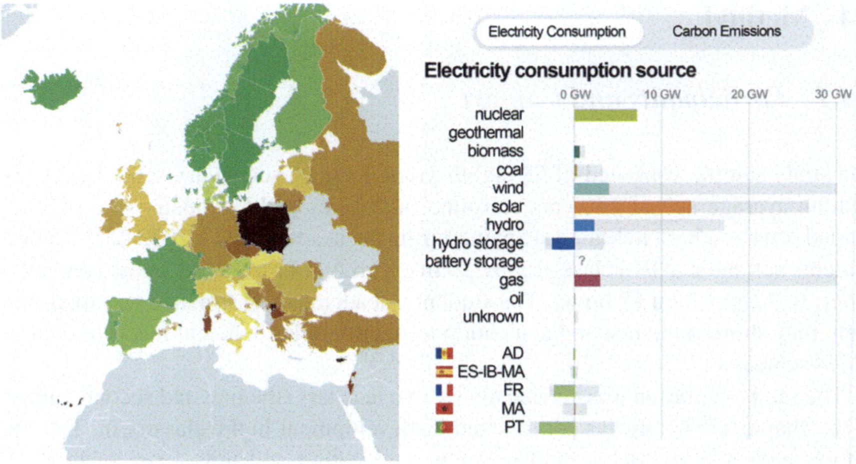

Fig. 7.1 *Electricity Maps* application

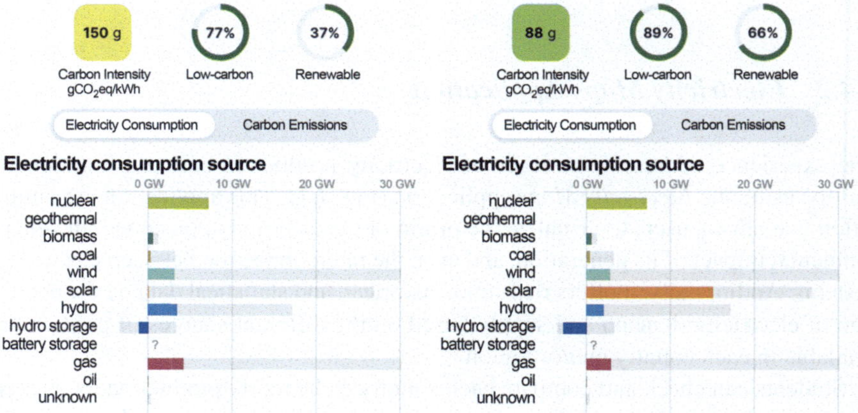

Fig. 7.2 Example of variation in consumption and carbon intensity with incorporating solar energy

7.4.3 Experience Description

The experience was carried out over seven sessions of 50 min each, with the development of the activity distributed in five of these sessions, while the remaining two were devoted to the presentation and discussion of each group's proposals. The tasks that were carried out are detailed below:

- *Task 1*. Reflection on the current state of energy in the world based on a video.

Prior to presenting the sequence to the students and as an introduction to the Energy unit in the Physics and Chemistry subject, we showed the video *Dreaming the Future: Energy in 2050* (https://www.youtube.com/watch?v=aOVR3R6aH0g) along with a "See Think Wonder" thinking routine (Fontichiaro, 2010). The video addresses what energy consumption will be like in the year 2050 and the challenges facing humanity with respect to energy generation for the entire world population. This task aimed to raise awareness of the global energy situation.

- *Task 2.* Presentation of the experience, individual reflection on energy production and consumption in various countries and sharing in small groups.

The students were introduced to the experience, which included using an application to compare electricity production and carbon dioxide emissions between Spain and two European countries. The teacher explained that the final product would be a group exhibition where they had to reflect on the comparison and propose improvements to reduce carbon dioxide emissions in Spain. The students showed interest and curiosity because they had not previously worked on these curricular contents with this methodology and had to use an application.

The teacher organised the class into groups of four students and assigned them to two countries according to size and economic situation to compare with Spain. Thus, for example, one group compared Spain with France and Germany, and another compared Spain with Ireland and Portugal. Individually, the students had to decide, without accessing any source of information, which of the three countries assigned to their group would have the most significant advantages and disadvantages from an energy point of view in terms of carbon dioxide emissions, renewable and non-renewable energy sources, energy prices, etc. (see the full statement in Figure 7.3).

Each group then pooled their ideas, looking for at least three similarities and differences between the contributions made by each component. Finally, the students were informed that, in groups, they had to establish a fictitious international body that would act as an inspector, in charge of carrying out and presenting a comparison between countries, emphasising potential improvements for Spain. To carry out this task, they had to analyse all the information in the application.

- *Task 3.* Group work with the *Electricity Maps* application.

This task consists of comparing the countries in the app. Before use, teachers explained the concepts of renewable energy, carbon intensity, low emissions or kilowatt-hours included in the app. Nevertheless, some groups required the teacher's attention to guide them in interpreting the data provided by the app until they became familiar with it. The students had to check possible variations in electricity production or consumption from the different energy sources available for each country and the carbon dioxide emissions generated during different time slots (e.g., offsetting solar energy during the night). Given that this process could be lengthy, the task was limited to two hours of their choice from those provided by the application. Once the comparison had been made and based on the data collected, they had to agree on at least two energy improvements for Spain, mainly related to reducing carbon dioxide emissions. As this was a group task, the teachers ensured that the distribution

Individual task

1. Different energy qualities are understood as energy sources in the three countries your group is assigned to, including Spain. Decide which of them you would have the most **advantages and disadvantages from an energy point of view** to the number of energy sources, price of energy, number of renewable and non-renewable energy sources, and carbon dioxide emissions into the atmosphere throughout the day and night. **Please justify your answer**.

Fig. 7.3 Example of an activity presented in the portfolio (image generated by ChatGPT, (AI-generated image), retrieved from ChatGPT by OpenAI)

of tasks was fair and that the sharing was appropriate. It was also stressed that data collection and interpretation should be done comparatively between countries.

- *Task 4*. Presentation and discussion of the work carried out. Presentation of opportunities for improvement in Spain.

This task consisted of an oral presentation of the comparisons made. The presentation format was free; they could refer to the fictitious international organisation they represented. After each presentation, there was a question-and-answer session in case the rest of the groups wished to make any comments or question the speakers.

- *Task 5*. Individual reflection on energy production and consumption in different countries.

The last task, similar to task 2, was carried out individually. It required them to indicate the advantages and disadvantages from an energy point of view of each country on which they had collected information.

In order to properly develop the sequence of activities, each student had a portfolio to respond in writing to the tasks set (Fig. 7.3) and record individual and group contributions.

To conclude this section, it is important to highlight that the presentation and debate process played a crucial role not only in promoting critical reflection, but also in enabling students to refine their communication strategies and deepen their conceptual understanding through feedback from both peers and teachers. This iterative process of evaluating the effectiveness of their arguments is pivotal to their metacognitive development, prompting them to adapt and refine their approaches based on peer interaction.

Furthermore, the concluding individual reflection enables students to critically evaluate their own learning progress. Indeed, this act of self-assessment, an essential component of the broader metacognitive process, enhances their capacity to assess their learning outcomes and facilitated the advancement of self-regulated learning. By reflecting on their evolving understanding of energy production and consumption, students are able to consolidate both their subject knowledge and their personal development.

Finally, the collaborative aspect of the activities encourages a critical review of students' thinking through the exchange of ideas within groups. This sharing of perspectives enhanced the learning experience and facilitates meaningful adjustments to students' initial assumptions. By engaging with their peers in a collaborative environment, students are empowered to challenge and refine their preconceived notions, thereby further enhancing the depth of the metacognitive learning process.

7.4.4 ENCIC-CT Model Application

Firstly, the ENCIC-CT model, with its multidimensional structure, is intentionally designed to cultivate CT within science education (Franco-Mariscal et al., 2025). This objective is reflected in its focus on global energy issues, which allows for the examination of complex, real-world issues from various perspectives. By focusing on a topic of significant societal and scientific relevance, the experience aligns with the core principles of the ENCIC-CT model, particularly its emphasis on using real-life problems to stimulate CT (Layer 1).

Moreover, the experience significantly contributes to the application of scientific knowledge regarding energy within the knowledge domain, as well as to the critical analysis of information, argumentation, and communicative skills within the skills domain, and to metacognitive development in the dispositions domain (Layer 2). This approach not only encourages students to engage in reflective consideration of the advantages and disadvantages inherent in various energy models but also deepens their understanding of the complexities involved.

Furthermore, Layer 3 of the ENCIC-CT model places a pronounced emphasis on the scientific practices of inquiry and argumentation. This includes the systematic gathering of information and data analysis, which not only enhances students' comprehension of energy-related issues but also boosts their capacity to critically evaluate information. Additionally, the process of formulating and debating solutions substantially improves students' ability to construct well-founded arguments.

Lastly, to support these efforts, a digital tool—specifically, a simulation for comparing electricity production and carbon dioxide emissions across various countries—served as the teaching strategy (Layer 4). This tool enabled students to apply scientific knowledge within a real-world context, thereby reinforcing the practical application of CT.

7.4.5 Data Collection and Analysis

The data collection instrument was the student's portfolio, where each student provided written responses to the assigned tasks and documented both individual and group reflections throughout the sequence of activities. The portfolio served as the primary tool for providing information about the integration and consolidation of knowledge, the critical analysis of information, and the metacognition acquired by students throughout the experience.

Concerning the application of scientific knowledge and critical analysis of information, two aspects were examined in the arguments presented in the initial and final individual productions in the country comparison (tasks 2 and 5, respectively). On the one hand, the use of energy sources and associated carbon dioxide emissions as evidence for the argument. On the other hand, the inclusion of terms relevant to understanding the problem, such as carbon intensity, low-carbon energy sources or renewable energies, all addressed in the application, was assessed. The use of other terms related to carbon dioxide emission or production and pollution, which were mentioned ambiguously, was also analysed.

The following arguments given by an eighth-grade student before and after using the application serve to illustrate this analysis:

> Ireland may have more energy advantages because of its green spaces and because it may have more renewable energy (Eighth-grade student response before using the application).
>
> Ireland emits the most CO_2 of the three countries compared, and the carbon intensity is three times higher than in Spain (Eighth-grade student response after using the application).

In this case, the initial argument is based solely on the student's experience or opinion. After using the app in the classroom, the student uses the terms carbon dioxide emission and carbon intensity to justify that Ireland produces more carbon dioxide than Spain, which the student incorporated into her comparison.

The percentage of students who used them at the two points in time was calculated for each evidence included in the arguments. Additionally, McNemar's test was applied to determine possible statistically significant differences between before and after, as these are nominal variables for related samples.

Regarding metacognition, the evolution of the arguments presented by each student in both individual and group tasks throughout the experience was analyzed qualitatively. This analysis focused on a sample of these tasks to assess how students' arguments developed over time, reflecting their growing understanding of the issue. The aim was to gain insights into the students' reflective processes and how their ability to evaluate and adjust their thinking evolved throughout the activities.

7.5 Results

7.5.1 Results About Knowledge Application and Critical Analysis of Information

Table 7.1 presents the percentage of students who mentioned each energy source or associated carbon dioxide emission before and after using the application to compare different countries' energy advantages and disadvantages. Table 7.2 presents the results obtained in the use of some relevant terms.

As seen in Tables 7.1 and 7.2, the application increased the inclusion of evidence in the arguments related to carbon dioxide sources or emissions associated with specific energy sources and energy-specific terms.

However, McNemar's test only found statistically significant differences in the sources coal ($p = 0.004$), natural gas ($p = 0.002$) and solar energy ($p = 0.031$), and in the use of the three terms worked out in the application (carbon intensity: $p = 0.008$, low-carbon energy: $p = 0.013$, renewable energy: $p = 0.000$), as well as in the ambiguous use of carbon dioxide emission/production ($p = 0.035$), in all cases in favour of the final moment.

Table 7.1 Students' percentages who mentioned the associated energy source or carbon dioxide emission before and after using the app

		Before using the app	After using the app	McNemar test (*p*)
Nuclear energy	Source	10.3	19.0	0.267
	Carbon dioxide	0.0	5.2	0.250
Geothermal energy	Source	0.0	0.0	–
	Carbon dioxide	0.0	0.0	0.063
Biomass energy	Source	3.4	1.7	1.000
	Carbon dioxide	3.4	0.0	1.000
Coal	Source	0.0	19.0	0.004
	Carbon dioxide	0.0	8.6	0.063
Wind energy	Source	0.0	22.4	0.167
	Carbon dioxide	0.0	1.7	1.000
Solar energy	Source	39.7	19.0	0.031
	Carbon dioxide	1.7	0.0	1.000
Hydropower	Source	10.3	15.5	0.549
	Carbon dioxide	0.0	1.7	1.000
Natural gas	Source	0.0	17.2	0.002
	Carbon dioxide	0.0	8.6	0.063
Oil	Source	12.1	5.2	0.289
	Carbon dioxide	0.0	0.0	–

Table 7.2 Percentages of students who mentioned different energy-related terms in their arguments

	Before using the app	After using the app	McNemar test (p)
Carbon intensity	0.0	13.8	0.008
Low-carbon energy	5.2	24.1	0.013
Renewable energy	3.9	75.9	0.000
It ambiguously mentions carbon dioxide emission or production	25.9	44.8	0.035
Mentions the term pollution in an ambiguous way	8.6	5.2	0.687

7.5.2 Results About Metacognition

This section presents selected evidence from students' portfolios to illustrate how metacognition is addressed throughout the learning process. To this end, the work of two students is shown as representative examples of metacognitive work carried out throughout the experience described above.

Firstly, the initial individual responses of student 2LH14 whose countries to compare were Ireland, Portugal and Spain and student 2MM23 whose countries were Hungary, Finland and Spain are shown, in which the students answer about the energy situation of the three countries based on their preconceptions only:

> Ireland may have more energy advantages because it has many green spaces and perhaps more renewable energy. Portugal seems to me that it could have hydroelectric energy because it has a lot of coastline. Although Spain may have a lot of renewable energy, I get the feeling that it emits more CO_2 into the atmosphere (Student 2LH14).

> I think that less energy is used in Spain because Finland and Hungary are much colder, and they have to use more energy to heat themselves. In Spain, prices are much higher than in Finland and Hungary. Finland uses more renewable energy than Spain and Hungary, and much more CO_2 is released into the atmosphere (Student 2MM23).

Below are the ideas generated by the two students following a class group discussion, conducted without using the app, where each member shared their responses to the first question. As previously mentioned, this question focused solely on their prior ideas. The students will be referred to as group A and B, respectively:

> Ireland is the most advanced country, and Spain not so much. The smaller the population, the lower the energy consumption. We cannot agree on which country has the highest CO_2 emissions, and we have not reached a consensus on which country has the most expensive energy (Group A, which includes the student 2LH14).

> We agree that Finland uses more energy due to the temperatures and that Finland is the "greenest" country. There is no consensus on which country is the richest in energy and in which country there is more pollution (Group B, which includes the student 2MM23).

Once the students have used the application, they return to the same group and individual tasks. This time, the responses of both groups are shown, showing the differences in the students' discourse as a result of the integration of real data and the comparison between countries:

> If there are more people, there is greater energy consumption. Depending on the region in which the country is located, it will have a greater or lesser number of natural resources, which affects energy production. On the other hand, the economic aspect, such as the GDP of each country, also has an influence, since a lower GDP may hinder the installation of renewable energy infrastructures (Group A, which includes the student 2LH14).

> Finland uses many low-emission and renewable energies, while in Hungary only solar energy is used as a renewable. Spain could use more renewable energies (Group B, which includes the student 2MM23).

Finally, their individual responses at the end of the experience are shown:

> Ireland is the country that emits the most CO_2 of the three, and its carbon intensity is approximately three times higher than in Spain. However, it is the most prepared to operate with renewable energies due to the resources it has. Portugal is a promising country in terms of renewable energies due to its use of solar energy, although it still uses natural gas. Since Spain does not have fossil fuel sources, it must acquire them commercially, and this usage emits a lot of CO_2 into the atmosphere, despite the use of renewables such as solar and hydroelectric energy (Student 2LH14).

> The worst situation is in Hungary because it heavily relies on nuclear energy, and its only renewable source is solar energy, while it also uses a lot of natural gas. Finland is in the best position as it uses many renewable and low-emission energies. Spain is in the middle because, although it uses a lot of renewable energies, it still relies heavily on natural gas (Student 2MM23).

Both students showed significant progress in their understanding of the factors that influence energy production and consumption in different countries. In their initial answers, both relied on general observations and vague assumptions, such as geographical features or climate, without delving into the analysis of energy types or national energy policies. 2LH14 focused on superficial environmental features, while 2MM23 mentioned climate and energy prices in a simplified way.

In their second answers, both students included more specific concepts. Student 2LH14 introduced key factors such as population, natural resources, and GDP, while 2MM23 analysed the use of nuclear energy, natural gas and renewable energy, demonstrating a deeper understanding of energy impacts and sustainability. This progress reflects a greater ability to handle complex variables and relate them to each country's energy situation.

Finally, in terms of metacognition, both students demonstrated improvement in their critical analysis. Student 2LH14 was more accurate in assessing how economic and geographical factors affect energy production, while student 2MM23 improved in structuring his argument by ranking countries according to their energy policies. Both students showed enhanced abilities to question their initial assumptions and to reflect critically on the factors influencing energy consumption and production.

7.6 Discussion and Educational Implications

The experience on energy generation and consumption has encouraged the students' interest and curiosity to delve deeper into a current problem, such as the energy reality of their country in comparison with different parts of the world, as well as the causes and consequences of this situation.

The experience facilitated the development of critical analysis of information, and the appropriate application of knowledge about energy, as dimensions of CT in the ENCIC-CT model (Chap. 1) (Franco-Mariscal et al., 2025), which was evident when students analysed, interpreted, evaluated and selected information from the application in a reflective, logical and informed manner (Bailin, 2002). This critical analysis was stimulated by comparing and contrasting electricity production and consumption data from different countries and identifying similarities and differences. The experience helped students to reflect on the variables that influence energy production and consumption in each country, such as climate, population, level of development, and energy policy, which helped them to develop their capacity for analysis and synthesis and to recognise the advantages and disadvantages of each energy model.

Applying energy knowledge to a real problem helped students better understand electricity-related processes, such as energy sources, generation forms, transport, storage, efficiency, sustainability, climate change, etc. In this sense, some improvements found in the final arguments are related to the use of coal, natural gas or solar energy sources, or more specific energy terms such as carbon intensity, low-carbon energy and renewable energy, to which their handling in the application undoubtedly contributed, and where McNemar's test found statistically significant differences. This may be because references to coal, natural gas or solar energy tend to be recurrent among students, as they are more familiar with these energy sources, as in the case of the use of fossil fuels such as petrol for cars, the use of natural gas at home for cooking or heating, or the presence of solar panels on buildings.

In short, critical analysis and subsequent reflection helped students develop their ability to argue and defend a position on a topical issue using evidence-based arguments (Jiménez-Aleixandre, 2010).

The final presentation also promoted metacognition and debate, as well as communicative skills, addressing the problems and challenges posed by the current energy model, along with possible solutions and alternatives to improve the situation. Students used the data to assess the environmental, social, and economic impacts of electricity production and consumption in Spain and to propose improvements in the country. Some of the ideas that emerged from these presentations were their concern about climate change, global warming, the security of energy supply, and energy catastrophes. These reasons have led some countries such as Denmark, Germany and the United States to restructure their energy systems (Araujo, 2014; Lee & Gloaguen, 2015).

In summary, the *Electricity Maps* application has the educational potential to develop CT in secondary school students in different dimensions. It is also an interdisciplinary tool that allows activities to be carried out jointly with other disciplines such as Geography, Biology, Mathematics or Technology, encouraging students' interest and motivation for science from an active, critical and ethical perspective, promoting a responsible and committed attitude towards the environment and society.

Acknowledgements This work is part of the Spanish National Plan R+D+i Project entitled "Citizens with critical thinking: A challenge for teachers in science education", reference PID2019-105765GA-I00, funded by MCIN/AEI/10.13039/501100011033.

References

Alderman, L., & Pronczuk, M. (2022, January 2). Europe plans to say nuclear power and natural gas are green investments. *The New York Times*.

Araujo, K. (2014). The emerging field of energy transitions: Progress, challenges, and opportunities. *Energy Research & Social Science, 1*, 112–121. https://doi.org/10.1016/j.erss.2014.03.002

Bailin, S. (2002). Critical thinking and science education. *Science & Education, 11*(4), 361–375. https://doi.org/10.1023/A:1016042608621

Batanero, C., & Díaz, C. (2011). *Estadística con proyectos* [Statistics with projects]. University of Granada.

Colucci-Gray, L., Camino, E., Barbiero, G., & Gray, D. (2006). From scientific literacy to sustainability literacy: An ecological framework for education. *Science Education, 90*(2), 227–252. https://doi.org/10.1002/sce.20109

Cruz-Lorite, I. M., Cebrián-Robles, D., Acebal-Expósito, M. C., & Blanco-López, A. (2022). Pre-service teachers' ideas about obtaining electricity in nuclear power stations in a role play context. *EURASIA Journal of Mathematics, Science and Technology Education, 18*(11), em2176. https://doi.org/10.29333/ejmste/12507

Eggert, S., Ostermeyer, F., Hasselhorn, M., & Bögeholz, S. (2013). Socioscientific decision making in the science classroom: The effect of embedded metacognitive instructions on students' learning outcomes. *Education Research International, 2013*, 309894. https://doi.org/10.1155/2013/309894

Ercan, O., Ural, E., & Tekbiyik, A. (2015). Pre-service teachers' attitudes towards nuclear energy and the effect of Fukushima nuclear disaster on their attitudes. *The International Journal of Social Sciences and Humanities Invention, 11*(2), 1669–1678. https://doi.org/10.18535/ijsshi/v2i11.01

Erduran, S., & Jiménez-Aleixandre, M. P. (Eds.) (2007). *Argumentation in science education: Perspectives from classroom-based research*. Springer. https://doi.org/10.1007/978-1-4020-6670-2

European Commission. (2018). *Action plan against disinformation*. European Commission. https://commission.europa.eu/document/download/b654235c-f5f1-452d-8a8c-367e603af841_en?filename=eu-communication-disinformation-euco-05122018_en.pdf

Feinstein, N. W., & Kirchgasler, K. L. (2015). Sustainability in science education? How the next generation science standards approach sustainability, and why it matters. *Science Education, 99*(1), 121–144. https://doi.org/10.1002/sce.21137

Fensham, P. (2011). Globalization of science education: Comment and a commentary. *Journal of Research in Science Teaching, 48*(6), 698–709. https://doi.org/10.1002/tea.20426

Fontichiaro, K. (2010). Cross curriculum: Awakening and building prior knowledge with primary sources: See, think, wonder. *School Library Monthly, 27*(1), 14–15.

Franco-Mariscal, A. J., Cano-Iglesias, M. J., España-Ramos, E., & Blanco-López, A. (2025). The ENCIC-CT model for the development of critical thinking. In A. J. Franco-Mariscal (Ed.),

Critical thinking in science education and teacher training (Chap. 1) (pp. 3–42). Springer. https://doi.org/10.1007/978-3-031-78578-8_1

Hodson, D. (2003). Time for action: Science education for an alternative future. *International Journal of Science Education, 25*(6), 645–670. https://doi.org/10.1080/09500690305021

International Atomic Energy Agency. (2022). *Near term and promising long term options for the deployment of thorium based nuclear energy*. IAEA-TECDOC-2009, IAEA. https://www.iaea.org/publications/15215/near-term-and-promising-long-term-options-for-the-deployment-of-thorium-based-nuclear-energy

Jiménez-Aleixandre, M. P. (2010). *10 ideas clave. Competencias en argumentación y uso de pruebas* [*10 Key Ideas: Competencies in Argumentation and Use of Evidence*]. Graó.

Kilinc, A., Boyes, E., & Stanisstreet, M. (2013). Exploring students' ideas about risks and benefits of nuclear power using perception theories. *Journal of Science Education and Technology, 22*, 252–266. https://doi.org/10.1007/s10956-012-9390-z

Lee, R. P. (2016). Misconceptions and biases in German students' perception of multiple energy sources: Implications for science education. *International Journal of Science Education, 38*(6), 1036–1056. https://doi.org/10.1080/09500693.2016.1176277

Lee, R. P., & Gloaguen, S. (2015). Path-dependence, lock-in, and student perceptions of nuclear energy in France: Implications from a pilot study. *Energy Research & Social Science, 8*, 86–99. https://doi.org/10.1016/j.erss.2015.05.001

Martín, C., Prieto, T., & Jiménez, A. (2013). El problema de la producción y el consume de energía: ¿Cómo es tratado en los libros de texto de educación secundaria? [The problem of production and consumption of energy. How is it treated in secondary education textbooks?]. *Enseñanza de las Ciencias, 31*(2), 153–171. https://doi.org/10.5565/rev/ec/v31n2.767

Park, M., & Liu, X. (2016). Assessing understanding of the energy concept in different science disciplines. *Science Education, 100*(3), 483–516. https://doi.org/10.1002/sce.21211

Sadler, T. D. (2009). Situated learning in science education: Socioscientific issues as contexts for practice. *Studies in Science Education, 45*(1), 1–42. https://doi.org/10.1080/03057260802681839

Spellman, F. R. (2014). *Environmental impacts of renewable energy*. CRC Press.

Sy, S. A., & Mokaddem, L. (2022). Energy poverty in developing countries: A review of the concept and its measurements. *Energy Research & Social Science, 89*, 102562. https://doi.org/10.1016/j.erss.2022.102562

Toulmin. S. E. (2003). *The uses of argument* (3rd ed.). Cambridge University Press. https://doi.org/10.1017/CBO9780511840005

Vila, L., Márquez, C., & Oliveras, B. (2023). Una propuesta para el diseño de actividades que desarrollen el pensamiento crítico en el aula de ciencias [A proposal for the design of activities that develop critical thinking in the science classroom]. *Revista Eureka sobre Enseñanza y Divulgación de las Ciencias, 20*(1), 1302. https://doi.org/10.25267/Rev_Eureka_ensen_divulg_cienc.2023.v20.i1.1302

Zeidler, D. L., Herman, B. C., & Sadler, T. D. (2019). New directions in socioscientific issues research. *Disciplinary and Interdisciplinary Science Education Research, 1*, 11. https://doi.org/10.1186/s43031-019-0008-7

Zeidler, D. L., Sadler, T. D., Simmons, M. L., & Howes, E. V. (2005). Beyond STS: A research-based framework for socioscientific issues education. *Science Education, 89*(3), 357–377. https://doi.org/10.1002/sce.20048

Chapter 8
Enhancing Argumentation and Decision-Making in Teacher Training Through Debates Within the Context of Formative Research in Uruguay

Daisy Imbert Romero, Claudia Cabrera Borges, Mariana Pérez Orlando, Silvia Carámbula, Gabriela Martínez, Marcela Pérez Salatto, Sair Aparicio, and Alicia Fagúndez

D. Imbert Romero (✉) · C. Cabrera Borges · M. Pérez Orlando · S. Carámbula · G. Martínez · M. Pérez Salatto · S. Aparicio · A. Fagúndez
National Public Education Administration (ANEP), Consejo de Formación en Educación (CFE), Montevideo, Uruguay
e-mail: dimbertromero@gmail.com

C. Cabrera Borges
e-mail: claudiaanahi@gmail.com

M. Pérez Orlando
e-mail: marianapodocente@gmail.com

S. Carámbula
e-mail: silvia.carambula@docente.ceibal.edu.uy

G. Martínez
e-mail: agabrielamm@gmail.com

M. Pérez Salatto
e-mail: micaelaperezsalatto@gmail.com

S. Aparicio
e-mail: sair.aparicio@docente.ceibal.edu.uy

A. Fagúndez
e-mail: maliciafd@gmail.com

D. Imbert Romero · C. Cabrera Borges
ORT University, Montevideo, Uruguay

A. J. Franco-Mariscal (ed.), *Critical Thinking in Science Education and Teacher Training*, Contemporary Trends and Issues in Science Education 64,
https://doi.org/10.1007/978-3-031-78578-8_8

8.1 Introduction

In order to contextualize the research conducted, it is essential to describe the setting in which the implementation of debates about socio-scientific issues within the framework of formative research takes place.

Formative research methodology has proven to be valuable in identifying ways to improve theories and models in all fields of education (Reigeluth & Frick, 1999). From a didactic point of view, it corresponds to proposals for activities that enable problem-solving learning (Turpo et al., 2020). In this process, students learn to investigate and develop research skills, thus allowing them to address the curriculum's contents comprehensively.

Chadha et al. (2023) refer to students as partners, indicating that this is an increasingly common notion in higher education. This author describes an experience of participatory action research in an introductory chemistry module in chemical engineering, whereby students were actively involved in various stages of the module's research, design and development.

In this context, since 2011, a group of teachers from the "Centro Regional de Profesores" (CeRP) of the Centre in Uruguay has been promoting that students in initial teacher training learn to solve problems that arise in practice through research by carrying out introductory research projects, now called formative research projects. Although research competence develops over the four years of initial teacher education and is expected to continue to improve during professional life, a significant obstacle to its development has been observed. This is none other than the difficulty students encounter in arguing when writing the discussion and conclusion of their reports. To overcome this difficulty, and through a collaboration with science education researchers at the University of Malaga (Spain), it was decided to replicate a debate training programme based on socio-scientific issues, which had already been implemented at that university with preservice science teachers (Cano-Iglesias, 2024).

The use of debates on socio-scientific issues can enhance formative research by addressing the challenge of developing students' argumentative skills, which are crucial for their research competency. Socio-scientific issues are inherently complex, requiring students to navigate the intersection of scientific facts with ethical, cultural, economic, and political considerations (Sadler & Zeidler, 2005). By engaging in structured debates on these issues, students are not only exposed to the multifaceted nature of real-world problems but are also encouraged to critically evaluate and articulate their perspectives.

Moreover, debates serve as an active learning strategy that aligns with the objectives of formative research. In this context, debates create a dynamic environment where students can practice their ability to construct well-reasoned arguments, critically evaluate evidence, and effectively communicate their conclusions (Fang et al., 2019)—skills that are also integral to formative research projects. By integrating these debates into the formative research methodology, preservice teachers can foster a more engaging and comprehensive learning experience, preparing students to tackle complex societal challenges with confidence and competence.

This chapter presents the findings of a debate programme with undergraduate students in Uruguay to promote argumentation and decision-making as critical thinking (hereinafter, CT) skills that will help them successfully develop their formative research.

8.2 Argumentation and Decision-Making as Indispensable Allies for Developing Critical Thinking in Formative Research Processes

This section shows the relationship between formative research and argumentation and decision-making as CT dimensions.

8.2.1 Formative Research

Formative research is a practice that brings educators closer to scientific knowledge, fostering innovative and high-quality professional development (Espinoza, 2020; Estrada & Estrada, 2019). From a pedagogical standpoint, it involves activities designed to facilitate learning through problem-solving (Turpo-Gebera et al., 2020). In this process, students not only learn how to conduct research but also develop essential investigative skills, enabling them to engage with curriculum content in a more holistic manner. Restrepo (2003) defines formative research as a teaching and learning strategy that begins with contextualized, relevant situations for future educators, as they identify the issues for project development through their pre-professional practice.

In the field of teacher training, formative research offers preservice teachers an opportunity to engage with knowledge production through the perspective of critical didactics. According to Imbert (2023), there is a link with curriculum theories:

> Critical metatheory frames critical didactics and implies: 1. Emancipation of the teacher, who, in addition to analysing the reality of the classroom, seeks to transform it. 2. Action research is a methodology that allows teachers to reflect on classroom learning and propose improvement actions. 3. Transformative practice is a type of practice that arises from reflection to change the existing situation. These three characteristics of critical didactics come together and become a reality through formative research (p. 51).

Formative research, which involves the relationship between teaching and research, stands as a cardinal scaffold in initial teacher education to motivate students to address topics directly linked to their disciplines and encourage the treatment of issues that have a greater affinity with the human condition.

In addition, it is essential to provide participants with opportunities to reflect on and evaluate the implementation of the design by identifying strengths and weaknesses. Participants are encouraged to point out any additional comments, with a particular focus on highlighting areas for improvement (Reigeluth & Frick, 1999).

Paucca et al. (2021) analysed the relationship between formative research and the achievement of research competences of university students in Peru, including teacher education students at different levels. They underline the importance of this relationship and conclude that there is a direct and significant relationship (p. 366). Valero (2021) stresses that the importance of formative research lies in being able to combine research methods with teaching and learning processes, especially in higher education, where the scholarly output of both students and educators defines the academic landscape. Along the same lines, Esquivel et al. (2023) identified the need to include this methodology in higher education. Sabariego et al. (2020), in their research aimed at developing research competence in preservice preschool teachers at the University of Barcelona (Spain), concluded that formative research enables the development of research competences in initial training and favours the achievement of three major types of research skills, following the typology of Moreno (2005): instrumental, methodological construction, as well as conceptual and social construction of knowledge. Postholm (2020) reflects on the importance of the role of the researcher in educational research, combining the role of the researcher in both action research and formative interventions. Formative research would be enhanced if researchers engaged with and supported teachers in their teaching, thus helping them apply new knowledge to contextual situations to develop their practice.

In formative research, knowledge is contextualised and transformed to solve a problem. In this sense, and taking into account that action research is a type of research within formative research, Hadfield (2012) analyses three fundamental points within the action research process: the nature of practitioners' practical reasoning, the bases of critical self-reflection and collective action, as fundamental aspects to avoid subjectivity in the interpretation of the results.

Reigeluth and Frick (1999) complement the above conceptualisations when they indicate that formative research corresponds to a type of developmental research or action research that aims to improve design theory and manage to design instructional practices that impact the learning process.

In the words of Pérez-Salatto and Pérez-Salatto (2023, p. 59), formative research is a didactic approach that generates multiple learning opportunities, both in the formation of research competences and in the construction of a critical perspective on teaching practice.

8.2.2 Argumentation and Decision-Making in Formative Research in the Context of Debates on Socio-scientific Issues

The literature shows that it is essential to teach students argumentation literacy (Horng et al., 2013; Koeneman et al., 2013) as it activates the articulation between theory and evidence (Braund et al., 2013) and also invites students to think critically before making a decision (Larrain et al., 2014). The study by Anisa et al. (2023) reaffirms this need by showing that, despite teachers' efforts to organise argumentative classes, the impact on improving this competence is limited. In their case, it was shown that students' argumentation level was mainly low, and only 9% changed their argumentative stance.

An alternative to address such weaknesses is incorporating formative research into teacher training so that teachers can develop research skills by researching problems identified from their practice (Carámbula et al., 2023). Carrying out research work of this nature requires the future teacher to make decisions to fulfil each of the stages of this process: "[...] constantly reflect on their practice, formulate hypotheses, define objectives, seek information on current knowledge, collect data, discuss them and draw conclusions (Carámbula et al., 2023, p. 208)."

Before delving into the relationship between argumentation and the various stages of the research process conducted by future teachers, it is essential to first clarify the concept of argumentation that underpins this work, along with its key components. Unlike the traditional approach, which focuses primarily on the structure or form of an argument, this study adopts Toulmin's Argument Pattern (TAP) (Toulmin, 1958), which places greater emphasis on the content of the argument. Notably, this model also encourages reflection on the structure of argumentative texts (Pinochet, 2015). According to Toulmin (1958), an argument moves from data to claim, where data refers to the information, background, or facts used to support the claim. TAP also incorporates the warrant, backing, modal qualifier, and rebuttal. Additionally, Toulmin (1958) asserts that a sound argument must include at least three essential components: data, claim, and warrant. Building on this model, Jimenez-Aleixandre (2010) presents a simplified version of the elements that constitute an argument, reducing them to evidence, warrant, and conclusion, while noting that additional elements can be incorporated when needed.

Although argumentation is used in different parts of the formative research process, some stages are more relevant. Problem justification is a stage in which prospective teachers must argue the importance of the problem identified and provide evidence to support this. Arguments should be based partly on experience in teaching practice, but above all; teachers should be able to support their claims about the relevance of research in a selected topic by using background information in that area. The research's discussion and conclusion are other stages where the ability to argue becomes relevant. In these stages, students must be able to critically decide what evidence emerges from the data collected that allows them to reach conclusions. This evidence will allow them to justify their findings and account for the degree of compliance with the proposed objectives. Conversely, discussion requires them to connect the results obtained with what the theoretical framework offers, bringing their evidence into dialogue with the knowledge that has already been built on the problem. In this way, the research approach allows future teachers to develop arguments based on methods and analysis procedures (Questa et al., 2022, p. 53).

These processes are of great relevance considering the multiplier effect of teacher training, which, in addition to preparing preservice teachers to develop coherent and evidence-based arguments, enables them to experience these processes and thus better prepare them to guide their students to argue better.

There is no doubt about the advantages of argumentation and decision-making applied in formative research to develop CT. However, to perform these skills well in formative research, it is necessary to train students in argumentation and decision-making beforehand. Since formative research starts from a problem, developing an activity that explicitly instructs in using argumentation and decision-making based on one of these problems before carrying out the research would be interesting. Different authors (Anisa et al., 2023; Bernal-Herrera et al., 2023) indicate that the use of debates around socio-scientific issues (see Chap. 1) (Franco-Mariscal et al., 2025) is a good alternative for promoting these CT dimensions and allows students to prepare for more complex processes such as those involved in formative research, which, in turn, enhances critical reflection on their practices.

8.3 Research Questions

This paper poses the research question: What improvements are produced in the argumentation and decision-making of Uruguayan preservice teachers who participate in a training programme of debates on socio-scientific issues?

After participating as debaters and listeners in several debates, preservice teachers are expected to improve their argumentative and decision-making skills, enabling them to better develop discussion and conclusion in the formative research projects they will undertake in the following years of their initial training.

8.4 Method

This section describes the participants and the context in which the implemented experience took place, as well as the data collection procedure and the analysis that was conducted.

8.4.1 Participants and Context

The experience was at the CeRP Centro, belonging to the Education Training Council of the National Public Education Administration of Uruguay. At this teacher training institute, students graduate as secondary school teachers.

The participants were 62 first-year preservice teachers from all majors currently offered by the centre (except Spanish Language): Biology (N = 12), Physics (N = 6), Chemistry (N = 2), Mathematics (N = 18), Law (N = 5), Sociology (N = 1), English (N = 2), Geography (N = 3), History (N = 12) and Philosophy (N = 1). This course includes a common professional pathway for all specialisations, where 62 preservice teachers take the Spanish Language/Language in four groups. Each group is made up of students from different specialisations.

The training programme of critical argumentation through debates was implemented in this subject for two reasons. Firstly, the subject is presented as an opportunity to strengthen communicative skills through argumentation, which is one of the course's main objectives. Secondly, from a logistical point of view, it was pertinent given the fact that the teacher in charge is part of the group of researchers.

8.4.2 Experience Description

The debate programme is a replication of the activity called Micro-debates that researchers from the Science Education and Competences (ENCIC) group at the University of Malaga (Málaga, Spain) are developing with preservice science teachers (Bernal-Herrera et al., 2023; Cano-Iglesias, 2024; Salcedo-Armijo et al., 2023).

Initially, without prior preparation, the preservice teachers took part in a debate (debate 0) whose subject matter was related to everyday life, namely holidays at the beach or in the countryside. This debate was chosen to familiarise them with the dynamics of persuasive exchange characteristic of debates. Although a debater and a moderator were designated for each position, the whole group participated in constructing arguments. This activity was the starting point to present the training experience and to begin the instruction on argumentation and CT. At the same time, questionnaires were designed in which the preservice teachers chose possible problems of social and scientific interest (socio-scientific issues) (Zeidler et al., 2019) that required informed decision-making based on scientific knowledge and ethical, social and cultural values. The issues presented were excessive consumption of plastic packaging, vaccination of children and adolescents with the Pfizer vaccine against coronavirus, packaged food identified with octagons, recycling, hunting animals for tourism, consumption of red meat, vegan food, reduction of fuel vehicles, renewable energies, production of artificial meat, gene therapy, cloning, buying glasses at fairs, cars without windscreen wipers and genetically modified food. Different subgroups were organised to participate in the micro-debates based on the preservice teachers' interests. Participants were informed about their role and the issue to be discussed a week before. The preservice teachers participated in 10 micro-debates during October and November 2022 (Table 8.1).

Each micro-debate consists of an initial stage where the whole group fills in a Google form (instrument 1) expressing their initial position for or against the issue and a justification. This was followed by the debate itself (lasting 10 min), after which the whole group again filled in a form (instrument 2) stating whether or not they had a change in their position and why. The same instrument also asks for the student's perceptions of the activity.

The participants in each micro-debate are three preservice teachers, one for and one against the problem, and a presenter, taking place with the following structure. In the first part, the presenter announces the issue to be debated in three minutes and may include a piece of fake news for dispersal. In the second part, each debater has one minute to present their position, while in the third part, arguments are exchanged for five minutes. Finally, each debater concludes their position in one minute. This part is the last instance where you can convince your opponent and the audience.

As previously mentioned, between debate 0 and the following debates, training sessions were held to instruct preservice teachers on the importance of CT. During these sessions, the structure and elements of the scientific argument were presented (Cebrián-Robles et al., 2021; Jiménez-Aleixandre, 2010; Toulmin, 1958), a specific debate was analysed (Franco-Mariscal & Cano-Iglesias, 2023), and case studies were presented as a basis for the construction of quality scientific arguments (Jiménez-Aleixandre, 2010). Instruction was given on the appropriate use of discursive connectors in constructing arguments (Calsamiglia & Tusón, 1999).

The total duration of the experience was six hours, of which two hours corresponded to the debate 0 and instruction, and four hours were dedicated to the micro-debates.

Table 8.1 Discussions carried out specifying the problem and the teaching staff's speciality

Debate	Group 1	Group 2	Group 3	Group 4
0	Holidays on the beach or in the countryside			
1	Gene therapy (Physics)	Reduce vehicles of fuel (Mathematics)	Purchase of lenses at trade fairs (Biology)	Cloning (Biology)
2	Vegan food (Biology, Physics, Mathematics, and Chemistry)			
3	Reduce plastic packaging (Visual Communication, Law, Geography, History and Sociology)			
4	Packaged with octagons (Law)	Cars without windscreen wipers (History, and Sociology)	COVID vaccines in children (Geography)	Renewable energies (Mathematics, and Philosophy)

8.4.3 ENCIC-CT Model Application

The research conducted utilizes the ENCIC-CT model (Franco-Mariscal et al., 2025), which is designed to enhance the development of CT in science education. This model is structured in several layers, each addressing different aspects. The first layer focuses on the use of socio-scientific issues, selected for their relevance to everyday life and their potential to foster CT. Preservice teachers engaged with a range of issues, as detailed in the experience description section.

The second layer of the model emphasizes the development of argumentation and decision-making within the skills domain. Students participated in activities that encouraged critical analysis and the construction of arguments based on evidence and sound reasoning in order to make informed decisions.

The third layer addresses the scientific practice of argumentation. A key aspect of this experience is that it not only strengthens preservice teachers' critical skills but also fosters the development of research competencies. These competencies are applied in formative research processes, enabling preservice teachers to incorporate CT into their professional practice.

In order to tackle the socio-scientific issues, the teaching strategy of micro-debates (Layer 4) was used, offering students opportunities to discuss and debate the issues from various perspectives, thus deepening their understanding of the topics explored.

8.4.4 Data Collection and Analysis

The data analysis on which this chapter focuses was based on the information gathered from the forms applied in the pre- (instrument 1) and post-debate (instrument 2).

During the programme, 242 and 223 responses were obtained for instruments 1 and 2. A total of 222 forms were taken as valid as they corresponded to students who completed both the initial and post-micro-debate moments, and those forms completed in only one of the instances were discarded, as the focus was on the comparison.

Each unit of analysis (UA) was coded with a number corresponding to the answer in the database made to analyse the forms. Each UA was assigned to one student. However, the same preservice teacher can provide multiple responses, as they completed both the pre-and post-discussion forms for each problem.

A first analysis quantified decision-making through the frequency and percentages of preservice teachers who were in favour, against or undecided, both before and after the intervention, as well as the changes observed. Pearson's Chi-square test for related samples was performed to find possible significant differences between the two points in time.

A second qualitative analysis was established for the preservice teachers who had experienced a change of position, different categories according to the type of argument were used between pre- and post-debate. For this purpose, we checked if these arguments contained conclusion, evidence and warrant as essential elements of an argument according to Toulmin's (1958) model. In this way, it was identified whether the arguments given provided evidence or data understood as observations, facts, experiments and both quantitative and qualitative pieces of information that are used to assess the veracity of claims; conclusion that corresponds to knowledge claims that are sought to confirm or disprove, and warrant that refers to the process of how a conclusion was reached, showing that the passage from the data to the claim is appropriate and valid (Jiménez-Aleixandre, 2010).

8.5 Results

This section first presents the decisions found before and after the micro-debates and the decision changes that took place. Subsequently, the arguments submitted by those students who changed their positions are analysed.

8.5.1 Decisions Before and After the Micro-debate and Decision Changes

The first aspect to consider is how the experience affects the preservice teachers' decision-making when faced with the socio-scientific issues proposed in the two moments of intervention. Figure 8.1 illustrates the percentages of preservice teachers in favour, against or undecided before and after each micro-debate for the different socio-scientific issues.

An overall view of the results indicates that the majority of the preservice teachers were in favour of the socio-scientific issues raised. Only the problem of buying glasses at a fair offered majority decisions against and in no case was the predominant position undecided.

Figure 8.2 illustrates, for each issue, the percentage of preservice teachers who maintained or changed their decision, identifying also in which direction the change occurs. In all micro-debates, more than 58% of the preservice teachers maintained their initial decision after the discussion. This percentage ranged from 58.33% for gene therapy to 92.0% for cars without wipers.

The Pearson's Chi-square statistical test only found statistically significant differences in the issues of vegan diet ($\chi^2 = 52.09$, $p = 0.00$), cloning ($\chi^2 = 11.96$, $p = 0.003$), fuel cars ($\chi^2 = 13.15$, $p = 0.011$), cars without windscreen wipers ($\chi^2 = 15.29$, $p = 0.00$), plastic containers ($\chi^2 = 32.50$, $p = 0.00$) and octagons in containers ($\chi^2 = 3.94$, $p = 0.047$).

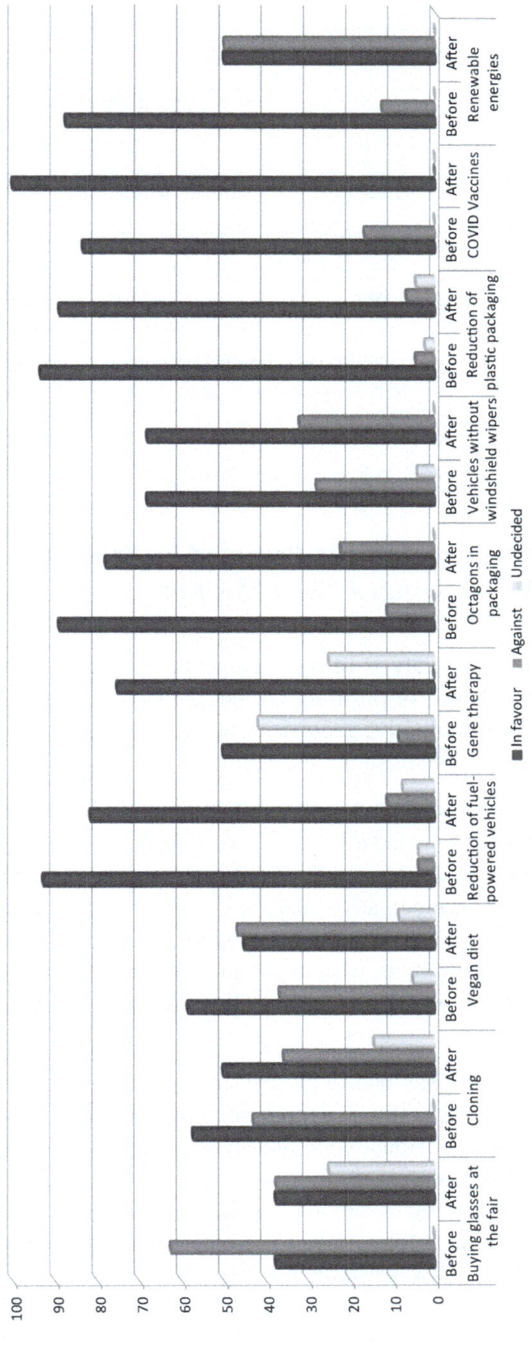

Fig. 8.1 Percentage of preservice teachers in each position on the different socio-scientific issues before and after the micro-debate

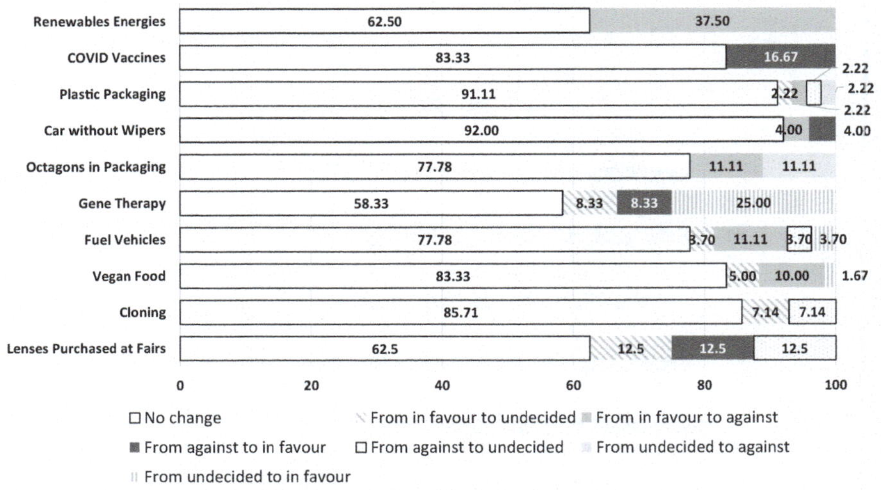

Fig. 8.2 Changes in decisions between before and after the micro-debate

8.5.2 Analysis of the Arguments Provided by Preservice Teachers Who Changed Their Position

This section delves into the analysis of the arguments offered by those preservice teachers who showed a change of position. Of the 222 valid forms, 40 participants changed their decision. A qualitative analysis of the arguments provided to support their decision allowed us to establish three categories:

- *Category A*. Preservice teachers who gave full arguments (providing conclusion, warrant and evidence) before and after the micro-debate (15 preservice teachers).
- *Category B*. Preservice teachers moved from a terse initial conclusion (yes or no) to more elaborate argumentative statements with conclusion, evidence and warrant in the post-debate (16 preservice teachers).
- *Category C*. Preservice teachers who maintained arguments that only included a short conclusion (yes or no) before and after the intervention (9 preservice teachers).

Table 8.2 illustrates some examples of preservice teachers' arguments in category B that they felt improved their ability to express their position on the socio-scientific issue posed, which could be attributable to their experience of the micro-debates.

Another evidence of the improvement of the critical positioning shown by some preservice teachers after the micro-debate was the identification that decisions do not have to respond to a dichotomy, given the complexity of the socio-scientific issues addressed (*I believe that every human being is free to make that decision and that neither of the two would be wrong,* UA51). On the other hand, UA55 made concrete proposals involving alternative paths that include contributions from both

Table 8.2 Examples of arguments shown by preservice teachers included in category B

Unit of analysis	Issue of the debate	Initial response	Final response	Warrant	Evidence	Conclusion
195	Reduce plastic packaging	Yes, I consider it important to reduce plastic consumption	No, because only 5% of the planet is polluted by plastics	Yes	Yes	Yes
216	Renewable energies	I think so	I do not think so because trees are cut down, flying species such as bats, birds and bees are harmed, as they can collide with the windmills. Moreover, it is also costly	Yes	Yes	Yes
237	COVID vaccine in children	No	The chances of the vaccine killing you are more significant than the chances of dying from COVID	No	Yes	No

the arguments for and against, in this case, on the issue of vegan food: *I am against a vegan diet. Yes, some vegan foods could be used and combined with meat.*

In addition, other preservice teachers argued from broader perspectives that included a more holistic view of the problems by integrating different aspects of the issue:

> My final decision is that veganism is a good choice for a healthy diet as it has multiple benefits for health, economy, and nature (UA60). (Vegan food issue showing health, economic, and environmental dimensions).
>
> I do not think so because trees are cut down, flying species such as bats, birds and bees are harmed, as they can collide with the mills. Moreover, it is also costly (UA216). (Renewable energies issue showing the environmental and economic dimensions).

Other decisions focused on the consequences of the problem:

> We have yet to think about what we would do with so many people and their commute to work if we reduced the number of vehicles. We would have to increase the number of local buses in Montevideo, but people would still complain about the driving times (UA102). (Decreasing fuel vehicles issue).

Alternatively, in the duality involved in identifying the cost–benefit associated with the problem:

> The chances of the vaccine killing you are more significant than the chances of dying from COVID-19 (UA 237). (COVID-19 vaccine issue in children).

8.6 Discussion and Educational Implications

Regarding decision-making, the micro-debates showed that at least 58% of the preservice teachers maintained their initial choice regarding the socio-scientific issue after participating in the activity. Depending on the situation, opinion changes ranged from 8 to 42% in their different variants. These findings exceed the observations of Anisa et al. (2023), who reported opinion changes in only 9% of the participants.

In addition, the Pearson's Chi-square test found statistically significant differences in some issues, revealing that the micro-debates impacted the preservice teachers. In this sense, the micro-debates helped some participants rethink their decision after hearing different points of view, which is evidence of the development of CT.

Despite the improvements, we share the perspective of Anisa et al. (2023) that its impact must be limited despite teachers' efforts to incorporate argumentation in the classroom. Therefore, it is imperative to continue to analyse the processes involved in decision-making in science education. Understanding preservice teachers' initial position is necessary to examine how they make decisions about the context and evaluate contradictions (Domènech & Márquez, 2014; Franco-Mariscal & Cano-Iglesias, 2023).

Another aspect to highlight regarding improving critical positioning is the understanding that decisions do not have to respond to a dichotomy, given the complexity of the problems addressed, and that both possibilities are possible. According to Morin (2002), "[…] the epistemology of complexity implies overcoming the subject/object dichotomy. On the principles of disjunctive and reductionist logic, the author states that the object is separated from its environment and isolated from the observer who observes it" (p. 64).

This idea is relevant for formative research since, in this case, the issues selected come from teaching practice and, therefore, respond to the complexity of the educational field, which requires responses in line with this complexity.

Regarding argumentation, using micro-debates led to some improvements, particularly in the quality of the arguments and the comprehensive analysis of the problem.

Concerning argumentative quality, the findings showed an evolution of many preservice teachers who initially offered incomplete arguments and moved after the intervention to arguments that included evidence, warrant and conclusion as essential elements. These argumentative changes are important because conclusions are invalid without evidence (Erduran & Jiménez-Aleixandre, 2007). The results obtained can be interpreted as evidence of improvements in the argumentation skills of some preservice teachers as identified by Jiménez-Aleixandre (2010), who indicates that argumentation implies the ability to evaluate claims based on evidence, as well as recognise that in science, conclusions must be supported by evidence that is stated through justification.

On the other hand, some preservice teachers evolved towards arguments, showing a multidisciplinary view of problems, thus showing progress in CT and integrated problem-solving. About this, education must be aligned with a perspective of teaching science in context (Occelli et al., 2020), allowing for "[...] understanding of science, its nature, analysis of everyday situations and the development of an active and critical role in social and environmental issues related to science and technology" (p. 1151).

This type of argument reveals that preservice teachers are aware that real problems cannot be tackled from a single discipline and require an interdisciplinary approach (Morin, 2002) as an indication of CT, as it is a rational process that develops from scientific knowledge and enables the construction and application of criteria (Bailin, 2002). Such criteria allow us to analyse, question and reflect on specific information and take a critical position concerning socio-scientific issues (Díaz & Jiménez-Liso, 2012).

In sum, the changes observed in both decision-making and argumentation can lead to improvements in the participants' future formative research projects. Specifically, it will discuss the justification of the research problem, the selection of evidence from their fieldwork, the discussion and conclusions, and a holistic view of the problems from different perspectives. From this research approach, it is hoped that future teachers will adopt a broader view of the problems that emerge from teaching practice and obtain the necessary tools to reflect on and transform their practices.

Closely related to this is the potential this type of practice offers for developing CT in preservice teachers. Jiménez-Aleixandre and Puig (2012), who define CT comprehensively, indicate that it combines elements associated with social emancipation and the evaluation of evidence. From their perspective, CT comprises the competence to propose ideas autonomously and the ability to participate in and reflect on the surrounding world (Franco-Mariscal et al., 2025). In the case of educators, the development of CT is particularly relevant to strengthen the processes of reflection on practices, which is fundamental in formative research.

Despite the improvements detected in the micro-debates experience, it is essential to identify the aspects that constitute weaknesses to obtain better results in a new cycle of micro-debates. These aspects refer to (1) the training programme should be planned with more time dedicated to its implementation, (2) between each micro-debate, the strengths and weaknesses of the arguments used should be reflected on with the preservice teachers using the recordings made, (3) it is essential to analyse the bibliographical sources and the arguments used in the preparation of the micro-debates to check if the sources used are reliable and credible. In summary, as an educational implication, developing the micro-debates programme through cycles of planning, debate, reflection, re-planning, and discussion would be advisable.

Acknowledgements This work is part of the Spanish National Plan R+D+i Project entitled "Citizens with critical thinking: A challenge for teachers in science education", reference PID2019-105765GA-I00, funded by MCIN/AEI/https://doi.org/10.13039/501100011033.

References

Anisa, A., Widodo, A., Riandi, R., & Muslim, M. (2023). Students' argumentation in science lessons: How effective is rebuttal analysis framework in representing the complexity of classroom argumentation? *Science & Education, 32*(3), 669–687. https://doi.org/10.1007/s11191-022-00320-8

Bailin, S. (2002). Critical thinking and science education. *Science & Education, 11*(4), 361–375. 1152. https://doi.org/10.1023/A:1016042608621

Bernal-Herrera, P., Cano-Iglesias, M. J., Franco-Mariscal, A. J., & Blanco-López, Á. (2023). Impacto de un debate sociocientífico en las habilidades argumentativas y en la toma de decisiones del profesorado de secundaria en formación inicial [Impact of a Socio-Scientific Debate on the Argumentative and Decision-Making Skills of Preservice Secondary Teachers]. *Enseñanza de las Ciencias, 41*(3), 113–132. https://doi.org/10.5565/rev/ensciencias.5861

Braund, M., Scholtz, Z., Sadeck, M., & Koopman, R. (2013). First steps in teaching argumentation: A South African study. *International Journal of Educational Development, 33*(2), 175–184. https://doi.org/10.33828/sei.v33.i1.7

Calsamiglia, H., & Tusón, A. (1999). *Las cosas del decir. Manual de análisis del discurso* [The things we say. Manual of discourse analysis]. Ariel.

Carámbula, S. B., Cabrera, C. A., Cabrera, A. R., Martínez, A. G., Pérez-Salatto, A. V. (2023). Evaluación de producciones de futuros profesores realizadas en procesos de investigación formativa. [Evaluation of Productions by Future Teachers Created During Formative Research Processes]. In C. Cabrera, D. Imbert, & C. Rebollo (Comp.), *Investigación formativa en profesorado: Aproximación a la didáctica crítica* [Training Research in Teachers: Approach to Critical Didactics] (pp.185–212). Magro Editores.

Cano-Iglesias, M. J. (2024). *Argumentación y toma de decisiones como habilidades del pensamiento crítico: Una investigación con estudiantes de ingenierías industriales y del Máster en Profesorado de especialidades científico-tecnológicas* [Argumentation and decision-making as critical thinking skills: A study with students in industrial engineering and the Master's in Teaching of scientific and technological specialties] [Doctoral Thesis, University of Malaga, Spain]. https://hdl.handle.net/10630/30996

Cebrián-Robles, D., Franco-Mariscal, A. J., & Blanco-López, A. (2021). Secuencia de tareas para enseñar argumentación en ciencias a profesorado en formación inicial a través de CoRubric. Ejemplificación en una actividad sobre una central salina [Sequence of tasks to teach argumentation in science to pre-service teachers through CoRubric. Exemplification in an activity on a saline power plant]. *Didáctica de las Ciencias Experimentales y Sociales, 40,* 149–168. https://doi.org/10.7203/DCES.40.18178

Chadha, D., Krishna, P., Bui, L., & Kogelbauer, A. (2023). How far do we go? Engaging students as partners in redesigning teaching. *Educational Action Research, 31*(4), 620–632. https://doi.org/10.1080/09650792.2022.2058974

Díaz, N., & Jiménez-Liso, M. R. (2012). Las controversias sociocientíficas: Temáticas e importancia para la educación científica [Socioscientific issues: Topics and importance to the scientific education]. *Revista Eureka sobre Enseñanza y Divulgación de las Ciencias, 9*(1), 54–70. https://doi.org/10.25267/Rev_Eureka_ensen_divulg_cienc.2012.v9.i1.04

Domènech, A. M., & Márquez, C. (2014). Which perspectives are referred in students' arguments about a socio-scientific issue? The case of bears' reintroduction in the pyrenees. In C. Bruguière, A. Tiberghien, & P. Clément (Eds.), *Topics and trends in current science education, 9th ESERA conference selected contributions* (pp. 71–84). Springer.

Erduran, S., & Jiménez-Aleixandre, M. P. (2007). *Argumentation in science education. Perspectives from classroom-based research.* Springer. https://doi.org/10.1007/978-1-4020-6670-2

Espinoza, E. E. (2020). La investigación formativa. Una reflexión teórica [Formative investigation. A theoretical reflection]. *Revista Conrado, 16*(74), 45–53.

Esquivel, J., Venegas, V., Venegas, C., Gonzáles, M., Bacón, N., & Valdivia, H. (2023). Formative research: Perceptions of communication science students at a Peruvian university. *Journal of Technology and Science Education, 13*(2), 565–582. https://doi.org/10.3926/jotse.1758

Estrada, A. D., & Estrada, J. (2019). La investigación formativa desde la transdisciplinariedad, para la reforma del pensamiento dentro de la formación profesional [The formative investigation from the transdisciplinarity, for the reform of the thought within the professional formation]. *Uniandes Episteme, Revista Digital de Ciencia, Tecnología e Innovación, 6*(2), 194–216.

Fang, S. C., Hsu, Y. S., & Lin, S. S. (2019). Conceptualizing socio-scientific decision making from a review of research in science education. *International Journal Science and Mathematics Education, 17*, 427–448. https://doi.org/10.1007/s10763-018-9890-2

Franco-Mariscal, A. J., & Cano-Iglesias, M. J. (2023). Desarrollo de habilidades de pensamiento crítico e iniciación a la investigación a través de debates en la formación inicial de profesores españoles de educación secundaria. Ejemplificación en un debate sobre las consecuencias de la erupción de un volcán [Development of critical thinking skills and initiation to research through debates in the initial training of Spanish secondary education teachers. Exemplification in a debate on the consequences of a volcano eruption]. In C. Cabrera, D. Imbert, & C. Rebollo (Comp.) *Investigación formativa en profesorado: Aproximación a la didáctica crítica* [*Training Research in Teachers: Approach to Critical Didactics*] (pp. 125–146). Magro Editores.

Franco-Mariscal, A. J., Cano-Iglesias, M. J., España-Ramos, E., & Blanco-López, A. (2025). The ENCIC-CT model for the development of critical thinking. In A. J. Franco-Mariscal (Ed.), *Critical thinking in science education and teacher training* (Chap. 1) (pp. 3–42). Springer. https://doi.org/10.1007/978-3-031-78578-8_1

Hadfield, M. (2012). Becoming critical again: Reconnecting critical social theory with action research practice. *Educational Action Research, 20*(4), 571–585. https://doi.org/10.1080/09650792.2012.727647

Horng, R. Y., Lu, P. H., Chen, P. H., & Hou, S. H. (2013). The effects of argument stance on scientific knowledge inquiry skills. *International Journal of Science Education, 35*(16), 2784–2800. https://doi.org/10.1080/09500693.2012.671558

Imbert, D. (2023). La didáctica crítica: Faro para las transformaciones en educación [The Critical Didactics: A Beacon for Transformations in Education]. In C. Cabrera, D. Imbert, & C. Rebollo (Comp.), *Investigación Formativa en Profesorado: Aproximación a la didáctica crítica* [*Training Research in Teachers: Approach to Critical Didactics*] (pp. 43–53). Magro Editores.

Jiménez-Aleixandre, M. P. (2010). *10 ideas claves. Competencias en argumentación y usos de pruebas* [10 key ideas. Competencies in argumentation and use of evidence]. Graó.

Jiménez-Aleixandre, M. P., & Puig, B. (2012). Argumentation, evidence evaluation and critical thinking. In *Second international handbook of science education* (pp. 1001–1015). Springer. https://doi.org/10.1007/978-1-4020-9041-7_66

Koeneman, M., Goedhart, M., & Ossevoort, M. (2013). Introducing pre-university students to primary scientific literature through argumentation analysis. *Research in Science Education, 43*(5), 2009–2034. https://doi.org/10.1007/s11165-012-9341-y

Larrain, A., Freire, P., & Howe, C. (2014). Science teaching and argumentation: One-sided versus dialectical argumentation in Chilean middle-school science lessons. *International Journal of Science Education, 36*(6), 1017–1036. https://doi.org/10.1080/09500693.2013.832005

Morin, E. (2002). The epistemology of complexity. In D. F. Schinitman, & J. Schnitman (Eds.), *New paradigms culture and subjectivity*. Hampton Press, Inc.

Moreno, M. G. (2005). Potenciar la educación. Un currículum transversal de formación para la investigación. [Enhancing education: A cross-curricular training program for research]. *REICE, Revista Electrónica Iberoamericana sobre Calidad, Eficacia y Cambio en Educación, 3*(1), 520–540.

Occelli, M., Pérez, E., García, L., Masullo, M., Grosso, P., Appendino, S., Galli, L., Ulloque, M., & Viarengo, M. (2020). Desarrollo de pensamiento crítico: Análisis de una experiencia interdisciplinaria centrada en una cuestión sociocientífica [Development of critical thinking: Analysis of an interdisciplinary experience focused on a socio-scientific issue]. *Revista Electrónica EDUCyT, 11*(extra), 1151–1162.

Paucca, N., Alfaro, M., García, J. O., Ramírez, J., & Rafayle, R. J. (2021). Investigación formativa y logro de competencias en estudiantes de una universidad pública—Lima] [Formative Research and Achievement in Students of a Public University—Lima. *PURIQ Revista de Investigación Científica, 3*(2), 227–235. https://doi.org/10.37073/puriq.3.349

Pérez-Salatto, M., & Pérez-Salatto, A. (2023). La investigación formativa. In C. Cabrera, D. Imbert, & C. Rebollo (Comp.), *Investigación Formativa en Profesorado: Aproximación a la Didáctica Crítica* [*Training Research in Teachers: approach to Critical Didactics*] (pp. 55–60). Magro Editores.

Pinochet, J. (2015). El modelo argumentativo de Toulmin y la educación en ciencias: una revisión argumentada [Toulmin's argumentative model and science education: An argued review]. *Ciência & Educação* (Bauru), *21*(2), 307–327. https://doi.org/10.1590/1516-731320150020004

Postholm, M. B. (2020). The complementarity of formative intervention research, action research and action learning. *Educational Research, 62*(3), 324–339. https://doi.org/10.1080/00131881.2020.1793684

Questa, M., Cabrera, C, & Tejera, A. (2022). ¿Qué es la investigación? In A. Tejera, & M. Questa (Coords.). *Competencias y herramientas de investigación aplicada con foco en la gestión educativa* [Skills and applied research tools with a focus on educational management] (pp. 9–37). University ORT Uruguay.

Restrepo, B. (2003). *Investigación formativa e investigación productiva de conocimiento en la universidad*. [Formative Research and Knowledge-Generating Research at the University.] Universidad Central. Nómadas.

Reigeluth, C. M., & Frick, T. W. (1999). Formative research: A methodology for creating and improving design theories. In C. M. Reigeluth (Ed.), *Instructional-design theories and models—A new paradigm of instructional theory* (pp. 633–652). Lawrence Erlbaum.

Sabariego, M., Hila, A., Gros, B., & Simón, B. (2020). Competencia investigadora e investigación formativa en la formación inicial del docente [Research competence and formative research in initial teacher training]. *Contextos Educativos Revista de Educación, 26*, 239–259.

Sadler, T. D., & Zeidler, D. (2005). Patterns of informal reasoning in the context of socioscientific decision making. *Journal of Research in Science Teaching, 42*(1), 112–138. https://doi.org/10.1002/tea.20042

Salcedo-Armijo, E., Cano-Iglesias, M. J., Blanco-López, A., & Franco-Mariscal, A. J. (2023). Uso de pruebas y decisiones de profesores en formación inicial en un debate sobre prohibición de plásticos [Use of evidence and decisions of pre-service teachers in a debate on plastics bans]. *Educación Química, 34*(4), 104–118. https://doi.org/10.22201/fq.18708404e.2023.4.84663

Toulmin, S. (1958). *The uses of argument*. Cambridge University Press.

Turpo, O., Quispe, P. M., Paz, L. C., & Gonzales, M. (2020). Formative research at the university: Meanings conferred by faculty at an Education Department. *Educação e Pesquisa, 46*, e215876. https://doi.org/10.1590/S1678-4634202046215876

Valero, V. N. (2021). La investigación formativa en la universidad [Formative research at the university]. *Revista Latinoamericana Ogmios, 1*(1), 7–8. https://doi.org/10.53595/rlo.2021.1.001

Zeidler, D. L., Herman, B. C., & Sadler, T. D. (2019). New directions in socioscientific issues research. *Disciplinary and Interdisciplinary Science Education Research, 1*(1), 1–9. https://doi.org/10.1186/s43031-019-0008-7

Chapter 9
Argumentation of Preservice Elementary Science Teachers on Breastfeeding in Public Places Using the Kialo Online Debate Platform

Miriam Palma-Jiménez, Daniel Cebrián-Robles, and Ángel Blanco-López

9.1 Introduction

One of the ways to learn science and develop critical thinking (hereinafter, CT) is through scientific argumentation (Blanco-López et al., 2015). According to Jiménez-Aleixandre (2010), argumentation is the ability to evaluate scientific statements based on evidence, recognising the importance that such statements must be justified, i.e. supported by evidence. Argumentation has the advantage of helping students learn to evaluate evidence and supports the learning processes of science in the classroom (Erduran & Jiménez-Aleixandre, 2007).

According to Osborne (2010), argumentation is essential to improve citizens' understanding of science and scientific literacy and develop collaboration and CT. Furthermore, argumentation allows, within CT, the use of criteria and evidence for evaluating knowledge, recognising the importance of considering alternatives and reviewing different points of view (Jiménez-Aleixandre & Puig, 2022). Therefore, argumentation has the advantage of helping students learn to evaluate evidence and supports science learning processes in the classroom (Erduran & Jiménez-Aleixandre, 2007).

Osborne et al. (2016) propose argument construction and argument critique as two key dimensions to develop scientific argumentation. Construction refers to students'

M. Palma-Jiménez (✉) · D. Cebrián-Robles · Á. Blanco-López
Science Education, Universidad de Málaga, Málaga, Spain
e-mail: miriampalma@uma.es

D. Cebrián-Robles
e-mail: dcebrian@uma.es

Á. Blanco-López
e-mail: ablancol@uma.es

A. J. Franco-Mariscal (ed.), *Critical Thinking in Science Education and Teacher Training*, Contemporary Trends and Issues in Science Education 64,
https://doi.org/10.1007/978-3-031-78578-8_9

ability to create coherent, evidence-based scientific arguments, which involves identifying a relevant claim, selecting evidence to support that claim, and constructing a warrant that explains how the evidence supports the claim (Osborne et al., 2016). Critique, on the other hand, refers to students' ability to evaluate and improve arguments, which involves identifying weaknesses in arguments, developing critical questions, and presenting evidence to support that critique (Osborne et al., 2016).

One way to develop CT through argumentation is by addressing socio-scientific issues (Sadler & Zeidler, 2005), as they are open problems, ill-structured, debatable and subject to multiple perspectives and solutions. These problems are suitable for the development of argumentation because they require not only scientific knowledge but also multidimensional knowledge (Jiménez-Tenorio et al., 2020) (economic, scientific, technological, political, ethical, etc.) to make decisions about these problems (Zeidler, 2014). Furthermore, these issues, due to their controversial nature, can lead to a distance between citizens and professional science, as there may be a gap between social needs and the solutions that end up being adopted (Puig & Evagorou, 2020). Several authors point out the importance of addressing socio-scientific issues in science classes (Colucci-Gray et al., 2006) and have been proposed as appropriate contexts for their learning (España & Prieto, 2010). Contextualisation of science teaching based on these problems differs from other approaches in that it integrates aspects related to science and its processes and those related to moral aspects (Bencze et al., 2020). One of the educational advantages of socio-scientific issues is that it helps students to improve their decisions and reasoning about them (Moreno-Fontiveros et al., 2022) based on scientific knowledge and a broader and more complex understanding of these problems (Bencze et al., 2020).

Human breastfeeding can be considered a socio-scientific issue of interest for preservice elementary science teachers (Palma-Jiménez et al., 2023), as it is related to the contents of nutrition and human relationships, which are necessary for their training (Illescas-Navarro et al., 2020). Among the scientific factors involved in this socio-scientific issue, we can highlight that breastfeeding has numerous benefits for the baby's health, as it offers protection against infections and diseases and improves the infant's cognitive development (Gartner et al., 2005). It also has a positive impact on the mother's health, as it prevents uterine and breast cancer (Stuebe, 2009). However, there are social factors that influence the continuity of human breastfeeding, such as women's return to work (Giménez et al., 2015), cultural aspects, such as breastfeeding in public (Morris et al., 2020) or beliefs detrimental to breast milk in popular culture, such as mothers with poor quality milk feeding less (Padró, 2019).

Controversies, such as breastfeeding in public places, constitute this socio-scientific issue. Morris et al. (2020) point out that mothers who breastfeed in public feel uncomfortable and vulnerable due to the perception of some people in society that this practice is selfish and repulsive. On the other hand, Komodiki et al. (2014) argue that increasing public acceptance of breastfeeding could improve mothers' confidence and normalise the practice, with potential health benefits for mothers and babies (Hauck et al., 2021). However, cultural and religious factors largely influence public acceptance of breastfeeding, making a global position complex (Komodiki et al., 2014).

The approach to socio-scientific issues and their controversies can be carried out using different teaching strategies, among which those involving argumentation practices predominate. In this sense, the debate has been highlighted as an effective educational tool for the development of argumentation (Bernal-Herrera et al., 2023; Brocos & Jiménez-Aleixandre, 2020; Capkinoglu et al., 2021; Cayci, 2020). Cayci (2020) considers it an educational resource that promotes scientific argumentation and literacy. On the other hand, Capkinoglu et al. (2021) used discussions on socio-scientific issues, such as nuclear energy, to highlight the importance of argumentation in science education. In addition, Brocos and Jiménez-Aleixandre (2020) have used debate in contexts related to everyday life, such as dietary choices, to promote the production of collectively written arguments.

9.2 Argumentation as a Critical Thinking Skill Through the Kialo Tool

The learning experience presented in this chapter mainly develops argumentation as one of the dimensions of the ENCIC-CT model (Chap. 1) (Franco-Mariscal et al., 2025). This work addresses the two dimensions of argumentation proposed by Osborne et al. (2016). On the one hand, the construction of arguments, as arguments, will be created using the context of the human breastfeeding controversy in public and, on the other hand, the critical dimension by evaluating arguments and conducting a counterargumentation and rebuttal exercise. Counterargumentation is identifying weaknesses in an argument and constructing an argument that contradicts the original argument; rebuttal would be an argument in response to a previous counterargument (Osborne et al., 2016).

Brocos and Jiménez-Aleixandre (2020) or Bernal-Herrera et al. (2023) used the debate as a resource for developing justifications and using evidence. Capkinoglu et al. (2021) used it to address the construction of conclusions, data, warrants, background and refutations. Cayci (2020) also used it to construct arguments, counterarguments, and rebuttals.

In order to facilitate the teaching–learning process of scientific argumentation through debates, different digital platforms allow for real-time discussion. One of these platforms is Kialo (https://www.kialo-edu.com/), which is open access and allows, based on an initial question, to create an online debate in real time where the different interventions are recorded in writing. When a user registers, he or she has the option of generating a public debate, to which any user can enter and intervene, or a private debate, to which only those invited can have access, with the option of making it public later. Once the topic to be debated is registered, a pro and a con thread is generated, and the same happens with each answer written in the debate. It allows different threads to be created on an initial topic with arguments and rebuttals. Each intervention in the debate allows for a maximum of 500 characters. As the debate develops, graphs are generated which visually represent the arguments, counterarguments and refutations of the different branches of the threads so that the user can see, in a visual way, which are the majority positions in the debate and the impact of the different arguments according to the number of responses. In addition, the graph allows direct access to any thread at any point in the debate, making it quicker to respond to a particular thread. Once a debate is over, it is saved in the cloud and can be re-accessed anytime.

In this way, Kialo is a tool that facilitates users to progressively move from building arguments to counterarguments and rebuttals. Chaudoin et al. (2017) describe Kialo as a visual reasoning platform where users collaborate to address specific questions by providing for and against arguments. Darmawansah et al. (2022) used Kialo with university students to improve argumentation, reflection and CT.

This chapter presents Kialo as a tool for developing competence in scientific argumentation through debate on the socio-scientific issue of human breastfeeding, specifically its controversy related to breastfeeding in public places. This work's novelty lies precisely in using the Kialo platform to promote argumentation.

9.3 Research Questions

This paper poses two research questions (RQ):

- RQ1: What are the results of the Kialo platform in the interplay of arguments, counterarguments and rebuttals in a debate on the breastfeeding controversy in public places conducted by preservice elementary teachers?
- RQ2: How do preservice elementary teachers rate using Kialo as an argumentation tool?

9.4 Method

9.4.1 Participants and Context

The sample of this study was 62 preservice elementary teachers from the 3rd year of the Degree in Primary Education at the University of Malaga (Malaga, Spain) during the academic year 2019–2020, who were taking the subject Science Teaching. This work is part of a broader formative programme in which preservice elementary teachers received training in scientific argumentation that lasted one semester (Palma-Jiménez et al., 2023). The learning experience presented here was one of the last activities in the formative programme, as it addresses one of the highest levels in the learning progression proposed by Osborne et al. (2016). The free-to-use Kialo platform was used as a teaching resource to design and implement the debate.

9.4.2 Experience Description

First, the professor explained the controversy about breastfeeding in public places to the preservice elementary teachers. Then, 14 groups of between four and five members were created to encourage the development of collaborative argumentation and teamwork. Seven groups were then randomly assigned to positions in favour of breastfeeding in public places and seven against. Thus, the preservice elementary teachers' position on the controversy did not have to coincide with the one assigned by the professor. This decision ensured that the debate could start from an equal position with the same number of groups in the two positions on the controversy. The groups were then given 30 min to search for information to prepare arguments for the debate (an opportunity for critical analysis of the information and communication within the group). After this time, a representative from each group was asked to register on the Kialo platform to avoid duplication of content by the different group members. Finally, the debate was held, which lasted approximately one hour (communication between groups).

The topic for discussion was presented through the statement: *for or against breastfeeding in public places*. Each group had to write an argument for or against the Kialo platform as assigned. In addition, to identify the different groups, each intervention in the debate had to be preceded by the group's name (e.g. group 1, group 2, etc.) Each group then had to make a counterargument to a previous argument. Then, the groups that had received counterarguments were to respond to them with rebuttals until the end of the debate.

9.4.3 ENCIC-CT Model Application

The experience is structured within the framework of the ENCIC-CT model (Franco-Mariscal et al., 2025). The first layer of this model addresses the socio-scientific issue of human breastfeeding in public spaces, a topic that resonates with both societal relevance and scientific inquiry. In the second layer, the focus shifts to the development of skills, with a particular emphasis on argumentation. This layer also encompasses the critical analysis of information and communicative skills, both of which are essential for informed scientific discourse. The third layer integrates the practice of scientific argumentation, where students engage in the construction, evaluation, and critique of arguments. This practice not only enhances their reasoning abilities but also aligns with real-world scientific processes, where evidence-based argumentation is crucial. Finally, in the fourth layer, the use of the Kialo platform is introduced as a teaching strategy to facilitate structured debates and discussions.

9.4.4 Data Collection and Analysis

In order to answer RQ1, the different interactions of arguments, counterarguments and rebuttals were collected on the Kialo platform. In addition, the results were analysed considering the number of interventions in the form of cumulative frequency and the graphs of interactions displayed by the Kialo platform to understand numerically and visually which interactions had taken place in the debate.

The user can view the findings in graphs as a semicircle (Fig. 9.1) or a tree (Fig. 9.2).

The circle in the blue centre (Fig. 9.1) represents the initial question. Around it is the first row, divided into green circular trapezoids, representing answers in favour

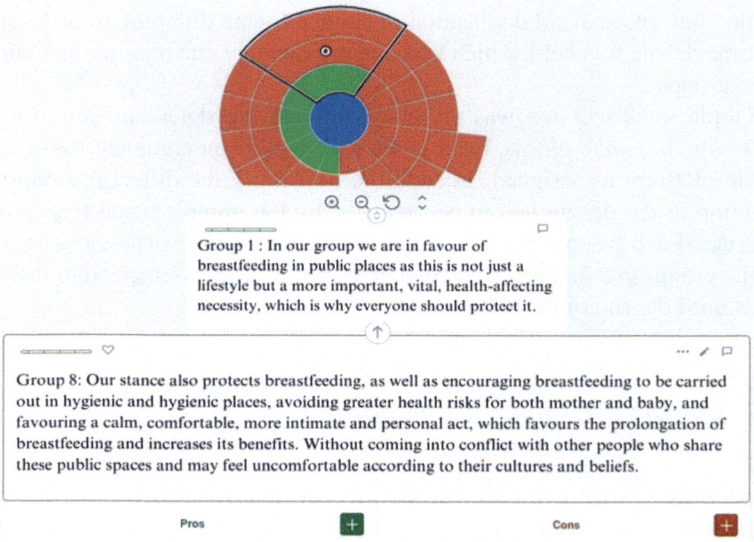

Fig. 9.1 Semicircle graph

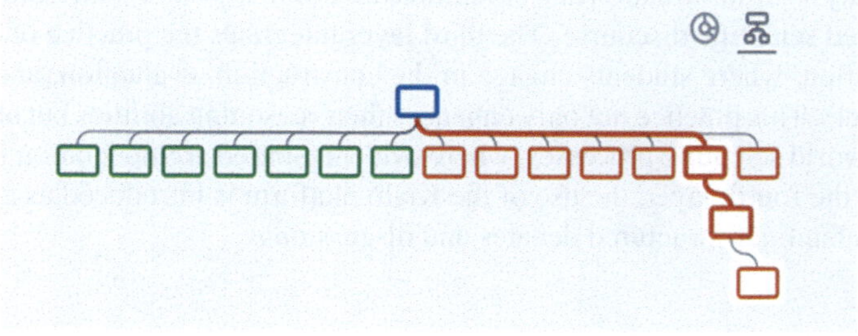

Fig. 9.2 Tree chart

of that question and red ones against it. The second row of red circular trapezoids corresponds to counterarguments in response to the arguments in row 1, while refutations to the counterarguments are shown from the third row onwards. Moreover, by clicking on each circular trapezoid, the content of each response can be visualised and is reflected in the graph with a small white circle with a black dot inside it, as we can see in Fig. 9.1 in the counterargument given by group 8 to the argument of group 1.

Figure 9.2 shows the tree graph, where the initial question is represented by a blue rectangle and the answers for and against are represented by green and red rectangles, respectively. However, unlike the previous graph, it only allows us to visualise the number of answers given to each argument when clicking on the rectangle to which we want to know the answer. Therefore, in this case, viewing the responses to each argument in isolation is only possible, and not globally.

Concerning the RQ2, an individual assessment was carried out with the following question: *What did you think about doing this activity with Kialo? Justify your answer*. From the answers to the preservice elementary teachers, a system of categories emerged based on the evaluation of Kialo as a teaching resource for developing the argumentation. These categories were then assigned to the responses to each preservice elementary teacher so that the same response could have several categories, but the same category could only be assigned to each response once. Finally, the cumulative frequency and the percentage of categories identified in the responses were calculated.

9.5 Results

9.5.1 Interactions of Arguments, Counterarguments and Rebuttals

The semicircle diagram (Fig. 9.3) allows for analysis of the different interactions in the debate concerning arguments, counterarguments and rebuttals.

As seen (Fig. 9.3), the preservice elementary teachers produced the same number of arguments for and against, as there were seven groups in each position, and each group had to formulate an argument. The arguments in the first row received one counterargument each, except for one of them, which did not receive any counterargument, perhaps because they did not make clear their position for or against. Concerning the rebuttals, the groups in favour made one rebuttal to each counterargument, except for one counterargument that received two rebuttals from the same group, perhaps because they wanted to express two different ideas and decided to make one rebuttal for each idea. However, one counter-thread received one rebuttal against and one rebuttal for, both of which were based on authoritative arguments,

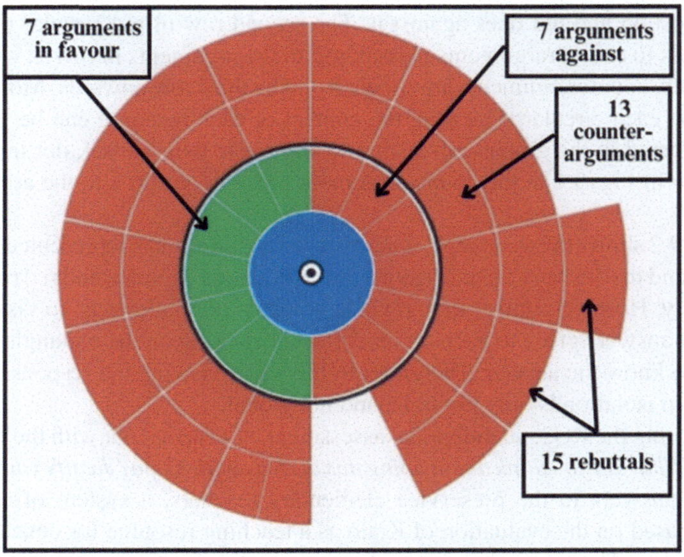

Fig. 9.3 Relationship of arguments, counterarguments and rebuttals in the debate

the first being based on legislation and the second on the World Health Organization. It may have been because the first group may have felt supported by relying on authority, and the other group wanted to respond using the same strategy.

Below are two intervention threads for and against, with their responses as counterarguments and rebuttals. At the end of each intervention, the name of the group to which it belongs is indicated in brackets.

Thread with the initial argument in favour of breastfeeding in public.

- Argument in favour: *We are in favour because we think that a baby does not understand places; it understands needs; if the mother has gone to the street for something she needs or simply for a walk, she should be able to breastfeed without any problem. There is no justification for discriminating against or insulting a mother for breastfeeding her child, as we are talking about a vital need that human beings have, which is to feed ourselves* (Group 6).
- Counterargument: *The baby is not aware, but the mother is, both of the place and her child's habits, and can, therefore, control breastfeeding schedules and prevent breastfeeding in public places as much as possible. Instead, breastfeeding can be done in places such as breastfeeding rooms. If these facilities are not available, you should try to find a more secluded place where you will not disturb anyone* (Group 14).
- Rebuttal in favour: *Breastfeeding rooms are practically scarce and close to public toilets, causing unpleasant smells. Babies do not decide when they are hungry, no matter how much you regulate their schedules, as they are unconscious. Breastfeeding a baby does not bother anyone as it is a natural act defended by the WHO,*

which supports women's breastfeeding by saying that they should breastfeed when and where they need to (Group 6).

Thread with initial argument against breastfeeding in public.

- Argument against: *We are against breastfeeding in public places because we consider that it is more convenient to breastfeed in adequate and conditioned spaces, quiet and hygienic, where both the baby and the mother feel comfortable, without shocks or reproachful looks, thus guaranteeing an adequate feeding process for the baby without putting the health of both of them at risk. Moreover, breastfeeding is so intimate and emotional that it belongs to the private sphere* (Group 8).
- Counterargument in favour: *We should make society aware that it is natural, no one should feel uncomfortable, and no mother should receive disapproving looks. It should be socially accepted and possible to do it wherever the mother and child see fit so that the child can fulfil its vital functions at the necessary time* (Group 1).
- Rebuttal against: *In a multicultural society such as ours, it is difficult to adopt the convictions of all the many cultures present. As we understand breastfeeding as natural and positive for mother and baby, we advocate that it should be carried out, but that it should be done in spaces set up for this purpose, such as the breastfeeding rooms, which are mandatory in public places reflected in the draft bill of the Junta de Andalucía of 2016 to promote healthy living and healthy eating* (Group 8).

9.5.2 Assessment of Preservice Elementary Teachers When Using Kialo for Argumentation

Table 9.1 shows the preservice elementary teachers' assessment of Kialo for each of the categories found, indicating their cumulative frequency and percentage.

Some examples of preservice elementary teacher responses, and the categories assigned in brackets, are shown below.

Example 1: *I found the support interesting and easy to use* [A1]. *I think it is very interesting because the answers are distributed in branches and by levels* [A8], making *it easier to see the argument, counterargument, and refutation* [A6], *which helps the student understand these terms. In addition, we will be working on informed opinion and the objectivity and subjectivity of children to be used with children in a classroom or as a home activity* [A2].

Example 2: *It is interesting* [A1], *as it helps us to learn to argue* [A7] *and to work in a debate with evidence* [A11], *as well as to put you in situations contrary to your thinking and have to defend it* [A12].

Table 9.1 Evaluation of Kialo as a resource for developing argumentation

Category	Frequency	Percentage
On argumentation (A)		
A1. Novel/interesting/engaging/attractive/dynamic ICT tool for practicing argumentation	38	25.3
A2. Use for teaching argumentation	23	15.4
A3. Helps to get multiple points of view	17	11.3
A4. Facilitates argumentation at a distance	15	10.0
A5. Improving argumentative interaction	13	8.7
A6. Facilitates the critical dimension (counter-argumentation and refutation)	12	8.0
A7. Helps to learn to argue	11	7.3
A8. Viewing the whole debate	3	2.0
A9. Complexity in arguing points of view contrary to one's own	3	2.0
A10 Asynchronous waiting to argue	2	1.3
A11. Allows the use of evidence	2	1.3
A12. Enrichment by arguing points of view contrary to one's own	2	1.3
On Kialo (K)		
K1. Limitation on the length of characters for arguments	5	3.3
K2. Difficulty in using the ICT tool	4	2.7
Total	150	100

9.6 Discussion and Educational Implications

The Kialo teaching strategy has made it possible to facilitate argumentation, counter-argumentation and refutation in a socio-scientific debate, specifically on the controversy of human breastfeeding in public places. Secondly, it has facilitated the visualisation of the impact of the different arguments in terms of the number of responses received. The arguments that received the highest number of counterarguments and rebuttals were those based on health and human nutrition, while the arguments that did not receive any counterarguments showed only personal opinions. Therefore, we think that one of the possible reasons for the lack of counterarguments in this argument may be because it did not make clear its position for or against the debate (Jiménez-Aleixandre, 2010).

Concerning the assessment of Kialo as a strategy for enhancing scientific argumentation, the categories with the highest percentages were those associated with its novelty, interest, attractiveness, and dynamic nature as an argumentative tool, as well as its utility for teaching argumentation. This aligns with the findings of Darmawansah et al. (2022) who stated in their study that Kialo could be an effective resource for improving students' argumentative expression skills. Also, it is interesting that the preservice elementary teachers considered that it facilitates the critical dimension (counter-argumentation and refutation), which was one of the intended purposes of the activity. Regarding how Kialo was implemented in this work, it is striking that two preservice elementary teachers considered it enriching to argue contrary to their own points of view. In contrast, three others indicated this as complex. Therefore, it would be interesting to investigate the effectiveness of arguing about a position contrary to one's own to develop scientific argumentation.

As difficulties of this resource, we consider that the limitation of the number of characters to 500 can be a factor affecting the development of a complete argument, but it can also help to synthesise the ideas and information used. In addition, the waiting time in Kialo responses was sometimes uncomfortable for preservice elementary teachers. However, it may have the advantage of favouring the analysis of the information provided since, this way, more time is available to construct counterarguments and rebuttals. Finally, it is important to consider that, before carrying out this activity, it is necessary to provide students with a scaffolding of how the Kialo platform works, which should be considered when scheduling the activity's development time.

Acknowledgements This work is part of the Spanish National Plan R+D+i Project entitled "Citizens with critical thinking: A challenge for teachers in science education", reference PID2019-105765GA-I00, funded by MCIN/AEI/10.13039/501100011033.

References

Bencze, L., Pouliot, C., Pedretti, E., Simonneaux, L., Simonneaux, J., & Zeidler, D. (2020). SAQ, SSI and STSE education: Defending and extending "science-in-context." *Cultural Studies of Science Education, 15*(3), 825–851. https://doi.org/10.1007/s11422-019-09962-7

Bernal-Herrera, P., Cano-Iglesias, M. J., Franco-Mariscal, A. J., & Blanco-López, A. (2023). Impacto de un debate sociocientífico en las habilidades argumentativas y en la toma de decisiones del profesorado de secundaria en formación inicial [Impact of a Socio-Scientific Debate on the Argumentative and Decision-Making Skills of Preservice Secondary Teachers]. *Enseñanza de las Ciencias, 41*(3), 113–132. https://doi.org/10.5565/rev/ensciencias.5861

Blanco-López, Á., España-Ramos, E., González-García, F. J., & Franco-Mariscal, A. J. (2015). Key aspects of scientific competence for citizenship: A Delphi study of the expert community in Spain. *Journal of Research in Science Teaching, 52*(2), 164–198. https://doi.org/10.1002/tea.21188

Brocos, P., & Jiménez-Aleixandre, M. P. (2020). El impacto ambiental de la alimentación: Argumentos de alumnado de Magisterio y Secundaria [Environmental impact of diet: Arguments of pre-service teachers and high school students]. *Enseñanza de las Ciencias, 38*(1), 127–145. https://doi.org/10.5565/rev/ensciencias.2802

Capkinoglu, E. Y., Cetin, P. S., & Metin Peten, D. (2021). How do pre-service science teachers evaluate the persuasiveness of a socioscientific argument? *International Journal of Science Education, 43*(4), 594–623. https://doi.org/10.1080/09500693.2021.1876273

Cayci, B. (2020). A study on the effectiveness of a teaching based on socio-scientific issues in the training of pre-service teachers. *Cypriot Journal of Educational Sciences, 15*(2), 220–231. https://orcid.org/0000-0003-3466-7033

Chaudoin, S., Shapiro, J. N., & Tingley, D. (2017). *Revolutionizing teaching and research with a structured debate platform*. https://scholar.princeton.edu/sites/default/files/jns/files/structureddebate.pdf

Colucci-Gray, L., Camino, E., Barbiero, G., & Gray, D. (2006). From scientific literacy to sustainability literacy: An ecological framework for education. *Science Education, 90*(2), 227–252. https://doi.org/10.1002/sce.20109

Darmawansah, D., Lin, C.-J., & Hwang, G.-J. (2022). Empowering the collective reflection-based argumentation mapping strategy to enhance students' argumentative speaking. *Computers & Education, 184*(1), 104516. https://doi.org/10.1016/j.compedu.2022.104516

Erduran, S., & Jiménez-Aleixandre, M. P. (2007). *Argumentation in science education: Perspectives from classroom-based research*. Springer. https://doi.org/10.1007/978-1-4020-6670-2

España, E., & Prieto, T. (2010). Problemas socio-científicos y enseñanza-aprendizaje de las ciencias [Socio-scientific issues and science teaching and learning]. *Investigación en la Escuela, 71*, 17–24.

Franco-Mariscal, A. J., Cano-Iglesias, M. J., España-Ramos, E., & Blanco-López, A. (2025). The ENCIC-CT model for the development of critical thinking. In A. J. Franco-Mariscal (Ed.), *Critical thinking in science education and teacher training* (Chap. 1) (pp. 3–42). Springer. https://doi.org/10.1007/978-3-031-78578-8_1

Gartner, L. M., Morton, J., Lawrence, R. A., Naylor, A. J., O'Hare, D., Schanler, R. J., Eidelman, A. I., et al. (2005). Breastfeeding and the use of human milk. *Pediatrics, 115*(2), 496–506. https://doi.org/10.1542/peds.2004-249

Giménez, V., Jimeno, B., Valles, M. D., & Sanz de Miguel, E. (2015). Prevalencia de la lactancia materna en un centro de salud de Zaragoza (España): Factores sociosanitarios que la condicionan [Breastfeeding prevalence in a Zaragoza primary health care centre. Socioeconomic factors related]. *Pediatría Atención Primaria, 17*(65), 17–26. https://doi.org/10.4321/S1139-76322015000100004

Hauck, Y. L., Bradfield, Z., & Kuliukas, L. (2021). Women's experiences with breastfeeding in public: An integrative review. *Women and Birth: Journal of the Australian College of Midwives, 34*(3), e217–e227. https://doi.org/10.1016/j.wombi.2020.04.008

Illescas-Navarro, M., Cruz-Guzmán, M., & Criado, A. M. (2020). El cuerpo humano y la salud en libros de texto de Educación Primaria (6–11 años), ¿qué aportan al problema socio-científico sobre la lactancia materna como parte de la alimentación en la primera infancia? [The human being and the health in the Spanish Primary textbooks (6–11 years old). What do they provide to the socio-scientific problem of the nutrition in the early childhood?] *Revista Eureka sobre Enseñanza y Divulgación de las Ciencias, 17*(1), 1–20. https://doi.org/10.25267/Rev_Eureka_ensen_divulg_cienc.2020.v17.i1.1302

Jiménez-Aleixandre, M. P. (2010). *10 Ideas Clave. Competencias en argumentación y uso de pruebas* (Vol. 12). Graó.

Jiménez-Aleixandre, M. P., & Puig, B. (2022). Educating critical citizens to face post-truth: The time is now. In M. P. Jiménez-Aleixandre & B. Puig (Eds.), *Critical thinking in biology and environmental education* (pp. 3–19). Springer. https://doi.org/10.1007/978-3-030-92006-7

Jiménez-Tenorio, N., Vicente, J. J., Yúñez, L. A., & Oliva, J. M. (2020). Fomentar la argumentación en clases de ciencias a través de una controversia sociocientífica en futuros docentes [Promoting argumentation in science classes through socio-scientific controversy among future teachers]. *Ápice, Revista de Educación Científica, 4*(1), 79–86. https://doi.org/10.17979/arec.2020.4.1.4639

Komodiki, E., Kontogeorgou, A., Papastavrou, M., & Volaki, P. (2014). Breastfeeding in public: A global review of different attitudes towards it. *Journal of Pediatrics & Neonatal Care, 1*(6), 1–4. https://doi.org/10.15406/jpnc.2014.01.00040

Moreno-Fontiveros, G., Cebrián-Robles, D., Blanco-López, Á., & España-Ramos, E. (2022). Decisiones de estudiantes de 14/15 años en una propuesta didáctica sobre la compra de un coche [Fourteen/fifteen-year-old students' decisions in a teaching proposal on the buying of a car]. *Enseñanza de las Ciencias, 40*(1), 199–219. https://doi.org/10.5565/rev/ensciencias.3292

Morris, C., Schofield, P., & Hirst, C. (2020). Exploration of the factors influencing attitudes to breastfeeding in public. *Journal of Human Lactation: Official Journal of International Lactation Consultant Association, 36*(4), 776–788. https://doi.org/10.1177/0890334419878119

Osborne, J. (2010). Arguing to learn in science: The role of collaborative, critical discourse. *Science, 328*(5977), 463–466. https://doi.org/10.1126/science.1183944

Osborne, J., Henderson, J. B., MacPherson, A., Szu, E., Wild, A., & Yao, S.-Y. (2016). The development and validation of a learning progression for argumentation in science. *Journal of Research in Science Teaching, 53*(6), 821–846. https://doi.org/10.1002/tea.21316

Padró, A. (2019). *Somos la leche: dudas, consejos y falsos mitos sobre la lactancia* [We're the Milk: Doubts, Advice, and False Myths about Breastfeeding]. Grijalbo Ilustrados.

Palma-Jiménez, M., Cebrián-Robles, D., & Blanco-López, Á. (2023). Impact of instruction based on a validated learning progression on the argumentation competence of preservice elementary science teachers. *Science & Education*. https://doi.org/10.1007/s11191-023-00468-x

Puig, B., & Evagorou, M. (2020). Design of a socioscientific issue unit with the use of modeling: The case of bees. *International Journal of Designs for Learning, 11*(1), 98–107. https://doi.org/10.14434/ijdl.v11i1.24142

Sadler, T. D., & Zeidler, D. L. (2005). Patterns of informal reasoning in the context of socioscientific decision making. *Journal of Research in Science Teaching, 42*(1), 112–138. https://doi.org/10.1002/tea.20042

Stuebe, A. (2009). The risks of not breastfeeding for mothers and infants. *Reviews in Obstetrics and Gynecology, 2*(4), 222–231.

Zeidler, D. L. (2014). Socioscientific issues as a curriculum emphasis: Theory, research, and practice. In N. G. Lederman & S. K. Abell (Eds.), *Handbook of Research on Science Education, Volume II* (pp. 711–740). Routledge.

Part IV
Inquiry and Critical Thinking

Chapter 10
Inquiring and Arguing About Yoghurt Making. A Study with Secondary School Students

Verónica Muñoz-Campos and Antonio Joaquín Franco-Mariscal

10.1 Introduction

For some time, different curricula have identified inquiry and argumentation as relevant scientific practices in science education (National Research Council, 2012). Furthermore, the Programme for International Student Assessment PISA test standards (OECD, 2017) include competences in epistemic knowledge, scientific explanations and inquiry, and the ability to interpret data and evidence scientifically (Allchin & Zemplén, 2020). Thus, students must learn to analyse and evaluate data, claims and arguments in various representations and draw appropriate scientific conclusions (OECD, 2017). This highlights the importance of integrating inquiry and argumentation in science teaching and learning.

Inquiry refers to teaching methods and strategies designed to help students improve their understanding of scientific content. A broader view, however, defines inquiry as the process of skills and abilities that students must understand and be able to perform. The first aspect denotes inquiry as a means to understand scientific content. In contrast, the second aspect indicates that inquiry should not be seen exclusively as a means but as an end (Teig, 2021).

Inquiry aims to enable students to make informed decisions without ignoring relevant scientific information, and argumentation plays a significant role in these processes. Inquiry involves carrying out activities that include observation, formulating questions and answers about the study problem, posing hypotheses, searching

V. Muñoz-Campos (✉) · A. J. Franco-Mariscal
Science Education, Universidad de Málaga, Málaga, Spain
e-mail: mucamv@gmail.com

A. J. Franco-Mariscal
e-mail: anjoa@uma.es

V. Muñoz-Campos
I.E.S. Portada Alta, Málaga, Spain

A. J. Franco-Mariscal (ed.), *Critical Thinking in Science Education and Teacher Training*, Contemporary Trends and Issues in Science Education 64,
https://doi.org/10.1007/978-3-031-78578-8_10

for information from various sources, designing the research, analyzing and interpreting data, and communicating results (Franco-Mariscal, 2015; Romero-Ariza et al., 2018).

In this context, the integration of argumentation becomes crucial. It provides the necessary structure and reasoning to navigate the complexities inherent in inquiry-based tasks (Sampson et al., 2011). While inquiry equips students with the tools to explore and understand scientific phenomena, argumentation empowers them to critically assess the validity of their findings, justify their conclusions, and engage in meaningful scientific discussions. This interplay between inquiry and argumentation ensures that students not only acquire scientific knowledge but also develop the capacity to apply it thoughtfully and effectively in real-world situations.

Moreover, argumentation requires specific intellectual and social skills (Muller & Perret-Clermont, 2009) that emphasize content knowledge and self-regulation to judge value in any situation. It makes it possible to differentiate between opinions and facts and to detect literal or judgmental meanings. An important function of argumentation is to investigate controversial or debated issues to arrive at reasoned judgments. Analyzing and critiquing individual arguments is one aspect of this process (Osborne et al., 2016) that helps to eliminate poorly reasoned arguments. Constructing arguments involves connecting evidence and conclusions by evaluating and reasoning about theoretical statements (Archila et al., 2022).

The integration of argumentation and inquiry is fundamental in science education and a challenge for teachers (Muñoz-Campos et al., 2020), closely related to developing critical thinking (hereinafter, CT).

10.2 Integrating Argumentation into Inquiry

Inquiry is characterised by questioning, hands-on activity and CT in science classrooms (Gerber et al., 2003). This practice encourages students to question, investigate, learn, report their effects and evaluate their knowledge (van Zee & Roberts, 2006). The benefits of using inquiry include excellent preservation of real information, increased flexibility and creativity in problem-solving and ensuring student motivation (Lord & Orkwiszewski, 2006). By incorporating socio-scientific and everyday life problems into this approach, as proposed by the ENCIC-CT model (Franco-Mariscal et al., 2025), teachers can further enhance the development of these skills.

The literature includes different approaches to inquiry in the classroom. Based on a review of various school inquiry proposals, Franco-Mariscal (2015) established seven dimensions for scientific competences linked to inquiry: (I1) problem formulation, (I2) information management, (I3) research planning and design, (I4) data collection and processing, (I5) data analysis and concluding, (I6) results communication, and (I7) attitude or critical reflection and teamwork.

Progress in these dimensions should enable students to strengthen their argumentation skills. For their initiation, it is advisable to use simple argumentation models such as Toulmin's model (2003) or the adaptation proposed by Jiménez-Aleixandre (2010), establishing conclusion, evidence and warrants as essential elements. For Jiménez-Aleixandre (2010), arguing consists of [...] *be able to evaluate statements based on evidence, i.e. to recognise that conclusions and statements must be justified, in other words, supported by evidence* (p.23).

Once argument construction has been mastered, it is possible to move on to the level of argument critique, as proposed by Osborne et al. (2016) in their learning progression model for scientific argumentation. These authors identify different difficulty levels for argumentation tasks in both dimensions (construction and critique), as shown in Table 10.1. In the table, these levels are correlated with the dimensions of Franco-Mariscal's (2015) inquiry approach.

Consequently, an important task in science education is to develop students' ability to argue coherently in the face of a given problem, such as those posed in a school inquiry.

Osborne et al.'s (2016) levels of argument construction and critique can be correlated with the different stages of the inquiry (Table 10.1), as exemplified below for an inquiry focused on how plastic degradation takes place (López-Fernández & Franco-Mariscal, 2021). The problem formulation (I1) involves the student making observations in their environment where they can identify evidence (0d) (e.g. the accumulation of plastics in their neighbourhood or the presence of microplastics in fish) that allows them to pose a problem. Through information seeking (I2), students provide and identify evidence (0c, 0d) that enables them to argue for the planning and design chosen for their investigation (I3) (e.g. photodegradable plastics are sensitive to light). Data collection and processing (I4) helps them obtain evidence (0c, 0d)

Table 10.1 Learning progression for scientific argumentation (Osborne et al., 2016) and its correlation with inquiry (Franco-Mariscal, 2015)

Osborne et al. (2016) learning progression			Franco-Mariscal's (2015) approach to inquiry
Level	Constructing	Critiquing	
0a	Constructing a claim		I5, I6
0b		Identifying a claim	I5
0c	Providing evidence		I2, I3, I4, I6
0d		Identifying evidence	I1, I2, I3, I4
1a	Constructing a warrant		I5, I6
1b		Identifying a warrant	I5
1c	Constructing a complete argument		I5, I6
1d	Providing an alternative counterargument		I7
2a	Providing a counter-critique		I7
2b	Constructing a one-sided comparative argument		I7
2c	Providing a two-sided comparative argument		I7
2d	Constructing a counterclaim with justification		I7

(e.g., after 42 days, colour changes were visible in a white plastic bag). Data analysis (I5) allows them to construct and identify justifications based on this evidence (1a, 1b) (e.g. the appearance of odour in plastic is related to the degradation of components of the material as an effect of the action of micro-organisms). In the conclusion (I5) stage, they first construct and identify a conclusion (0a, 0b) and then construct a complete argument (1c) (e.g. exposure of plastics to ambient conditions for 100 days is insufficient for complete degradation of the plastic to be observed since the samples studied have remained practically unchanged during this time. Therefore, much more time is required to monitor chemical changes). The communication of results (I6) constitutes in itself the complete process of argumentation since students must show their results by constructing complete arguments (1c) in which they provide a conclusion (0a), warrants (0c) and evidence (1a). Finally, to reach an adequate conclusion, students must engage in critical reflection (I7) in which other peers may intervene. This stage offers opportunities for proposing alternative counter-arguments (1d) (e.g. different types of plastics can degrade in less time than those studied, such as biodegradable plastics), constructing new arguments (2a), providing counter-critiques (2b), providing two-sided comparative argument (2c), and constructing a counterclaim with justification (2d). In addition, students should be able to recognise the strengths and weaknesses of their arguments.

Finally, to effectively integrate both scientific practices, the teacher must provide opportunities to develop these levels of argumentation. This can be achieved by including specific tasks within the inquiry process, where the student is encouraged to address various aspects of argumentation, such as elaborating arguments and counter-arguments and indicating evidence and warrants, among others.

10.3 Research Questions

This research aims to determine the argumentation skills of secondary school students when conducting an inquiry on the production of homemade yoghurt. More specifically, the inquiry is guided to answer the following research questions (RQ):

- RQ1: What evidence do secondary school students (grades 8, 9 and 10) use to argue whether or not the product obtained is yoghurt?
- RQ2: Is each evidence used in your arguments adequately justified?
- RQ3: Are there differences between the number of evidence and warrants given by students in different grades?

10.4 Method

10.4.1 Participants and Context

The inquiry discussed in this chapter is part of a broader teaching–learning sequence for secondary school students on yoghurt consumption and making (Muñoz-Campos et al., 2020). The design and classroom implementation of the sequence aim to provide opportunities for the student to develop skills related to inquiry and argumentation. The latter practice is introduced to make an informed decision about the problem.

The activity was carried out with 73 students aged between 13 and 15 from two secondary schools in Malaga (Spain), one in the capital city and one in a coastal area of the province. 65% of the participants were girls, and 35% were boys. All of them were taking Chemistry in compulsory secondary education in grades 8 (N = 19, 13 years old), 9 (N = 39, 14 years old) and 10 (N = 15, 15 years old), being compulsory in grades 8 and 9, and optional in grade 10. All the students had yet to receive training on inquiry and argumentation before this experience. The students' names in the textual quotations illustrating the experience are pseudonyms. These quotations were taken from their portfolios or classroom comments.

10.4.2 Experience Description

The inquiry involves making homemade yoghurt in the school laboratory. This experience is framed within the ENCIC-CT model (Franco-Mariscal et al., 2025), presenting yoghurt as a daily-life problem that students can relate to and solve. Food chemistry is a context that secondary students find interesting, encompassing several scientific disciplines such as chemistry, biology, physics, and technology (Hultin, 1997). Cooking recipes serve as a powerful didactic tool to motivate students, as they are familiar with the preparation and elaboration of recipes at home. By comparing the kitchen and its utensils with a laboratory and its instruments, students can draw meaningful parallels between everyday activities and scientific practices.

The inquiry was conducted in four sessions of one hour each through different tasks that posed leading questions to guide the experience (Fig. 10.1).

These tasks are described below with some examples from grade 10 students.

Task 1. What does it take to make yoghurt?

The inquiry begins with an activity that aims to familiarise students with the ingredients of milk and yoghurt, essential foodstuffs for producing yoghurt. Students were presented with a label of both foods to analyse their ingredients. Together with these labels and others they had brought from home, they were asked to make a list of the ingredients of each food.

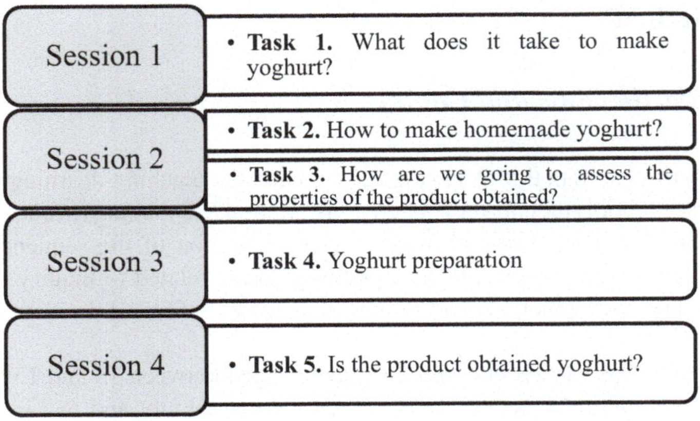

Fig. 10.1 Tasks set out in the experience

Students were very struck by the fact that some labels showed the name of the product itself (milk) as an ingredient and, in many cases, the presence of a small number of ingredients. Some students appreciated this lack of information on the labels:

> The label of my yoghurt says very few things. On the labels of my classmates, I have seen more ingredients, which my yoghurt probably also contains (Eva, S04).

This was a real difficulty, as some students needed help to identify the ingredients on the labels, with a short list of ingredients on the one hand and more detailed nutritional information on the other, usually in a larger size (Table 10.2). This nutritional information refers to the calories the carbohydrates, fats, proteins and vitamins in the milk contribute to the body. Most students thought that the ingredients of the food were the components of this nutrition information.

In addition, some students thought manufacturers did not include all ingredients because of patent issues.

Several students needed to learn the name or function of some of the terms on the labels (lactic ferments, stabilisers, etc.) (*The student Marina asked me what stabilisers were used for,* teacher-researcher's diary).

Bacteria as ingredients also captured their interest (*When we consume yoghurt, we are ingesting bacteria*; Ibrahim, S01), and the question arose whether all bacteria that exist in nature are harmful to human health or whether, on the contrary, some are necessary for the proper functioning of our organism.

They were also struck by the presence of certain ingredients that they did not commonly associate with dairy products, such as vegetable oils (*I would not say I like milk, but if it has oil in it, that is one more reason not to drink it*; Nerea, S05).

After analysing the labels, they were asked to list the ingredients they considered indispensable for making yoghurt. The students found it challenging to reach a consensus on the ingredients selection, concluding on four elements (milk, cream, bacteria and lactic ferments). This choice revealed a need for knowledge that milk

Table 10.2 Example of information contained on a milk label

INGREDIENTS	
Whole milk, cream, skimmed milk powder and starter cultures. This product may contain traces of gluten, soya, soya products and nuts	
NUTRITIONAL INFORMATION Average value per 100 g	**Portion (125 g)**
Energy value 476 kJ / 115 kcal	592 kJ / 143 kcal
Fat 9.7 g Of which saturated 6.6 g	12 g 8.3 g
Carbohydrate 3.7 g Of which sugars 3.7 g	4.6 g 4.6 g
Dietary fibre 0 g	0 g
Protein 3.3 g	4.1 g
Salt 0.13 g	0.16 g
Calcium 96 mg (12% NRV*) *NRV: Nutrient Reference Values	120 mg (15% NRV*)
One portion = 1 cup = 125 g. The pack contains four portions	
CONSERVATION	
Keep refrigerated between 1 °C and 8 °C	

and cream contribute similarly to the final product and used bacteria and starter cultures as synonyms (*Starter cultures are the bacteria necessary to form yoghurt,* Carlos' portfolio, S11).

The answer was no when asked if they thought any essential ingredients for yoghurt production were missing from the labels, as many did not mention bacteria.

By carrying out this task, the students learned that many of the ingredients on the labels are not necessary for producing yoghurt but important for its preservation.

> I believe that yoghurt should contain only milk and ferment without adding other ingredients (Ignacio, S07).
>
> Yoghurt has a lot of unnecessary ingredients (Juan Carlos, S09).
>
> We do not have to add vitamins when preparing yoghurt because the milk already contains them (Nerea's portfolio, S05).
>
> I have learned that yoghurt contains stabilisers (Sergio, S12).

The degree of participation of the students in the activity was high (*The best thing was the dynamism of the class, learning all together*; Ignacio, S07).

Task 2. How to make homemade yoghurt?

The students investigate the variables involved in the yoghurt production process. To do so, the teacher provided a table in which they had to include the ingredients involved, the variables to control related to the production conditions, and possible aspects to change. For instance, the teacher mentioned the quantity of milk as a variable associated with the ingredients and the temperature as a variable related to

Variable	I can use...
Amount of milk	1L, 2L, 3L
Amount of sugar	5gr, 10gr
Number of bacteria	20, 30, 40
Kind of milk	skimmed, whole
Time	24h, 48h
Temperature	5°, 20°, 40°, 60°
Place of conservation	Refrigerator

Fig. 10.2 Variables involved in yoghurt production

the production process. Figure 10.2 shows an example of a table completed by a grade 10 student.

The students showed interest in doing the task because they had to prepare the yoghurt for the next session. One student proposed to make a label for the prepared yoghurt.

Once the variables had been studied, they were asked to develop several recipes for making yoghurt, considering the selected variables. An example recipe was:

> 150 grams of whole milk, 10 grams of natural yoghurt without sugar, temperature 20 °C and processing time of 24 hours. The natural yoghurt is added to the milk and stirred for 30 seconds (Claudia, S02).

Doubts emerged when determining the required quantities and processing conditions for the procedure (*I need to know how much yoghurt, milk, or the temperature to make a good yoghurt*; Marta, S10).

From the recipes, each student chose one to prepare their yoghurt. Some students found it difficult to justify their choice:

> I have chosen this recipe because of the proportion and the temperature (Maria, S14).
>
> These are the right quantities to make yoghurt with the variables of time and temperature, which must also be controlled (Claudia, S02).

Some students were surprised to discover that yoghurt could be made at a temperature different from that of the refrigerator, as some labels indicated for its preservation. For others, the surprise came when they realised the need to use one yoghurt to make another, which raised questions about how the initial yoghurt was produced.

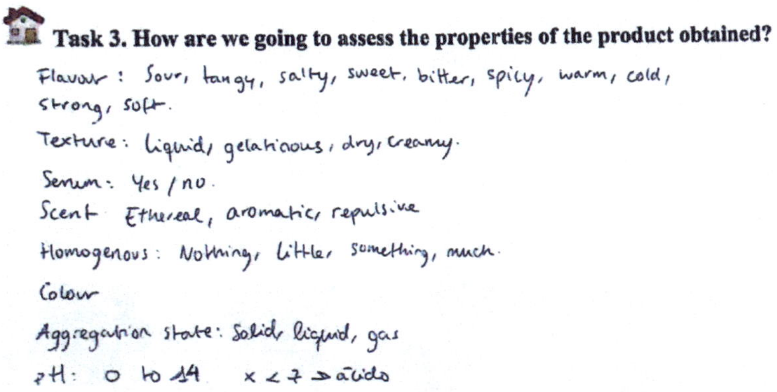

Fig. 10.3 Proposed properties to be measured in yoghurt

Task 3. How are we going to assess the properties of the product obtained?

This task explains to students that, in any process, not only those related to food, it is necessary to assess whether the product obtained is the desired one. To do so, they must measure different properties depending on the product.

They were asked to list the properties they thought were necessary to know whether the product obtained was yoghurt. The most frequently cited properties were texture, odour, flavour, colour, density, state of aggregation and homogeneity. Subsequently, they had to look for information on measuring these properties to carry out the measurements the next day. Examples of some scales proposed by a student for various properties are shown in Fig. 10.3.

pH was one of the least mentioned properties. Initially, the students did not know its meaning, but after the teacher's explanation, they related it to beauty products.

> Shampoo manufacturers measure the pH of their products (Ana's portfolio, S08).
>
> My sister's gel has a pH of 7, which means it is neutral (Alessandra's portfolio, S03).

Several students found it difficult to find a white colour scale. The teacher brought the scale in a toothpaste box into the classroom to solve this.

Task 4. Yoghurt Making

Before preparing their recipes, the students evaluated the properties of the starter mix and a commercial plain yoghurt, recording their observations in Table 10.3.

The teacher provided several yoghurt pots in case some students wished to use them in their experiments. This suggested to some students the size of the pot as a new study variable. The students were highly motivated. Proof of this is that they had all brought the necessary ingredients to make their recipes, including some hard-to-find compounds such as preservatives or thickeners, and they were quick to start making their recipes (Fig. 10.4).

Each student prepared two recipes. Collaborative work was necessary, so several students lent others some of the missing ingredients. During the preparation, some

Table 10.3 Properties of yoghurts

Products	Properties					
Commercial plain yoghurt						
Initial mix						
Final product 1						
Final product 2						

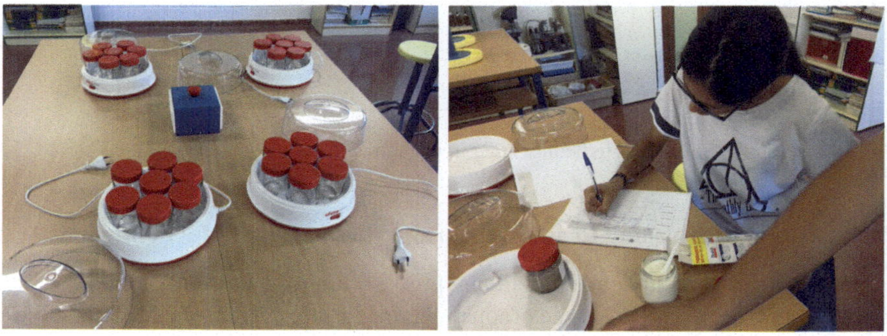

Fig. 10.4 Student making yoghurt in the classroom

students realised they would probably not get yoghurt because of the quantities, e.g., too much milk and not enough yoghurt. At the end of the session, many students expressed their concern to the teacher that they did not know if their recipes were adequate.

The temperature of the experiment plays a key factor in obtaining the yoghurt. About this variable, students chose to use a fridge, leave it in the room, or use the yoghurt maker's specific setting (40 °C). One student expressed concern about the low probability of success for yoghurt placed in the fridge (*They think that by just mixing milk and yoghurt and putting it in the fridge, yoghurt will come out. It is impossible*; Ibrahim, S01).

Task 5. Is the product obtained yoghurt?

The next day, the students tested the prepared yoghurts using the same procedures to measure the properties of the commercial yoghurt and the starter mix.

Each student tasted their yoghurt using the scales agreed upon in the sharing. In this case, a qualitative scale for taste (*I would not say I like it at all, I like it a little, I like it a lot, I like it a lot*). Yoghurts in which the milk was cut were not tasted.

All the students were excited to try their recipes, but some were disappointed to find that the result did not meet the characteristics of a yoghurt:

> It is liquid but tastes like yoghurt (Carlos, S11).
>
> It has a very strong, acidic taste (Marta, S10).

PRODUCTS	PROPERTIES					
	Texture	Scent	Flavour	Colour	Density	Aggregation State
Test Yogurt	Creamy	Sweet	Acid and salty	White 2	Homogeneous	Solid
Initial Mix	Liquid	Bitter	Acid and Salty	White 4	Heterogeneous	Liquid
Final product 1	Thick	Bitter	Acid and salty	White 4	Homogeneous	Liquid
Final product 2	Creamy	Sweet	Sweet	White 5	Homogeneous	Solid

Fig. 10.5 Properties of various yoghurts filled in by a student

After tasting the yoghurts, the students exchanged comments on the processing and discussed possible reasons why their products did not look like yoghurt. They even shared their products so that others could also try them. Figure 10.5 illustrates the table of properties filled in by one student.

Some students who had not obtained yoghurt proposed improvements to their experiment, such as leaving it in the yoghurt maker longer.

10.4.3 ENCIC-CT Model Application

The experience was conducted using the ENCIC-CT model (Franco-Mariscal et al., 2025), focusing on yoghurt production as an everyday problem (Layer 1), which is relevant for several reasons. Yoghurt is widely consumed by both children and adolescents (Parra, 2012) and is a staple in many diets worldwide. Understanding its production process has significant implications for health, nutrition, and responsible consumption. Numerous advertising campaigns highlight the health benefits of yoghurt (Cifelli et al., 2020), making it valuable to explore its production and the associated chemical and biological processes. Although yoghurt is currently produced industrially, it is important to remember that homemade yoghurt was common until a few decades ago. Similarly, it can also be made in the classroom using a simple and safe procedure (Muñoz-Campos et al., 2020). Moreover, from a sustainability perspective, learning to make yoghurt at home offers an alternative to industrial products, helping to reduce waste, control ingredients, and lower the carbon footprint related to production and transportation.

Moreover, the experience supports the development of various dimensions of CT. Within the knowledge domain, it encompasses the application of knowledge across several disciplines: the chemical reaction of fermentation and food chemistry (chemistry), the properties of matter (physics), and microorganisms, bacterial growth, nutrition, health, and healthy living habits (biology). Within the skills domain, the experience centers on argumentation (Layer 2), which is the main focus of this chapter.

The experience also relies on inquiry and argumentation as core scientific practices (Layer 3). Argumentation plays a key role in this context because it allows students to develop the skills needed to formulate and evaluate evidence-based scientific arguments, thereby helping them to build and defend their conclusions in a logical and coherent manner. The objective is for students to not only collect and analyze data but also to substantiate their findings with well-reasoned arguments. Additionally, specific teaching strategies (Layer 4) were not necessary for implementing this experience, as the effective integration of inquiry and argumentation is inherent in the design of the activity itself.

10.4.4 Data Collection and Analysis

This study focuses on the argumentation expressed by the students in the inquiry. The data collection instruments included the students' portfolio, specifically three questions from task 5, and the teacher-researcher's diary.

The written questions administered in task 5 were: *Considering the properties of the three products (commercial yoghurt, initial mixture and final product) recorded in task 5, argue whether or not your final product obtained is yoghurt. To do so, answer the following questions: (a) what has changed, (b) what has remained the same, (c) is the final product yoghurt?* Table 10.4 shows some answers to the three questions for one case where yoghurt was obtained and one where it was not.

The three questions were analysed together as a single argument for the presence of evidence, warrants and conclusions, according to Toulmin's (2003) simplified model. To achieve this, the researchers independently analysed the arguments and reached a consensus in cases of disagreement. Conclusions were categorized into two groups: "yoghurt obtained" or "yoghurt not obtained." Only evidence related to physical or chemical properties that could confirm whether the product was indeed yoghurt were considered. For warrants, those supported by laboratory evidence—such as changes in specific parameters like pH—or those that involved comparing the product's properties with those of other substances were deemed valid.

Table 10.4 Example of answers in task 5

Student	What has changed?	What remains the same?	Is the final product yoghurt?
15	The bacteria feed on the glucose in the milk, and we have obtained solid yoghurt with more consistency—a more sour taste	The colour is the only thing that has stayed the same	Yes, it is yoghurt because the bacteria have acted
6	The taste has changed; at first, it was sweeter and more acidic. The texture has also changed; before, it was more liquid and became more consistent	The colour remains the same; it is still white	It is not yoghurt because we leave it at room temperature for 24 h and add cream and sugar

Frequencies and percentages of students were determined for each element of the argument. The number of evidence showing warrants was also studied. Finally, the quality of the argumentative elements was analysed according to the following rubric for evidence (0: no evidence, 1: evidence correct but imprecise, 2: all evidence correct and accurate) and warrants (0: no justification, 1: justification that does not relate evidence to conclusion, 2: justification that links evidence to conclusion).

The teacher-researcher's diary served as an instrument to record observations, behaviors, and reactions in the classroom, offering a formal and systematic method for capturing insights and reflections. Each teacher used this instrument independently to track the progress of each session. The results section presents a qualitative overview of some of these observations and reflections.

10.5 Results

Table 10.5 presents the findings of the analysis of the arguments given by the students.

All students expressed a conclusion in their arguments indicating whether or not they had obtained yoghurt. The majority conclusion in all grades was that they had yet to get yoghurt. It was observed that the higher the grade, the more critical the students became when they realised that the product produced did not meet the characteristics of yoghurt.

Regarding the evidence, most students used the properties outlined in task 3 to evaluate the yoghurt. Among the most frequently cited properties, mentioned by more than 50% of the students in all grades, the texture (*Yes, it is yoghurt, because when it was put in the fridge, it became creamier*) and the taste (*the taste changed, at first it tasted sweeter and then it became more acidic*) stood out. Furthermore, in grades 9 and 10, the colour was also the subject of attention in high percentage (*the colour remains the same, it is still white*), while in grade 10, the smell was also a relevant evidence (*the scent remains the same*). Similarly, this is reflected in the teacher-researcher's diary (*almost all the students rely on evidence from the previous activity to determine whether the product obtained is yoghurt or not*).

Moreover, several students in grades 9 and 10 used the yoghurt composition and processing conditions worked on in task 2 as evidence. Even though ingredients are a determining factor in obtaining yoghurt, only 20% of students in grade 10 and 12.8% in grade 9 could argue whether their product was yoghurt by mentioning its composition or the presence of bacteria. One grade 9 student expressed it as *I have obtained milk with fruit,* while another grade 10 student justified her answer by stating that *yoghurt was not formed because we added too much milk.* Only one student in grade 9 used the bacteria as evidence to argue that her product was yoghurt (*The bacteria feed on the glucose in the milk, and we have obtained solid yoghurt with more consistency*). Notably, only 15.3% of the students in grade 9 used the processing temperature as evidence. In contrast, they all used it to elaborate their recipe (*It is not yoghurt, as I made a mistake in the mixing and temperature*).

Table 10.5 Frequency (f) and percentages (%) of students in each grade for each element of the argument

	Grade 8 (N = 19)		Grade 9 (N = 39)		Grade 10 (N = 15)	
	f	%	f	%	f	%
Conclusion						
Yes, I have obtained yoghurt	7	36.8	13	33.3	2	13.3
I did not get yoghurt	12	63.2	26	66.6	13	86.6
Evidence						
Taste	11	57.9	22	56.4	9	60.0
Texture	10	52.6	24	61.5	8	53.3
Colour	6	31.6	20	51.3	10	66.7
Smell	8	42.1	10	25.6	8	53.3
Consistency	7	36.8	12	30.8	4	26.7
Density	1	5.3	1	2.6	2	13.3
Aggregation state	5	26.3	11	28.2	5	33.3
Yoghurt composition	0	0.0	7	12.8	3	20.0
Processing temperature	0	0.0	6	15.3	0	0.0
Average number of evidence per student	2.5		2.8		3.3	
Warrants						
Taste	0	0.0	5	12.8	0	0.0
Texture	0	0.0	3	7.7	2	13.3
Colour	0	0.0	1	2.6	0	0.0
Smell	0	0.0	3	7.7	0	0.0
Consistency	1	5.3	11	28.2	3	20.0
Density	0	0.0	0	0	0	0.0
Aggregation state	0	0.0	4	10.2	0	0.0
Yoghurt composition	1	5.3	7	17.9	4	26.6
Processing temperature	0	0.0	2	5.1	0	0.0
Average number of warrants per student	0.1		0.9		0.6	

Note The percentages of evidence and warrants refer to the total number of students, and each student may indicate more than one evidence or warrant

In most cases, students used the evidence provided to strengthen the conclusion with warrants. Although there were differences in the three grades in the number of evidence used to support their arguments, students in all grades agreed to link the warrants to the composition and consistency of the yoghurt. Grade 9 students provided the greatest wealth of justifications, relating them to a greater diversity of evidence. The teacher-researcher's diary reflects this observation with the following insight: *Students tend to offer more evidence than justifications in their responses, suggesting that explanations or reasoning are often lacking in their classroom contributions.*

As seen (Table 10.5.), the average number of evidence per student increases with increasing grade, which shows that the arguments are more complete regarding the number of evidence. However, the same is not valid for the warrants where the maximum value is observed in grade 9. In this sense, the warrants should be improved since the average per student is less than 1 in all cases.

Table 10.6 shows the frequency and percentages of students who included a certain number of evidence and warrants in their arguments. Sixty-three per cent of students in grades 8 and 9 gave more than four pieces of evidence in their arguments, while only 50% of students in grade 10 gave more than four pieces. In all grades, the amount of evidence identified exceeded the number of warrants, and this difference became more pronounced as the quantity of evidence increased. As shown in Table 10.6, warrants are provided for all the evidence shown in very few cases (1 evidence/1 warrant, and 3 evidence/3 warrants).

Table 10.6 Evidence and warrants mentioned by students in each grade

Number of evidence	Number of warrants	Students' frequency					
		Grade 8 (N = 19)		Grade 9 (N = 39)		Grade 10 (N = 15)	
		f	%	f	%	f	%
0	0	3	15.8	3	7.7	2	13.3
	1	0	0.0	4	10.2	1	6.6
1	0	1	5.3	0	0.0	0	0.0
	1	1	5.3	3	7.7	2	13.3
2	0	4	21.0	0	0.0	0	0.0
	1	1	5.3	5	12.8	0	0.0
3	0	2	10.5	4	10.2	3	20.0
	1	0	0	2	5.1	0	0.0
	2	0	0	2	5.1	0	0.0
	3	0	0	1	2.6	0	0.0
4	0	5	26.3	0	0.0	2	13.3
	1	0	0	5	12.8	0	0.0
	2	0	0	2	5.1	0	0.0
5	0	2	10.5	1	2.6	1	6.6
	1	0	0	2	5.1	0	0.0
6	0	0	0	1	2.6	0	0.0
	1	0	0	0	0.0	1	6.6
	2	0	0	2	5.1	2	13.3
7	1	0	0	0	0.0	1	6.6

Table 10.7 Quality of evidence and warrants of students in each grade

	Grade 8 (N = 19)		Grade 9 (N = 39)		Grade 10 (N = 15)	
	f	%	f	%	f	%
Evidence						
No evidence	17	89.5	11	28.2	8	53.3
Correct but inaccurate evidence	0	0.0	0	0.0	0	0.0
Correct and accurate evidence	2	10.5	28	71.8	7	46.7
Warrants						
No warrant	17	89.5	11	23.4	8	47.1
Warrant does not link evidence to conclusion	0	0.0	13	27.7	4	23.5
Warrant linking evidence to conclusion	2	10.5	23	48.9	5	29.4

There is a significant difference in the number of warrants provided by students in the different grades. Thus, if we focus on students who express at least one warrant, we find 10% of students in grade 8, 76% in grade 8 and 47% in grade 10. This finding was noted by the teacher in their classroom observation, recorded in their diary as follows: *Grade 10 students face greater challenges in articulating their justifications compared to grade 9 students.*

Table 10.7 analyses the quality of the evidence and warrants shown. The quality of the evidence used in all grades is considered satisfactory since all students who included evidence in their arguments were correct and accurate, and in no case was evidence found to be inaccurate. However, the need to improve the warrants was identified as significant percentages of students in grades 9 and 10 did not connect adequately with the evidence presented and their conclusions.

10.6 Discussion and Educational Implications

Limited student argumentation in educational settings when addressing socio-scientific issues is a widespread challenge (Kolstø, 2006). The literature shows that some of the difficulties encountered when using argumentation in the science classroom are that students are not able to relate cause-result, are not able to show evidence, not unable to determine the changes occurred, not being able to identify the elements related to the discussion, not being able to create alternative thoughts and common epistemic points of view (Hogan, 2000; Jiménez-Aleixandre et al., 2000; Kuhn, 1991). Conversely, inadequate evaluation of evidence can lead to incorrect conclusions, highlighting the importance of teaching students how to critically analyse information. Furthermore, the complexity of scientific concepts can impede their full understanding, which in turn affects their ability to argue effectively. Additionally, improper use of value-based reasoning may impact how students interpret and debate topics (Khishfe, 2021).

Some of these ideas have been found in our study. In this regard, although most of the students could provide a good amount of evidence based on the properties of the yoghurt to argue for the transformation of milk into yoghurt (RQ1), this evidence was not sufficiently justified to support the conclusion (RQ2). In other words, students were unaware of the conclusions they could draw from the evidence, which calls for revising their arguments (Sandoval & Reiser, 2004) as it affects their quality, particularly in grade 8, where the absence of warrants is very high (RQ3).

This study shows that the inquiry seemed to help construct arguments (Osborne et al., 2016) mainly in obtaining evidence but needed more opportunities to provide the necessary warrants. As an improvement, there is a need to include new questions in the tasks that allow them to justify the evidence they are obtaining.

The development of inquiry in the classroom, specifically the making of yoghurt in the laboratory, revealed that students compared the products obtained. This particular point can be used to implement an activity in which students can carry out argumentative critiques, as proposed by Osborne et al. (2016).

In summary, the findings indicate that it is feasible to integrate argumentation into inquiry. Still, they also point to the need for careful design of argumentation tasks, especially when students need a more substantial background in this practice. In such circumstances, presenting these tasks following a sequential progression in argumentation is suggested, as Osborne et al. (2016) proposed. In this context, initial inquiries with students could focus exclusively on argument construction, progressing to investigations that include argument critique once this skill has been consolidated.

Acknowledgements This work is part of the Spanish National Plan R+D+i Project entitled "Citizens with critical thinking: A challenge for teachers in science education", reference PID2019-105765GA-I00, funded by MCIN/AEI/10.13039/501100011033.

References

Allchin, D., & Zemplén, G. (2020). Finding the place of argumentation in science education: Epistemics and whole science. *Science Education, 104*(5), 907–933. https://doi.org/10.1002/sce.21589

Archila, P. A., Molina, J., Danies, G., Truscott de Mejía, A. M., & Restrepo, S. (2022). Using the controversy over human race to introduce students to the identification and the evaluation of arguments. *Science & Education, 31*, 861–892. https://doi.org/10.1007/s11191-021-00299-8

Cifelli, C. J., Agarwal, S., & Fulgoni, V. L. (2020). Association of yogurt consumption with nutrient intakes, nutrient adequacy, and diet quality in American children and adults. *Nutrients, 12*(11), 3435. https://doi.org/10.3390/nu12113435

Franco-Mariscal, A. J. (2015). Competencias científicas en la enseñanza y el aprendizaje por investigación. Un estudio de caso sobre corrosión de metales en secundaria [Scientific competences in teaching and learning through research. A case study about the corrosion of metals in secondary education]. *Enseñanza de las Ciencias, 33*(2), 231–252. https://doi.org/10.5565/rev/ensciencias.1645

Franco-Mariscal, A. J., Cano-Iglesias, M. J., España-Ramos, E., & Blanco-López, A. (2025). The ENCIC-CT model for the development of critical thinking. In A. J. Franco-Mariscal (Ed.), *Critical thinking in science education and teacher training* (Chap. 1) (pp. 3–42). Springer. https://doi.org/10.1007/978-3-031-78578-8_1

Gerber, B. L., Price, C., Barnes, M., Hinkle, V., Barnes, L., Gordon, P., et al. (2003). Excellence in rural science teaching: Examining elements of professional development models. Paper presented at the Annual Meeting of the National Association for Research in Science Teaching. Philadelphia.

Hogan, K. (2000). Exploring a process view of students' knowledge about the nature of science. *Science Education, 84*(1), 51–70. https://doi.org/10.1002/(SICI)1098-237X(200001)84:1%3c51::AID-SCE5%3e3.0.CO;2-H

Hultin, H. O. (1997). Future directions for Food Chemistry. *Food Technology, 51*(6), 42–46.

Jiménez-Aleixandre, M. P. (2010). *10 ideas clave. Competencias en argumentación y uso de pruebas.* [10 key ideas. Competences in argumentation and use of evidence.]. Graó.

Jiménez-Aleixandre, M. P., Rodríguez, A. B., & Duschl, R. A. (2000). "Doing the lesson" or" doing science": Argument in high school genetics. *Science Education, 84*(6), 757–792. https://doi.org/10.1002/1098-237X(200011)84:63.0.CO;2-F

Khishfe, R. (2021). Explicit instruction and student learning of argumentation and nature of science. *Journal of Science Teacher Education, 32*(3), 325–349. https://doi.org/10.1080/1046560X.2020.1822652

Kolstø, S. D. (2006). Patterns in students' argumentation confronted with a risk-focused socio-scientific issue. *International Journal of Science Education, 28*(14), 1689–1716. https://doi.org/10.1080/09500690600560878

Kuhn, D. (1991). *The skills of argument.* Cambridge University Press. https://doi.org/10.1017/CBO9780511571350

López-Fernández, M. M., & Franco-Mariscal, A. J. (2021). Indagación sobre la degradación de plásticos con estudiantes de secundaria [Inquiry about the degradation of plastics with high school students]. *Educación Química, 32*(2), 21–36. https://doi.org/10.22201/fq.18708404e.2021.2.76553

Lord, T., & Orkwiszewski, T. (2006). Moving from didactic to inquiry-based instruction in a science laboratory. *American Biology Teacher, 68*(6), 342–345. https://doi.org/10.1111/10.2307/4452009

Muller, N., & Perret-Clermont, A. N. (2009). *Argumentation and education: Theoretical foundations and practices.* Springer. https://doi.org/10.1007/978-0-387-98125-3

Muñoz-Campos, V., Franco-Mariscal, A. J., & Blanco-López, A. (2020). Integration of scientific practices into daily living contexts: A framework for the design of teaching-learning sequences. *International Journal of Science Education, 42*(15), 2574–2600. https://doi.org/10.1080/09500693.2020.1821932

National Research Council (NRC) (2012). *A framework for K-12 science education: Practices crosscutting concepts, and core ideas.* National Academy Press.

OECD, Organisation for Economic Co-operation and Development. (2017). *PISA 2015. Assessment and analytical framework: Science, reading, mathematic and financial literacy.* OECD Publishing.

Osborne, J. F., Henderson, J. B., MacPherson, A., Szu, E., Wild, A., & Yao, S. (2016). The development and validation of a learning progression for argumentation in science. *Journal of Research in Science Teaching, 53*(6), 821–846. https://doi.org/10.1002/tea.21316

Parra, R. A. (2012). Yogur en la salud humana [Yoghurt in human health]. *Revista Lasallista de Investigación, 9*(2), 162–177.

Romero-Ariza, M., Abril, A. M., & Quesada, A. (2018). Empowering teachers to bring authenticity and responsive action into the science classroom. *School Science Review, 371*, 40–45.

Sampson, V., Grooms, J., & Walker, J. P. (2011). Argument-driven inquiry as a way to help students learn how to participate in scientific argumentation and craft written arguments: An exploratory study. *Science Education, 95*(2), 217–257. https://doi.org/10.1002/sce.20421

Sandoval, W. A., & Reiser, B. J. (2004). Explanation-driven inquiry: Integrating conceptual and epistemic scaffolds for scientific inquiry. *Science Education, 88*(3), 345–372. https://doi.org/10.1002/sce.10130

Teig, N. (2021). Inquiry in science education. In T. Nilsen, A. Stancel-Piatak, & J. E. Gustafsson (Eds.), *International handbook of comparative large-scale studies in education*. Springer.

Toulmin, S. (2003). *The uses of argument* (3rd ed.). Cambridge University Press.

van Zee, E. H., & Roberts, D. (2006). Making science teaching and learning visible through web-based "snapshots of practice." *Journal of Science Teacher Education, 17*(4), 367–388. https://doi.org/10.1007/s10972-006-9027-2

Chapter 11
Analysis of Critical Thinking Through Scientific Thinking in the Design of Inquiries by Preservice Preschool Teachers

María Marta Alarcón-Orozco, Antonio Joaquín Franco-Mariscal, and Ángel Blanco-López

11.1 Introduction

Although the development of critical thinking (hereinafter, CT) has been a constant in education throughout history (Jiménez-Aleixandre & Puig, 2022), the contemporary challenges facing society today, such as the recent pandemic, climate change and the proliferation of fake news, make it particularly relevant. In science education, works such as those by Bargiela et al. (2022) and Blanco-López et al. (2017) are already available that propose concrete activities for implementation in the classroom.

In Spain, CT is highlighted as an essential skill in the minimum teaching requirements for the infant education stage (Government of Spain, 2022). Promoting scientific thinking is prescribed in this context, as well as the subject of didactics of natural sciences in preschool degree (Government of Spain, 2007). Both types of thinking have been presented in the literature indistinctly and interchangeably, which can lead to confusion (García-Carmona, 2023), hence the need to establish the delimitations between the two.

The works of García-Carmona (2023) and Vázquez-Alonso and Manassero-Mas (2018) show the parallelism between scientific thinking and CT by highlighting the skills of one and the other and also pointing out their differences. The first authors

M. M. Alarcón-Orozco (✉) · A. J. Franco-Mariscal · Á. Blanco-López
Science Education, Universidad de Málaga, Málaga, Spain
e-mail: mmartaalarcon@uma.es

A. J. Franco-Mariscal
e-mail: anjoa@uma.es

Á. Blanco-López
e-mail: ablancol@uma.es

A. J. Franco-Mariscal (ed.), *Critical Thinking in Science Education and Teacher Training*, Contemporary Trends and Issues in Science Education 64,
https://doi.org/10.1007/978-3-031-78578-8_11

point out that one of the reasons for teaching students to think through CT skills lies in the fact that it teaches scientific thinking at the same time. García-Carmona (2023), meanwhile, distinguishes the purpose of both thoughts, understanding that scientific thinking has as its objective "finding the best rational explanation of a phenomenon among different possible explanatory proposals (i.e., generating scientific knowledge)" and the purpose of CT is "choosing the most defensible idea/position among others that are also defensible". That is, while CT shares numerous skills and practices with scientific thinking, unlike scientific thinking, as Ennis (2018) states, the presence of two or more defensible ideas is not a drawback. Table 11.1 presents the skills of scientific thinking and CT and those that are common to both.

Table 11.1 Skills characteristic of scientific thinking and CT and common to both

Scientific thinking	Critical thinking	Common skills
Manage scientific knowledge related to the issue to be investigated	Make moral/ethical, political, etc., and scientific assessments/analyses related to the issue to construct complex arguments	Have respect for evidence
Formulate researchable questions and hypotheses (where appropriate) to initiate scientific inquiry	Make decisions or take one's position on the issue based on arguments	Possess an informed understanding of epistemic and non-epistemic aspects of NOS
Obtain data or evidence through the planning and development of scientific inquiries	Maintain a sceptical attitude towards new data, information or ideas	Assess the reliability of data or information
Look for patterns of behaviour and generalisations to elaborate explanations (models, causal relationships, etc.)	Emotionally manage frustrations in the face of errors or unexpected results	Interpret data and information (inference)
Apply logical reasoning	Respect ideas or opinions different from one's own if they are argued and do not incur illegalities	Construct arguments from evidence
Elaborate explanations from evidence		Understandably communicate ideas or opinions
Discard those ideas that have no logical/rational or empirical support		Discuss different ideas and opinions
		Being aware of the strengths and weaknesses of one's ideas or opinions and those of others (metacognition)

Source García-Carmona (2023)

Vázquez-Alonso and Manassero-Mas (2018) consider that the operability of CT models can be exemplified in inquiry-based science education (IBSE) approaches. This approach suggests the idea of basing science teaching on inquiry methods that encourage the active involvement of students in their learning process by asking questions, designing and conducting experiments, and analysing and obtaining results (Sjøberg, 2019). Previous studies, such as Croner's (2003) on laboratory practices, already offered a similarity between these stages and CT.

Lederman et al.'s (2013) proposal for IBSE allows students to approach problems similarly to scientists, enabling them to become familiar with scientific work, develop a procedural understanding of science, and apply scientific thinking skills. These aspects are detailed in Table 11.2, drawn from the studies by García-Carmona (2023) and Vázquez-Alonso and Manassero-Mas (2018), where the stages of the inquiry sequence proposed by Alarcón-Orozco et al. (2022) are integrated.

Table 11.2 Aspects of scientific thinking associated with basic skills and stages of inquiry

Aspects of scientific thinking	Associated basic skills	Stages of the inquiry
Create and test hypotheses	Ask questions, identify problems, formulate hypotheses, plan and develop research, apply mathematical thinking and statistical analyses, etc.	Formulation of the research question
		Hypothesis formulation
		Design of the experience
		Data collection and analysis
Think about the causes and effects	Control for effects of multiple variables; attribute causality; use logic; make valid and reliable interpretations; etc.	Choice of variables
Construct explanations from evidence	Issue critical judgments on what is observed or measured; accept and reject explanations; use available scientific evidence and knowledge (consult the literature); justify and validate ideas; present arguments; develop representative and explanatory models, etc.	Drawing conclusions
Be aware of and control for one's thinking (meta-processes)	Have a sceptical attitude; show open-mindedness; challenge knowledge with alternatives; evaluate assumptions; predict; seek new ideas and expertise with creativity and imagination; etc.	Decision-making
Communicate, evaluate, share, collaborate, and think about the information	Communicate and share knowledge; work cooperatively in teams; debate with colleagues on theories and solutions; evaluate others' results and conclusions, etc.	Communication

Source Based on Alarcón-Orozco et al. (2022) and García-Carmona (2023)

CT, according to Jiménez-Aleixandre and Puig's approach (2022), presents a series of components related to the use of criteria and evidence for the evaluation of knowledge, the willingness to consider alternatives and revise points of view that connect with argumentation as a scientific practice, the formation of an independent opinion questioning one's group and authority, and a critical analysis of discourses that justify social inequalities that connect with education for citizenship and social practice. For these authors, almost all definitions of CT consider the practice of argumentation. Bargiela et al. (2022), while sharing this view, consider that, at the preschool, CT is promoted through dialogical interaction in environments that awaken curiosity, encourage the formulation of questions and allow the search for answers through experiments, as well as the construction of valid justifications by discussing possible alternatives.

11.2 Critical Thinking in Inquiry in Preschool

Research, exemplified by Samarapungavan et al. (2008), has shown how children in preschool who receive specific science training can demonstrate a functional understanding of scientific inquiry processes and important science concepts during their investigations, mainly when these are intentionally structured by a trained adult (Gropen et al., 2017). At these levels of cognitive development, a study by Monteira and Jiménez-Aleixandre (2016) indicates that tapping into the natural tendency of children at this age to enjoy exploring and observing their environment (Eshach & Fried, 2005; French, 2004; Ramanathan et al., 2022) can help preschoolers develop scientific practices and enjoy observing, thinking about, and questioning the natural world (Worth, 2010). In addition, research on the development of cognitive flexibility suggests that, by age four, children can revise their predictions and, ultimately, their ideas when they have more information (Fischer, 2012; Gropen et al., 2011). Akerson and Volrich (2006) previously demonstrated that K-1 students can engage in meaningful inquiry activities if these are adapted to their developmental level.

Inquiry promotes active learning and enables the development of the basic scientific thinking skills described in Table 11.2. Thus, children can create and formulate hypotheses by observing and categorising observations, discovering regularities, asking questions and anticipating answers. Also, when guided by an adult, they can plan and carry out investigations by controlling the effects of possible variables and use logic to make valid and reliable interpretations through a dialogical process (Bargiela et al., 2022) in which scientific knowledge is constructed between teacher and pupil. The questions the teacher poses during inquiry-based activities facilitate preschoolers' construction of explanations through explicit reflection on scientific processes and the nature of scientific knowledge. This approach, used by Akerson and Volrich (2006), helps preschoolers grasp how scientific knowledge is developed, which is crucial for fostering their CT. In this way, children can think about the causes of a phenomenon and the effects of possible changes in variables. These questions can help them make critical judgements about what is observed or measured, accept and/or reject scientific evidence and available knowledge, and construct explanations from the evidence. In addition, they can be helped to maintain a sceptical attitude that allows them to question knowledge with alternatives, evaluate assumptions, and seek new ideas, enabling them to become aware of their learning and favouring the processes of metacognition.

On the other hand, the methodological strategies characteristic of the preschool stage based on cooperation and teamwork (Government of Spain, 2022), together with circle time central activities in the daily dynamics of the classroom, allow knowledge to be shared, ideas to be put forward, learning to be communicated and agreements to be reached on the subject of study (Bustamante et al., 2018). This approach allows science education to be integrated with other areas of the curriculum, such as reading and writing. Such integration not only strengthens scientific learning but also fosters critical skills across various contexts (Akerson & Volrich, 2006).

IBSE is consolidated as one of the most appropriate approaches in preschool due to its potential to develop CT skills (Bargiela et al., 2022). To this end, following the proposal of Evagorou et al. (2020), it is necessary to encourage dialogue and introduce argumentation practices that help students to give evidence on key aspects in the development of inquiry processes such as the choice of variables, the procedure used to measure the variables, taking the data and representing them.

In short, considering these premises on the similarities and differences between scientific and CT, the former is a preliminary step towards developing the latter. At the preschool, inquiry practices can promote the development of scientific thinking and lay the foundations for CT through a collaborative dialogue between teachers and students. As preschool is a key time to begin an approach to science, preschool teachers must develop the CT skills necessary to effectively implement IBSE in the classroom to promote CT in students. Since scientific thinking is considered a precursor to CT, it is essential to assess the level of development of these skills during initial teacher training at this stage of education. This analysis is the central objective of this chapter.

11.3 Research Question

This study addresses the following research question: What level of performance do preservice preschool teachers show in scientific thinking skills when designing inquiry activities?

11.4 Method

11.4.1 Participants and Context

The participants in this study were 121 Spanish preservice preschool teachers who were studying the third year of the preschool degree in two different centres (Faculty of Education Sciences in Malaga and the University Centre of Antequera) of the University of Malaga (Spain). All but two of the participants were female, with an average age of 23 years. For the vast majority, their previous scientific training ended with what they received during the 3rd year of compulsory secondary education in Spain (ages 14–15). This data is significant because it confirms the poor pre-university scientific training of students in this degree who usually come from humanities or social science backgrounds and show an attitude of rejection towards science, perhaps promoted by negative experiences with these subjects at school (Alarcón-Orozco et al., 2021). The preservice teachers participated in a training programme on IBSE implemented during the 2018–2019 academic year in the subject Teaching of Natural Sciences of the preschool degree, which included a specific competence, knowledge of scientific methodology and promoting scientific thinking and experimentation.

The implementation was carried out by two different trainers (the first two authors of the chapter), one in each university centre. Both were science education researchers with extensive teaching experience in the initial training of preschool teachers. The two trainers were involved in the design of the training programme and co-ordinated the implementation.

11.4.2 Experience Description

The scheme followed to introduce preservice teachers to inquiry experiences consisted of three stages:

- *Stage 1*: The preservice teachers conducted an inquiry related to a daily-life problem, such as whether or not to consume yoghurt, in the role of science learners (Muñoz-Campos et al., 2020). They then visualised and analysed, identifying the inquiry processes, an experience developed with pupils in the preschool (De la Calle, 2005).
- *Stage 2*: Working in small groups, the preservice teachers designed an inquiry aided by a template in which they assumed the role of trainee preschool teachers.
- *Stage 3*: The preservice teachers participated in a fair of inquiry experiences in which they disseminated their work. As a final task, they were asked to write a report on the designed inquiry sequence and make any proposals for improvement that they considered necessary for its development and implementation in the preschool stage.

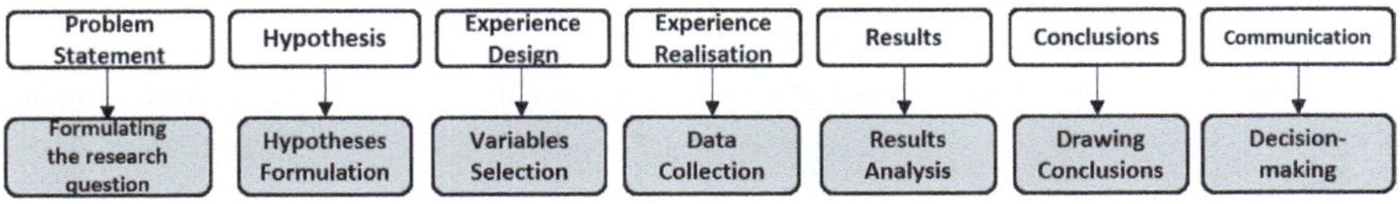

Fig. 11.1 Complete inquiry sequence. *Source* Alarcón-Orozco et al. (2022)

The participants used the inquiry sequence diagram in Fig. 11.1 during the training programme. The scientific thinking skills are highlighted at the top of the figure, and the stages of inquiry are at the bottom.

11.4.3 ENCIC-CT Model Application

We consider that this training programme significantly contributes to the development of CT among preservice preschool teachers, in line with the ENCIC-CT model outlined in Chap. 1 (Franco-Mariscal et al., 2025). In this study, the everyday problem addressed (the first layer of the model) is how to make yoghurt, which the preservice teachers approached as science learners. Later, as trainee preschool teachers, they explored other daily-life problems as contexts for implementing IBSE with preschoolers. The programme supports the development of all dimensions included in the second layer of the ENCIC-CT model. However, the experience examined in this chapter, where preservice preschool teachers design their own inquiries, primarily focuses on applying scientific knowledge, argumentation, and decision-making regarding inquiry implementation in preschool settings, as well as on communicating their ideas with peers. This experience emphasizes the scientific practice of inquiry (the third layer of the model). For the fourth layer (the teaching strategy), the experience involves preservice preschool teachers taking on the role of teachers who design their own lessons.

11.4.4 Data Collection and Analysis

The reports submitted supported identifying the aspects of scientific thinking studied and analysing them according to the rubric (Table 11.3) based on the designs made by the preservice teachers. This rubric, constructed based on Croner (2003), includes three performance levels: excellent, good, and needs improvement, valued at 3, 2, and 1, respectively. For the analysis, the different stages of the inquiry were taken from the reports, and it was verified whether their presentation contained the aspects shown in the rubric. This analysis was carried out by consensus among the authors of this chapter. To obtain a representative sample of the two groups of preservice teachers that included the different topics freely selected by the participants, we worked with a sample of 20 of the 44 reports.

Table 11.3 Evaluation rubric from Croner (2003)

Aspects of scientific thinking	Inquiry stage	Excellent (3)	Good (2)	Needs improvement (1)
Creating and testing hypotheses	Formulation of the research question	The problem is explicitly stated as a question and indicates a direct causal relationship between the independent and dependent variables	The problem must be clearly stated as a question but show a causal relationship between the student's understanding of the independent and dependent variables	The problem is either not posed at all or in a way that shows little or no relationship between the independent and dependent variables
	Formulation of the hypothesis	The hypothesis clearly explains how the change in the independent variable affects the dependent variable	The prediction shows a relationship between the independent variable and the dependent variable	The prediction does not appear or shows no relationship between the independent and dependent variables
Thinking about causes and effects	Choice of the independent variable	Clearly identified independent variable	An independent variable was identified, but there may be an element of vagueness	Independent variable needs to be correctly identified
	Choice of the dependent variable	The dependent variable was clearly identified	A dependent variable was identified, but there may be an element of vagueness	The dependent variable needed to be correctly identified
	Choice of the controlled variable	Controlled variables are clearly identified	Controlled variable identified, but there may be an element of vagueness	Controlled variable needs to be correctly identified
Creating and testing hypotheses	Data collection	The data are collected and organised in a table that displays them clearly and logically in an easily understandable way	The data is collected in a table, but the table can be organised in a way that is easier to understand	Data are collected but show little organisation and must be presented more comprehensibly

(continued)

Table 11.3 (continued)

Aspects of scientific thinking	Inquiry stage	Excellent (3)	Good (2)	Needs improvement (1)
	Analysis of results	A graph is drawn that clearly represents the data in the table and is easily understandable. It has a descriptive title, identifies vertical and horizontal axes, and properly labelled units	A graph is drawn that represents the data in the table and is legible but could be more easily understood using another format. It provides a descriptive title, but there may be some improvements, such as identifying axes and labelling appropriate units	A graph is not drawn or is drawn in such a way that the descriptive title needs to be included; axes are not straight; units need to be correctly labelled; intervals on the axes are not evenly spaced
Building explanations from evidence	Drawing conclusions	The conclusion clearly shows whether the data obtained supports or refutes the prediction and offers an explanation	The conclusion indicates whether the prediction is correct but must clearly show how the data supports or refutes it	The conclusion has little or no relation to the prediction. It needs to show how the evidence supports the conclusion

In order to analyse participants' performance in their scientific thinking skills, the following relationship was established with the stages of inquiry by the proposals of Alarcón-Orozco et al. (2022) and García-Carmona (2023):

- The *Create and test hypotheses* aspect of scientific thinking corresponds to the median of the stages corresponding to the formulation of the research question, formulation of the hypothesis, data collection and analysis of results that are interpreted based on graphic representations.
- *Think about causes and effects,* which represent the median of the choice of variables stage distinguishing between independent, dependent, and controlled variables.
- *Construct explanations from evidence* is linked to the median of the conclusion development stage.

Once the responses had been categorised, a statistical analysis was carried out to determine possible significant differences in the performance of the different stages of inquiry and between the other aspects of scientific thinking. Given the small sample size (20 cases) and the ordinal nature of the variables, non-parametric tests were chosen. Parametric tests assume a normal distribution and may be inappropriate for small samples or when data do not meet these assumptions. Since ordinal variables represent categories with a meaningful order but not a continuous scale, non-parametric methods are more appropriate. Therefore, the Friedman test (χ^2) was used to assess differences among the set of variables, while the Wilcoxon test (Z) was applied to examine differences between pairs of specific variables. All analyses were performed with SPSS v. 29.0.2.0 software.

11.5 Results

Table 11.4 shows the rubric value assigned to each stage for each inquiry and the medians.

As can be seen, the preservice teachers selected topics attractive to the preschoolers, mostly centred on food and whose end products could be tasted. This preference could be due, on the one hand, to the yoghurt-making inquiry sequence that they carried out assuming the role of science learner and, on the other hand, to the qualitative nature of the dependent variables by obtaining products that could be evaluated with the senses. The ease of measuring these variables may have influenced this choice.

There are no significant differences in the medians obtained for the different inquiries, which range from 1 to 3 (Table 11.4). These findings suggest that the choice of topic only significantly affects the participants' overall performance in their role as trainee preschool teachers in designing the inquiry.

However, when analysing the stages of the inquiry, differences between the medians were observed, with the lowest value (1) being found only for the problem statement and the highest value (3) for choosing the different types of variables and data collection. The Friedman test detected statistically significant differences between the different stages (χ^2 [df = 7] = 48.821, $p < 0.001$). Furthermore, the Wilcoxon test revealed that these differences were found between the problem statement and other stages, namely formulating hypotheses (Z = – 2.640, $p = 0.008$), the choice of independent variables (Z = – 3.365, $p < 0.001$), the choice of dependent variables (Z = – 3.490, $p < 0.001$), the choice of controlled variables (Z = – 3.769, $p < 0.001$) and data collection (Z = – 2.910, $p = 0.004$), and in all cases in favour of the latter. Table 11.5 shows the results of the Friedman test for the stages of each inquiry.

Table 11.6 shows, for each inquiry, the medians for each aspect of scientific and global thinking.

The aspect of scientific thinking with the highest median (3) was considering causes and effects. The Friedman test detected the existence of statistically significant differences between the aspects considered (χ^2 [df = 2] = 14.528, $p < 0.001$), with the Wilcoxon test showing that these differences were between thinking about causes and effects and the other aspects (creating and testing hypotheses: Z = – 3.506, $p < 0.001$; constructing explanations from evidence: Z = – 2.937, $p = 0.003$). In all cases, the differences favoured thinking about causes and effects. Table 11.7 shows the findings of the Friedman test for the aspects of scientific thinking.

Table 11.4 Findings obtained in the stages of each inquiry and median values

	Problem statement	Hypothesis	Design of the experience			Realisation experience	Results	Conclusions	
Thematic of inquiry	formulate research question	Formulate Hypotheses	Independent variable	Dependent variable	Controlled variable	Data collection	Analysis of results	Conclusions	Median
Heat resistance of school waxes	2	2	2	1	3	1	1	3	2
Soap making	1	1	1	1	3	1	2	2	1
Plant growth	3	3	3	3			1	1	3
Chemical reaction between eggshell and vinegar	1	3	3	3	3	3	1	3	3
Non-Newtonian fluids	1	1	3	3	3	3	3	2	3
Blowing up a balloon with vinegar and baking soda	2	3	3	3	3	3	2	2	3
Changes of status	1	2	3	3	3	3	3	2	3
Building a thermometer	1	1	3	3	3	3	2	3	3
Slime making	1	1	2	3	3	3	1	1	2
Generating electricity with food	1	1	3	3	3	1	2	1	2
Milkshake making	1	2	3	3	3	1	1	2	2

(continued)

Table 11.4 (continued)

	Problem statement	Hypothesis	Design of the experience			Realisation experience	Results	Conclusions	
Thematic of inquiry	formulate research question	Formulate Hypotheses	Independent variable	Dependent variable	Controlled variable	Data collection	Analysis of results	Conclusions	Median
Preparation of sponge cake	1	1	3	3	3	1	1	1	1
Making pancakes	1	2	3	3	3	2	2	1	2
Flan making	3	3	3	3	3	3	3	2	3
Making cupcakes	2	3	3	3	3	3	3	3	3
Biscuit making	1	3	3	3	3	3	2	3	3
Bread making	1	1	3	3	3	2	1	2	2
Making fruit smoothies	2	2	1	1	1	2	1	1	1
Making pancakes	2	3	2	3	3	2	2	1	2
Making slush	1	1	2	3	3	3	2	3	3
Median	1	2	3	3	3	3	2	2	2

Table 11.5 Friedman test results for stages of inquiry

Inquiry stage	Average range	Friedman test statistics
Problem statement	2.68	N = 20
Formulating hypotheses	4.00	$\chi^2 = 48.821$
Choice of independent variables	5.43	df = 7
Choice of dependent variables	5.70	$p = 0.000$
Choice of controlled variables	6.23	
Data collection	4.78	
Analysis of results	3.45	
Drawing conclusions	3.75	

Table 11.6 Findings in the aspects of scientific thinking analysed

Themes of the inquiry	Creating and testing hypotheses	Thinking about causes and effects	Building explanations from evidence	Median
Heat resistance of school waxes	2	2	3	2
Soap making	1	2	2	2
Plant growth	3	3	1	3
Chemical reaction between eggshell and vinegar	2	3	3	3
Non-Newtonian fluids	2	3	2	2
Blowing up a balloon with vinegar and baking soda	3	3	2	3
Changes of status	2	3	2	2
Building a thermometer	2	3	3	3
Slime making	2	2	1	2
Generating electricity with food	1	3	1	1
Milkshake making	1	3	2	2
Preparation of sponge cake	1	3	1	1
Making pancakes	2	3	1	2
Flan making	3	3	2	3
Making cupcakes	3	3	3	3
Biscuit making	2	3	3	3
Bread making	1	3	2	2
Making fruit smoothies	2	1	1	1
Making pancakes	2	3	1	2
Making slush	2	3	3	3
Median	2	3	2	2

Table 11.7 Friedman test results for aspects of scientific thinking

Inquiry stage	Average range	Friedman test statistics
Creating and testing hypotheses	1.58	N = 20
Thinking about causes and effects	2.65	$\chi^2 = 14.528$
Building explanations from evidence	1.78	df = 2
		$p = 0.001$

11.6 Discussion and Educational Implications

The results offer an in-depth insight into the relationship between the stages of inquiry and scientific thinking skills in the sample analysed. These findings contribute to theoretical understanding and have practical implications for the design of science education strategies.

The problem statement appears to be the inquiry stage, where preservice teachers scored the lowest. This may be due to the difficulty of formulating questions that can be addressed by IBSE, the preservice teachers' lack of experience in these matters, and the need for more development of specific critical skills to define the inquiry problem clearly. On the contrary, the high median values observed in the steps related to choosing variables and taking data suggest that preservice teachers are influential in these aspects and show a solid understanding of how to select and define variables in an experimental design or take data, thus demonstrating an advanced ability in planning and executing research.

The lack of significant statistical differences between problem posing and data analysis, and between problem posing and drawing conclusions may suggest that preservice teachers perform more uniformly in these areas. These facts may indicate that these stages pose specific challenges that must be considered in designing IBSE strategies. The significant differences detected between the stages of the problem statement and hypothesis formulation, and the different types of variables and data collection indicate where participants found it most difficult to make the logical connection between these phases. In order to address this situation, the teacher should provide guidance and support to students, encourage participation in the question formulation process and provide concrete examples to guide them. In this way, and in line with Akerson and Volrich (2006) and Bargiela et al. (2022), an environment can be established that stimulates curiosity, generates questions, and encourages seeking answers through experimentation while building strong justifications through the search for evidence to support or disprove possible alternatives. The study on the aspects of scientific thinking showed that creating and testing hypotheses and constructing explanations from evidence were the aspects with the lowest median values. We understand that these findings could result from the procedure used to calculate their value since the stages of inquiry related to these aspects also obtained relatively low values. These results suggest that preservice teachers encounter difficulties generating predictions and designing experiments to test them, as well as in the associated explanations. The statistically significant differences in these aspects of scientific thinking indicate that these areas require further attention and development in the teaching process.

On the other hand, before the inquiry design, preservice teachers should be invited to identify possible gaps related to the topic of study. This preliminary step lays the foundation for formulating solid hypotheses and favours a deeper understanding of the problem to be investigated. Consequently, the teacher's task goes beyond teaching procedures and involves cultivating in students the ability to identify gaps in knowledge and formulate questions that lead to meaningful exploration.

Furthermore, taking into account the characteristics of CT (Table 11.1), we understand that, during teacher training, it is necessary to pay attention to comprehensively approaching problems, taking into account their multiple dimensions, which encompass scientific, technical, ethical, environmental and economic aspects, among others (Franco-Mariscal et al., 2025). During this process, preservice preschool teachers must foster respect for evidence and personal autonomy, allowing them to develop independent opinions. Likewise, a sceptical attitude towards new data or information should be instilled, along with respect for divergent ideas (García-Carmona, 2023). Additionally, attention must be paid to managing emotions, especially the frustrations that may arise in the face of mistakes or unexpected results (García-Carmona, 2023).

Considering the results, the need to propose measures to enhance scientific thinking more critically is evident. A key strategy would be to enrich inquiry proposals by introducing elements that encourage argumentation and decision-making, thus facilitating a deeper approach to CT. For instance, in addition to receiving pedagogical support during the design phase, preservice teachers could be encouraged to base the choice of subject matter not only on curricular terms but also by considering assessments and analyses that include moral, ethical, political and other aspects, in addition to scientific-technological ones (García-Carmona, 2023).

Preservice preschool teachers are expected to argue the reasons for their statements and choices throughout the process, from the problem statement to the selection of variables. For instance, in the proposed inquiry on soap making, the student should adequately justify the choice of certain dependent variables, such as hardness or odour, when expressing evidence on the effectiveness of the soap as a final product.

As a limitation of the study, communication and decision-making are two aspects of scientific thinking that could not be assessed with the instrument used. These aspects will be analysed in future work, based on the study of the initial teacher trainees' productions after the science fair, where they presented their research.

Acknowledgements This work is part of the Spanish National Plan R+D+i Project entitled "Citizens with critical thinking: A challenge for teachers in science education", reference PID2019-105765GA-I00, funded by MCIN/AEI/10.13039/501100011033.

References

Alarcón-Orozco, M. M., Díez-Ojeda, M., & Franco-Mariscal, A.J. (2021). ¿Cuál es la percepción sobre el origen de una investigación en los maestros en formación? [What is the perception of pre-service teachers regarding the origin of a research project?] In *Actas de los 29 Encuentros de Didáctica de las Ciencias Experimentales* (pp. 711-719). University of Córdoba.

Alarcón-Orozco, M. M., Franco-Mariscal, A. J., & Blanco-López, A. (2022). Ayudando a maestros en formación inicial a desarrollar indagaciones en la Educación Infantil. [Assisting preservice teachers in developing inquiries in preschool.] *Revista Eureka sobre Enseñanza y Divulgación de las Ciencias, 19*(1), 1601. https://doi.org/10.25267/Rev_Eureka_ensen_divulg_cienc.2022.v19.i1.1601

Akerson, V. L., & Volrich, M. L. (2006). Teaching nature of science explicitly in a first-grade internship setting. *Journal of Research in Science Teaching, 43*(4), 377–394. https://doi.org/10.1002/tea.20132

Bargiela, I. M., Blanco-Anaya, P., & Puig, B. (2022). Las preguntas para la indagación y activación de pensamiento crítico en educación infantil. [Inquiry and Critical Thinking Activation Questions in Preschool.] *Enseñanza de las Ciencias, 40*(3), 11–28. https://doi.org/10.5565/rev/ensciencias.5470

Blanco-López, A., España, E., & Franco-Mariscal, A. J. (2017). Estrategias didácticas para el desarrollo del pensamiento crítico en el aula de ciencias. [Teaching strategies for the development of critical thinking in the teaching of science]. *Ápice, Revista de Educación Científica, 1*(1), 107–115. https://doi.org/10.17979/arec.2017.1.1.2004

Bustamante, A. S., Hindman, A. H., Champagne, C. R., & Wasik, B. A. (2018). Circle time revisited. How do preschool classrooms use this part of the day? *The Elementary School Journal, 18*(4), 610–631.

Croner, P. (2003). Developing critical thinking skills through the use of guided laboratory activities. *The Science Education Review, 2*(2), 46:1–46:13.

De la Calle, C. (2005). *Proyecto el agua. 3 años.* [Project The Water. 3 years.] https://bit.ly/2tX2lPa

Ennis, R. H. (2018). Critical thinking across the curriculum: A vision. *Topoi, 37*(1), 165–184. https://doi.org/10.1007/s11245-016-9401-4

Eshach, H., & Fried, M. N. (2005). Should science be taught in early childhood? *Journal of Science Education and Technology, 14*(3), 315–336. https://doi.org/10.1007/s10956-005-7198-9

Evagorou, M., Nicolau, C., & Lymbouridou, C. (2020). Modelling and argumentation with elementary school students. *Canadian Journal of Science, Mathematics and Technology Education, 20*, 58–73. https://doi.org/10.1007/s42330-020-00076-9

Fischer, K. W. (2012). Starting well: Connecting research with practice in preschool learning. *Early Education and Development, 23*(1), 131–137. https://doi.org/10.1080/10409289.2012.63787

Franco-Mariscal, A. J., Cano-Iglesias, M. J., España-Ramos, E., & Blanco-López, A. (2025). The ENCIC-CT model for the development of critical thinking. In A. J. Franco-Mariscal (Ed.), *Critical thinking in science education and teacher training* (Chap. 1) (pp. 3–42). Springer. https://doi.org/10.1007/978-3-031-78578-8_1

French, L. (2004). Science as the center of a coherent, integrated early childhood curriculum. *Early Childhood Research Quarterly, 19*(1), 138–149. https://doi.org/10.1016/j.ecresq.2004.01.004

García-Carmona, A. (2023). Scientific thinking and critical thinking in science education. Two distinct symbiotically related intellectual processes. *Science & Education.* https://doi.org/10.1007/s11191-023-00460-5

Gobierno de España [Government of Spain]. (2007). Orden ECI/3854/2007 de 27 de diciembre por la que se establecen los requisitos para la verificación de los títulos universitarios oficiales que habiliten para el ejercicio de la profesión de Maestro en Educación Infantil. [Order ECI/3854/2007 of December 27, establishing the requirements for the verification of official university degrees that qualify for the practice of the profession of preschool teacher.] *Boletín Oficial del Estado*, 312, 29 de diciembre. https://www.boe.es/eli/es/o/2007/12/27/eci3854

Gobierno de España [Government of Spain]. (2022). Real Decreto 95/2022 de 1 de febrero, por el que se establece la ordenación y las enseñanzas mínimas de la Educación Infantil. [Royal Decree 95/2022 of February 1, establishing the organization and minimum curriculum for preschool.] *Boletín Oficial del Estado*, 28, 2 de febrero. https://www.boe.es/eli/es/rd/2022/02/01/95/con

Gropen, J., Clark-Chiarelli, N., Hoisington, C., & Ehrlich, S. B. (2011). The importance of executive function in early science education. *Child Development Perspectives, 5*(4), 298–304. https://doi.org/10.1111/j.1750-8606.2011.00201.x

Gropen, J., Kook, J. F., Hoisington, C., & Clark-Chiarelli, N. (2017). Foundations of science literacy: Efficacy of a preschool professional development program in science on classroom instruction, teachers' pedagogical content knowledge, and children's observations and predictions. *Early Education and Development, 28*(5), 607–631. https://doi.org/10.1080/10409289.2017.1279527

Jiménez-Aleixandre, M. P., & Puig, B. (2022). Educating critical citizens to face post-truth: the time is now. In B. Puig & M. P. Jiménez-Aleixandre (Eds.), *Critical thinking in biology and environmental education. facing challenges in a post-truth world* (pp. 3–20). Springer. https://doi.org/10.1007/978-3-030-92006-7

Lederman, N. G., Lederman, J. S., & Antink, A. (2013). Nature of science and scientific inquiry as contexts for learning of science and achievement of scientific literacy. *International Journal of Education in Mathematics, Science and Technology, 1*(3), 138–147.

Monteira, S. F., & Jiménez-Aleixandre, M. P. (2016). The practice of using evidence in kindergarten: The role of purposeful observation. *Journal of Research in Science Teaching, 53*(8), 1232–1258. https://doi.org/10.1002/tea.21259

Muñoz-Campos, V., Franco-Mariscal, A. J., & Blanco-López, Á. (2020). Integration of scientific practices into daily living contexts: A framework for the design of teaching-learning sequences. *International Journal of Science Education, 42*(15), 2574–3260. https://doi.org/10.1080/09500693.2020.1821932

Ramanathan, G., Carter, D., & Wenner, J. A. (2022). Framework for scientific inquiry in preschool. *Early Childhood Education Journal, 50*, 1263–1277. https://doi.org/10.1007/s10643-021-01259-1

Samarapungavan, A., Mantzicopoulos, P., & Patrick, H. (2008). Learning science through inquiry in kindergarten. *Science Education, 92*(5), 868–908. https://doi.org/10.1002/sce.20275

Sjøberg, S. (2019). *Critical perspectives on inquiry-based science education (IBSE) in Europe. Position paper written for EUN partnership, European schoolnet.* https://www.academia.edu/38491775\h

Vázquez-Alonso, A., & Manassero-Mas, M. A. (2018). Más allá de la comprensión científica: educación científca para desarrollar el pensamiento [Beyond understanding of science: science education for teaching fair thinking]. *Revista Electrónica de Enseñanza de las Ciencias, 17*(2), 309–336. http://reec.uvigo.es/volumenes/volumen17/REEC_17_2_02_ex1065.pdf

Worth, K. (2010). Science in early childhood classrooms: Content and process. *Early Childhood Research & Practice (ECRP), 12*(2), 1–17.

Chapter 12
Comprehensive Analysis of the Environmental Awareness Problem. A Service-Learning Experience Based on Inquiry

Fabianna Otárola Benavides, Jessica Soto-Martínez, Manuel Martínez-Lobos, Sylvia Moraga-Toledo, Teresa Lupión-Cobos, and Cristina García-Ruiz

12.1 Introduction

In an increasingly volatile and uncertain world characterized by a growing variety of complex social issues, the 2030 agenda (United Nations, 2015) established by the Organization for Economic Co-operation and Development (OECD) in its report on the future of education, represents a fundamental generational change, the central element of which is the imperative need to instil and emphasize the concept of sustainability among our students, preparing them to be part of a responsible and empowered citizenship. This transformative effort prioritizes collaboration over division and the enduring pursuit of sustainability over short-term profits, as Schleicher (2018) highlights.

F. Otárola Benavides · M. Martínez-Lobos
Universidad de Playa Ancha, Valparaíso, Chile
e-mail: fabiana.otarola@upla.cl

M. Martínez-Lobos
e-mail: manuel.martinez@upla.cl

J. Soto-Martínez · S. Moraga-Toledo
Liceo Artístico Guillermo Gronemeyer Zamorano, Valparaíso, Chile
e-mail: sylvia.moraga@upla.cl

T. Lupión-Cobos · C. García-Ruiz (✉)
Science Education, Universidad de Málaga, Málaga, Spain
e-mail: crisgarcia@uma.es

T. Lupión-Cobos
e-mail: teluco@uma.es

A. J. Franco-Mariscal (ed.), *Critical Thinking in Science Education and Teacher Training*, Contemporary Trends and Issues in Science Education 64,
https://doi.org/10.1007/978-3-031-78578-8_12

Consequently, scientific education must assume a fundamental role in this paradigm shift, actively contributing to developing knowledge, skills, attitudes and values that empower students to emerge as authentic agents of change, devising the ability to exert a positive influence in their environment and defending the cause of creating an inclusive and sustainable future. In fact, since the Earth Summit in Rio de Janeiro in 1992, different initiatives and international reports have been highlighting the need for teachers in all educational fields to contribute to citizens acquiring an adequate vision of the problems and challenges of current society, necessary for decision making (Li et al., 2021).

This scientific education displayed in the educational curriculum aims to ensure that students acquire the basic concepts and ideas of science that allow them to understand nearby experiences and situations and generate creative solutions to everyday problems through scientific knowledge (Ministry of Education of Chile [MINEDUC], 2015). In this way, it contributes to informed and responsible decision-making regarding issues of social interest, mainly about the environment, such as global warming and climate change (Cohen et al., 2020). Thus, students are expected to feel like an active part of a scientifically informed community, capable of making good decisions and using science and its resources to coexist and improve socially.

Under these premises, the design and incorporation into the practice of educational experiences, based on scientific practices and aligned with the Sustainable Development Goals (SDG), is presented as an effective strategy to promote scientific literacy and promote scientific-technological identities (Grimalt-Álvaro & Couso, 2021). These experiences, focused on contexts and problems of daily life of interest to students, facilitate the development of critical thinking (hereinafter, CT) (Franco-Mariscal et al., 2025). In addition, they contribute to promoting social justice and sustainable development by addressing socially relevant issues for students (Reynante et al., 2020). Despite its importance, this strategy is still in development, with little research on promoting CT and using scientific practices in the classroom (Puig & Jiménez-Aleixandre, 2022).

One of these scientific practices, inquiry, has been widely recognized in the literature as a promotor of CT (Verawati et al., 2020; Wahyudi et al., 2019), as it fosters student understanding through scientific research, posing problems and proposing solutions and reasoned conclusions through processes such as analysis, evaluation or inference.

On the other hand, the study of service-learning—a methodology that combines learning and community action goals through structured educational projects in response to community- identified needs (Armstrong et al., 2021)—has become consolidated as a strategy, model, philosophy, and pedagogical orientation, being recognised as a key teaching practice to increase students' engagement (Kuh, 2008). Service-learning contributes to social justice due to its influence at different levels, contexts, and social dimensions (Chiva-Bartoll et al., 2020), considered from the constructivist approach. It links to basic, generic and transversal competencies declared by institutional training models (Lebovits & Bharath, 2019). Therefore, the relevance of exploring the link between service-learning and teaching based on socially responsible methodologies is evident in encouraging meaningful learning spaces to promote a scientific-technological education of students that is motivating, close, and valuable in their lives.

Consequently, this chapter presents the design and implementation of a teaching experience based on the scientific practice of inquiry, focused on SDG 12 on responsible production and consumption and aimed at secondary education students. Thus, through the service-learning project entitled "Analysing industrial pollution in our community", aimed at evaluating the effects of pollution produced by human action in a coastal area, the CT is integrated into a comprehensive approach to addressing issues.

12.2 The Comprehensive Analysis of the Problem as a Critical Thinking Skill Through Inquiry and Service-Learning

CT is a concept that has been extended to education in general, constituting a focus of interest by including transversal and functional skills for personal, social and work life and structured through different higher-order skills constitutive of thinking that can act as facilitators or obstacles to teaching and learning (Vieira & Tenreiro-Vieira, 2016). Thus, for Solbes and Torres (2012), the development of thinking requires the person to acquire a set of skills to address and discuss socio-scientific issues, in addition to the need to understand science as a human activity that has relationships with technology-environment-society, its promotion reaching great importance in science education and its projection in citizen education (Blanco-López et al., 2017).

With different meanings regarding the dimensions that compose CT (Franco-Mariscal et al., 2025; Puig & Jiménez-Aleixandre, 2022), this chapter will focus on the dimension related to the comprehensive analysis of problems. In the interconnected society we find ourselves in, the ability to address complex problems holistically has become paramount. Hence, the adoption of a holistic approach, considering both scientific and technical aspects, as well as social issues (ethical, cultural, economic or philosophical) and emphasising environmental sustainability, results in a catalyst for developing CT.

Integral understanding of a science issue by secondary education students is essential to foster their reasoning, problem-solving, and informed decision-making abilities. In addition, it encourages them to explore, question and analyse scientific concepts in depth through problems and challenges that require applying their knowledge to find solutions, which leads them to investigate scientific topics of interest and present their findings to their educational environment and community.

Furthermore, the combination of inquiry and service-learning contributes to this development, as it generates a greater appreciation of the complexity of scientific problems, applying knowledge to a practical and community-oriented context.

Inquiry, which fosters creativity and curiosity, is intrinsically linked to problem-solving, creative thinking, and divergent thinking (Ying, 2022), thus contributing to a comprehensive approach to research problems.

This synergy of approaches promotes in students the ability to identify the multi-dimensional nature of issues, as well as the development of creative, ethical and culturally sensitive solutions, which impacts citizens and social practice (Puig & Jiménez-Aleixandre, 2022).

12.3 Research Questions

Considering the need to create environmental awareness about the actions that impact the students' closest environment, as well as the CT skill described, we pose the following research question: How can we promote in students the comprehensive analysis of problems as a dimension of CT in the face of environmental problems in their community, using inquiry and service-learning to achieve positive awareness about their environment?

For doing so, a service-learning experience has been developed based on the approach of scientific inquiry practice, aimed at second-year high school students, and implemented in an educational centre in the commune of Quilpué (Chile).

12.4 Method

12.4.1 Participants and Context

We designed a teaching–learning experience aimed at promoting SDG 12 related to responsible production and consumption. The experience was implemented in the Natural Sciences subject of a local High School and had the participation of 19 second-grade students, aged between 15 and 16 years, distributed in small workgroups.

Its design and development in the classroom were addressed by a teacher from the high school itself, who was advised by researchers in science education from the University of Playa Ancha (Chile) and the University of Málaga (Spain), also counting on the support of an expert researcher in environmental pollution, all co-authors of this chapter. This collaboration resulted from the INCE Network (www.redince.com) for the promotion of inquiry-based science education between Chile and Spain, covered by a framework agreement (reference 8.07/5.12.5574) between both universities.

The context focused on a latent problem, environmental pollution, in the area where the educational centre was located (areas of Quintero and Puchuncaví, in Chile). These two coastal communes, located in the province of Valparaíso, are characterized by a large industrial park, the largest in the entire country, with more than 15 industries, including thermoelectric plants, copper and oil refineries, and chemical industries, among others (Saravia et al., 2016). Both communes of Quintero and Puchuncaví had great relevance for the region due to their wheat and legume crops, one of the area's main economic activities. However, with the passage of time and the installation of the industrial park, agricultural activity decreased abruptly, with a drop of more than 70% in the case of wheat and 50% in the case of legumes (Arellano-Escudero, 2017). Consequently, in 1993, the bay was declared an area saturated by sulphur dioxide and particulate matter, generating various problems, affecting not only the agricultural and fishing industries, but also the health of its inhabitants (Tume et al., 2020), causing physical, emotional and social discomfort in the population, which is suspicious of the ecosystem that was once fertile and prosperous (Graham, 2019; Moreno-Jiménez & Fuenzalida, 2017).

Politically, the situation highlights the challenges and controversies in balancing industrial development with environmental protection (Espinoza, 2015). Socially, the contamination has adversely affected the local communities, leading to health problems and impacting their livelihoods, particularly in agriculture and fishing. Ethically, the situation raises questions about the responsibility of industries and the government to protect the environment and ensure public health and the justice of imposing environmental burdens on specific communities. As it can be seen, the issue addressed includes environmental, social, economic and ethical aspects.

12.4.2 Experience Description

Under this scenario, we designed an experience that would allow evaluation of the effects of pollution produced by human action, using the cultivation of legumes, and with the ultimate objective of putting what was learned during the sequence at the community service, disseminating the results obtained to the citizens.

The educational objectives addressed, differentiated into learning (LG) and skills development (SG) goals according to the Chilean curriculum (MINEDUC, 2015), are shown in Table 12.1.

The experience begins with the students through the leading question, *how to demonstrate the presence of pollutants produced by industrial activity in the Quintero and Puchuncaví area?* from which the phases described in Fig. 12.1 are developed.

In phase I (*Targeting and exploration*), awareness of the issue was carried out through inquiry questions that allowed the focus to be developed, which led the students to a literature review to identify formal and non-formal sources of information, analysing audiovisual material, as well as current news about pollution in the study area. Once the material was analysed, each group identified the primary contaminants present in the area, their environmental effects, and the comprehensive health of people exposed daily to said contamination.

After the identification of the possible variables, in phase II (*Introduction to new knowledge*), through dialogue, reflection, and decision-making regarding the problem

Table 12.1 Learning (LG) and skills development (SG) goals of the teaching–learning experience and their relationship with CT dimensions, according to the ENCIC-CT model (Chap. 1) (Franco-Mariscal et al., 2025)

Learning goals (LG)	Skills development goals (SG)	CT dimensions and descriptors (Franco-Mariscal et al., 2025)
LG1. Explain, through models and experimentation, the properties of solutions in close examples LG2. Investigate and explain the applications that have arisen as a result of genetic manipulation to generate foods, detergents, clothing, drugs or others, and evaluate their ethical and social implications, considering the physical state (solid, liquid and gaseous), its components (solute and solvent) and the amount of dissolved solute (concentration) LG3. Explain and evaluate the effects of human actions (environmental conservation, crops, forestation and deforestation, among others) and natural phenomena (droughts, volcanic eruptions, among others) about the balance of ecosystems, the availability of renewable natural resources and non-renewable, and possible measures for sustainable development	SG1. Plan and develop research that allows collecting evidence and contrasting hypotheses using technological and mathematical tools SG2. Critically analyse social, economic, ethical and environmental implications of problems related to public controversies involving science and technology	*Comprehensive analysis of the problem.* Address problems comprehensively, in their complexity, considering the scientific, technical, ethical, cultural, philosophical, social, environmental, economic and other related dimensions
	SG3. Construct, use and communicate scientific arguments	*Argumentation.* Create solid arguments and question the validity of the arguments, rejecting conclusions not based on evidence, and detecting argumentative fallacies *Comunicative skills.* Communicate decisions using appropriate language according to the context and goals or intentions

*Although, taking into account the SG3 of the Chilean curriculum, this directly relates to the dimensions of argumentation and communicative skills, this chapter will focus, as has been specified, on the development of the comprehensive analysis of the problem dimension

posed, students established research questions, raising the working hypotheses, as well as the goals, which served to guide the work in the subsequent phases.

In phase III (*Structuring and synthesis of new knowledge*), students of each group built their experimental design, once again through dialogue and reflection on the most appropriate methodology to respond to the guidelines proposed in the previous phase. This design focused on sowing beans (*Phaseolus vulgaris L.*) and lentils (*Lens culinaris*) in different substrates, with each group agreeing on the growing conditions according to the questions and hypotheses posed. At this time, the need to make a

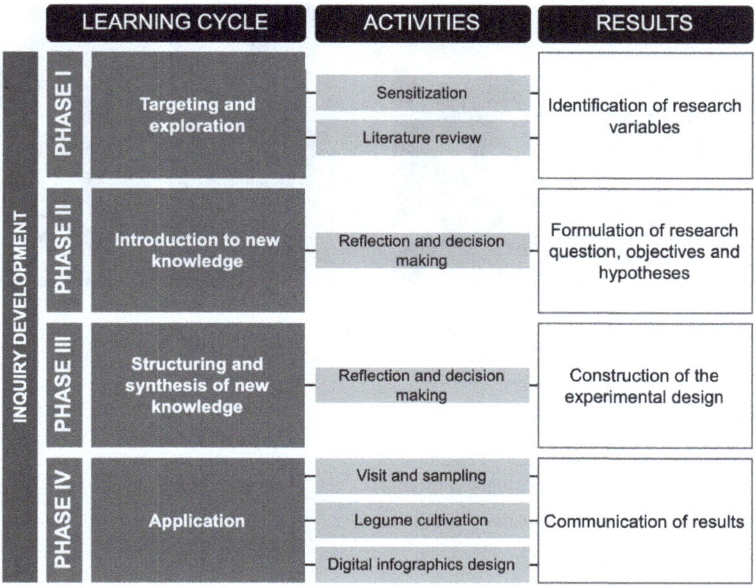

Fig. 12.1 Learning cycle diagram with a focus on scientific inquiry

daily record of what was observed in the crops was also indicated, so each group created an observation record table.

Finally, phase IV (*Application*), temporarily located during the development of phase II, focused on a guided tour to collect water and soil samples from the contaminated sector to carry out subsequent legume crops. Furthermore, for its development, the group of students received advice from an expert researcher, co-author of this work, on the protocols for collecting soil and water samples, paying particular attention and care in handling and hygiene during the process (Fig. 12.2). After that, several activities took place to analyse the results obtained and communicate them to the rest of the community through disseminating the experience and conclusions on the social networks available to the educational centre (*Instagram* and similar), thus calling for reflection regarding the causes and effects of environmental pollution.

12.4.3 Implementation of the Inquiry

After completing the phase I, the working groups decided on the study variables. Thus, they determined the dependent variable to be the growth of the legume and the independent variables to be the soil type for planting and the irrigation water used. Thus, each group agreed on a standard design, using one of the two types of seeds (*Phaseolus vulgaris L.* or *Lens culinaris*) (controlled variable), which would allow for a comparison of the results obtained by each group (Table 12.2).

Fig. 12.2 Photographs of the students during the guided tour of the study area

Table 12.2 Variables of the study

Culture	Soil type	Water type
1	Campiche estuary (polluted)	Potable water
2	Campiche estuary (polluted)	Campiche estuary (polluted)
3 (control)	Organic	Potable water
4	Organic	Campiche estuary (polluted)

After the guided tour and sample collection, each group proceeded with the experimental design, placing all the crops in the same place within the classroom and watering them with a frequency of two or three days, depending on the humidity observed in the substrates.

The observations were recorded in logs for approximately four weeks, in which, in addition to the experimental design, the changes observed each day were recorded and compared with the control culture (3).

From the observations, the groups showed that the seeds of both *Phaseolus vulgaris* L. and *Lens culinaris* only managed to germinate and grow a few centimetres when using the conditions described in cultures 1 and 3 (Fig. 12.3). Those legume seeds planted in the Campiche estuary, using potable water for irrigation, germinated and grew similarly to the control groups. Among the hypotheses for this, the class group determined that the presence in irrigation water of heavy metals from industrial pollution, such as lead, nickel, cadmium and manganese, could affect the physiological processes of plants, reducing growth and presenting a lower biomass production. This hypothesis was contrasted with the literature on heavy metals' effect on plants (Pokorska-Kiewiada et al., 2018) and with the evidence from experimentation that revealed that irrigation water quality is decisive for the cultivation of legumes used in this research.

Fig. 12.3 Photographs of the observation process

12.4.4 ENCIC-CT Model Application

This chapter applies the ENCIC-CT model (Franco-Mariscal et al., 2025) to the development of CT in science education, addressing a socio-scientific issue related to industrial pollution in a coastal area (Layer 1). It focuses on the CT dimension concerning the comprehensive analysis of complex issues within the skills domain and addresses social engagement as a disposition (Layer 2). The described experience employs the scientific practice of inquiry (Layer 3), allowing students to investigate the problem in depth. Additionally, the teaching strategy of service-learning (Layer 4) is incorporated, including elements of community engagement and reflective practice. This strategy enhances the relevance and impact of the learning experience by connecting students to the social realities of their surroundings, while also promoting a sense of civic responsibility. By integrating these interconnected layers, the ENCIC-CT model provides a structured framework for students to develop CT through engagement with real-world problems.

12.4.5 Data Collection and Analysis

Students recorded all the documentation and development of the project through group logbooks, which were evaluated using a rubric designed to assess all the stages of the scientific inquiry.

At the end of the project, and after disseminating the results through social networks, a day of reflection was held to evaluate the level of satisfaction of each group regarding the investigation. In this reflection, guiding questions were posed to determine the perception of student's learning experience and their assessment of the project (questions Q1 and Q2) and, on the other, to qualitatively analyse the dimension of CT addressed, estimating whether students had dealt with the pollution problem holistically, considering the environmental, social, political, economic and ethical factors that were dealt with throughout the project (questions Q3 and

Q4). Hence, students collected their written responses to questions Q1 to Q4 in their logbook and a classroom discussion was held.

- Q1. How do you rate the service-learning project on the most contaminated area in your region?
- Q2. How was your experience during the execution of the project, which also allowed you to take a guided tour of the study area?
- Q3. To what extent did the project allow you to analyse pollution and its effects in the Quintero and Puchuncaví area?
- Q4. To what extent has this project motivated you to develop a sustainable attitude towards the environment?

Subsequently, we selected the information of interest to analyse the content of the students' responses registered in the logbooks and established around questions Q1 to Q4, categorising all the responses, using the ATLAS.ti software (version 8.4.4). We organised the information according to the dimensions and codes that arose, which were submitted for discussion by the project researchers and are specified in Table 12.3.

12.5 Results

Table 12.3 shows the codes that arose per question, their frequency in the students' responses and the relative percentage.

Regarding the assessment of the project and its contextualization in the most contaminated area of the region (Q1), about 40.0% expressed having developed an awareness at an environmental level (*It clearly makes you realize how bad the environment is in terms of how we harm it or contaminate it, and the industry especially is the one that affects it the most*) and to a lesser extent (33.3%) at a social level (*I had not been aware that it was so serious because of the people who get sick due to the heavy metals*). Although only 13.3% stated they were completely unaware of the issue in the Quintero area, only 6.7% declared they had prior knowledge of it.

Concerning the experience of the project (Q2), 39.0% expressed that they found it exciting and motivating (*It was a pleasant experience, and in fact, it was pretty fun to be able to see the growth process of legumes*), finding it educational in a lower percentage (17.0%) and innovative (5.5%). The interest in the different stages of the project is distributed relatively equally between the visit to the study area (11.0%), the taking of samples (11.0%) and the monitoring of the legume cultivation (11.0%), followed by communication of the results obtained (5.5%).

Q3 and Q4, aimed at developing a comprehensive analysis of the problem as a dimension of CT, show equally disparate results. Thus, the answers about the analysis of pollution and its effects (Q3) focused exclusively on environmental aspects related to the type of soil (30%) (*I was able to see the process and changes using the different lands*) and the type of water (30%) (*We realized that with contaminated water it was more challenging to germinate*). At the same time, regarding the development of

Table 12.3 Findings of the analysis of the leading questions Q1–Q4

Code	Frequency	Relative percentage (%)
Q1. How do you rate the service-learning project on the most contaminated area in your region?		
Awareness at a social level	5	33.3
Environmental awareness	6	40.0
Ignorance of the environmental problem	2	13.3
Prior knowledge of the environmental problem	1	6.7
No answer	1	6.7
Total	15	100.0
Q2. How was your experience during the execution of the project, which also allowed you to take a guided tour of the study area?		
Educational	3	17.0
Exciting and motivating	7	39.0
Innovative	1	5.5
Interest in communicating results	1	5.5
Interest in legumes culture	2	11.0
Interest in sample collection	2	11.0
Interest in visiting study area	2	11.0
Total	18	100.0
Q3. To what extent did the project allow you to analyse pollution and its effects in the Quintero and Puchuncaví area?		
Analysing environmental aspects (soil type)	3	30.0
Analysing environmental aspects (water type)	4	30.0
No answer	4	40.0
Total	11	100.0
Q4 To what extent has this project motivated you to develop a sustainable attitude towards the environment?		
Actions towards sustainability		
Saving in water consumption	1	10.0
Reduction of plastic consumption	1	10.0
Pollution awareness		
Contamination by other solid waste	2	20.0
Microplastic pollution	1	10.0
Pollution from industrial emissions	1	10.0
No answer	4	40.0
Total	10	100.0

sustainable attitudes towards the environment (Q4), awareness about contamination by other waste stands out, such as solids (20.0%), industrial emissions (10.0%) and microplastics (10.0%). However, only two sustainability actions were proposed, such as saving water consumption (10.0%) and reducing plastic consumption (10.0%).

12.6 Discussion and Educational Implications

Implementing the inquiry-based service-learning experience has generated some positive results, although there is still a wide range of improvements. In general, the group of students valued the experience positively, demonstrating a better predisposition towards the use of this type of teaching–learning approach and positively valuing the fact of having been able to visit the area to investigate since it encouraged them even more with the development of the investigation and allowed them to observe in situ the environmental contamination present in the place.

Students highlighted that the approach through inquiry contributed to a motivating learning experience, especially about collecting actual samples for analysis (*very educational when taking the samples and carrying out the project by growing the legumes*) (Fig. 12.4). In addition, he expressed feeling supported to face the decision-making, even if it entailed some errors (*The work was not accomplished very well, but I believe that we were still able to meet the objective of knowing and understanding why this happens and its causes*). Although to a large extent, they stated that they had become environmentally aware of the effect of pollutants on their environment, thus contributing to positive attitudes towards carrying out sustainable actions (*I learned a lot about pollution and that we can avoid it in some way*), these did not materialize into specific actions they could do.

The service-learning application allowed the students to share the results obtained in their research on industrial pollution, not only with their educational community but with their close social environment, thus promoting awareness and citizen actions among students. This awareness also promoted attitudes and values in students

Fig. 12.4 Pollution emitted by the study area's industrial cordon and examples of project dissemination on social networks

directed towards social justice and sustainable development, motivating them to take action measures to mitigate industrial pollution in their region (Tume et al., 2020).

In addressing the problem comprehensively, as outlined in Fig. 12.1, our project incorporated stages such as awareness-raising and an extensive review of the problem's variables and its multi-level consequences. Despite these efforts, students fell short in grasping the issue. This shortfall is particularly evident in their inability to fully engage with and reflect on the broader implications of the problem (Puig & Jiménez-Aleixandre, 2022; Verawati et al., 2020). In fact, most students only prioritised the environmental aspect of the problem, ignoring economic, social, political or ethical aspects. Only a minority of students considered social aspects in addition to environmental ones. Thus, although the inquiry allowed students to delve deeper into a real and pressing environmental problem in the central area of Chile, the development of CT skills linked to the comprehensive analysis of the problem was not satisfactorily achieved.

In response to the research question posed, we consider that, although the project contributed to promoting active participation of students in their community, it did so exclusively at an environmental level, without considering other aspects.

This difficulty in integrating the multifactorial analysis is added to others that arose during the execution of the project, such as the management of resources by the teachers involved or the complexity expressed by students in the search for reliable sources of information, as well as the sampling procedure. These obstacles encourage us to incorporate actions and new tasks for future implementations. In this way, we consider it essential to delve into the social aspects of the problem, including the task of collecting evidence at a social level, for example, through interviews with residents of the Quintero and Puchuncaví area, either with authorities or with members of NGOs involved in the environment.

Regarding the economic dimensions of the issue, one potential task involves a detailed analysis of economic data at both local and national levels. This approach aims to understand the economic impact more thoroughly. In order to explore the political aspects, a critical analysis of the electoral promises made by various parties concerning the affected area could be conducted, offering insights into the political response and commitments. Lastly, to delve into the ethical aspects, an in-depth reading and subsequent class discussion of selected UNESCO reports, particularly *Environmental Ethics and International Policy Report* (Have, 2006), would be beneficial, framing the discussion within a global ethical context.

In conclusion, we find that the educational implications of this experience are significant. However, there is still a wide range of improvement since it highlights the importance of designing educational experiences that integrate inquiry and service-learning approaches in science teaching, addressing problems that are real and close to the student community and ultimately developing CT.

Acknowledgements M. G. Otárola Benavides and M. G. Soto Martínez thanks the INCE Network (www.redince.com) and the Pedagogy in Chemistry and Sciences program at the University of Playa Ancha for accompanying them in preparing and implementing the project. Dr García-Ruiz thanks the RYC 2020 program, funded by the State Research Agency and the European Social Fund (reference: RYC2020-029033-I/AEI/10.13039/501100011033). This work is part of the Spanish National Plan R+D+i Project entitled "Citizens with critical thinking: A challenge for teachers in science education", reference PID2019-105765GA-I00, funded by MCIN/AEI/10.13039/501100011033.

References

Arellano-Escudero, N. (2017). Arsénico sobre Puchuncaví: Metabolismo de la minería y sufrimiento ambiental [Arsenic upon Puchuncaví: Metabolism of mining and environmental suffering]. *Revista Iberoamericana De Viticultura, Agroindustria y Ruralidad, 3*(10), 71–91.

Armstrong, D., Murck, B., & Poe, J. C. (2021). Service learning opportunity for undergraduate science students. Integrating problem-based learning in the high school science curriculum. *Journal of Chemical Education, 98*(4), 1275–1282. https://doi.org/10.1021/acs.jchemed.0c01275

Blanco-López, Á., España-Ramos, E., & Franco-Mariscal, A. J. (2017). Estrategias didácticas para el desarrollo del pensamiento crítico en el aula de ciencias [Teaching strategies for the development of critical thinking in the teaching of science]. *Ápice, Revista de Educación Científica, 1*(1), 107–115. https://doi.org/10.17979/arec.2017.1.1.2004

Chiva-Bartoll, O., Ruiz-Montero, P., Capella-Peris, C., & Salvador-García, C. (2020). Effects of service learning on physical education teacher educaction students' subjetive happiness, prosocial behavior, and professional learning. *Frontiers in Psychology, 11*, 331. https://doi.org/10.3389/fpsyg.2020.00331

Cohen, R., Zafrarni, E., & Yarden, A. (2020). Science teachers as proponents of socio-scientific inquiry-based learning: From professional development to classroom enactment. In M. Evagorou, J. A. Nielsen, & J. Dillon (Eds.), *Science teacher education for responsible citizenship. Towards a pedagogy for relevance through socioscientific issues* (pp. 117–132). Springer. https://doi.org/10.1007/978-3-030-40229-7_8

Espinoza, L. A. (2015). El polo industrial Quintero-Ventanas. ¿Hacia dónde fue el desarrollo? [Industrial site Quintero-Ventanas. Consequences of development]. *MILLCAYAC—Revista Digital de Ciencias Sociales, 11*(3), 245–269.

Franco-Mariscal, A. J., Cano-Iglesias, M. J., España-Ramos, E., & Blanco-López, A. (2025). The ENCIC-CT model for the development of critical thinking. In A. J. Franco-Mariscal (Ed.), *Critical thinking in science education and teacher training* (Chap. 1) (pp. 3–42). Springer. https://doi.org/10.1007/978-3-031-78578-8_1

Graham, M. (2019). *Diario de mi residencia en Chile: en el año 1822.* [Journal of My Residence in Chile: In the Year 1822.] Penguin clásicos.

Grimalt-Álvaro, C., & Couso, D. (2021). Developing STEM identities in students in the "big middle". connections between identity and socioeconomic level. In E. Carlton, R. S. Schwartz, D. Bressler, N. Makki, & J. Plummer (Eds.), *NARST 94th annual international conference developing STEM identities in students in the "big middle". Connections between identity and socioeconomic level* (p. 119). NARST.

Have, H. A. M. J. (Ed.). (2006). *Environmental ethics and international policy*. UNESCO.

Kuh, G. D. (2008). *High-Impact Educational Practices: What they are, who has access to them, and why they matter*. Association of American Colleges and Universities.

Lebovits, H., & Bharath, M. N. (2019). Service-learning as a tool to cultivate democratically minded students: A conceptual framework. *Journal of Public and Nonprofit Affairs, 5*(3), 277–292. https://doi.org/10.20899/jpna.5.3.277-292

Li, S., Liu, X., Yang, Y., & Tripp, J. (2021). Effects of teacher professional development and science classroom learning environment on students' science achievement. *Research in Science Education, 52*, 1031–1053. https://doi.org/10.1007/s11165-020-09979

Ministerio de Educación de Chile [MINEDUC]. (2015). *Bases curriculares 7º básico a 2º medio.* [Curriculum Framework for 7th Grade to 12th Grade.] Ministerio de Educación, Gobierno de Chile. https://www.curriculumnacional.cl/614/articles-37136_bases.pdf

Moreno-Jiménez, A., & Fuenzalida, M. (2017). Medición y cartografía del impacto socialmente percibido ante industrias contaminantes con un estimador de densidad núcleo (kernel) y SIG: aplicación al caso del complejo industrial Ventanas, Chile. [Measurement and mapping of socially perceived impact related to pollutant industries using a kernel density estimator and GIS. The case study of Ventanas industrial site (Chile)]. *Estudios Geográficos, 78*(282), 225–256. https://doi.org/10.3989/estgeogr.201708

Pokorska-Kiewiada, K., Rajkowska-Mysliwiec, M., & Protasowicki, M. (2018). Acute lethal toxicity of heavy metals to the seeds of plants of high importance to humans. *Bulletin of Environmental Contamination and Toxicology, 101*, 222–228. https://doi.org/10.1007/s00128-018-2382-9

Puig, B., & Jiménez-Aleixandre, M. P. (Eds.). (2022). *Critical thinking in biology and environmental education. Facing challenges in a post-truth world*. Springer. https://doi.org/10.1007/978-3-030-92006-7

Reynante, B. M., Selbach-Allen, M. E., & Pimentel, D. R. (2020). Exploring the promises and perils of integrated STEM through disciplinary practices and epistemologies. *Science & Education, 29*, 785–803. https://doi.org/10.1007/s11191-020-00121-x

Saravia, P., Armingol, K., & Garland, B. (2016). El derrame de petróleo en Quintero, V región de Chile: Una mirada desde las organizaciones sociales [Oil spill in Quintero bay—Chile's V región. A look from social organizations]. *Población y Sociedad, 23*(2), 179–206.

Schleicher, A. (2018). *The future of education and skills: Education 2030.* OECD Education Policy Perspectives. OECD.

Solbes, J., & Torres, N. (2012). Análisis de las competencias de pensamiento crítico desde el aborde de las cuestiones sociocientíficas: Un estudio en el ámbito universitario. [Analysis of critical thinking competencies from the perspective of socio-scientific issues: A study in the university context]. *Didáctica De Las Ciencias Experimentales y Sociales, 26*, 247–269. https://doi.org/10.7203/DCES.26.1928

Tume, P., Barrueto, K., Olguin, M., Torres, J., Cifuentes, J., Ferraro, F. X., Roca, N., Bech, J., & Cornejo, O. (2020). The influence of the industrial area on the pollution outside its borders: A case study from Quintero and Puchuncaví districs, Chile. *Environmental Geochemistry and Health, 42*(8), 2557–2572. https://doi.org/10.1007/s10653-019-00423-2

United Nations. (2015). *General assembly. Resolution 70/1. Transforming our world: The 2030 agenda for sustainable development.* United Nations.

Verawati, N. N. S. P., Hikmawati, H., & Prayogi, S. (2020). The effectiveness of inquiry learning models intervened by reflective processes to promote critical thinking ability in terms of cognitive style. *International Journal of Emerging Technologies in Learning (iJET), 15*(16), 212–219. https://doi.org/10.3991/ijet.v15i16.14687

Vieira, R. M., & Tenreiro-Vieira, C. (2016). Fostering scientific literacy and critical thinking in elementary science education. *International Journal of Science and Mathematics Education, 14*(4), 659–680. https://doi.org/10.1007/s10763-014-9605-2

Wahyudi, N. N., Verawati, S. P., Ayub, S., & Prayogi, S. (2019). Effectiveness of inquiry-creative-process learning model to promote critical thinking ability of prospective physics teachers. *Journal of Physics: Conference Series, 1417*, 1–6. https://doi.org/10.1088/1742-6596/1417/1/012071

Ying, H. (2022). Effectiveness of inquiry-based science laboratories for improving teamwork and problem-solving skills and attitudes. *Journal of Research in Science Teaching, 59*(3), 329–357. https://doi.org/10.1002/tea.21729

Part V
Role-Playing Games, Technologies, Multimedia Resources, and Critical Thinking

Chapter 13
Role-Playing as a Strategy for Developing a Critical Analysis of Information and Decision-Making Through the Socio-scientific Issue of Russian Gas

María del Mar López-Fernández

13.1 Introduction

We live in an ever-changing world where it is challenging to foresee with certainty the knowledge needed for the future society. As Kilag et al. (2024) stated, it is necessary to foster skills in our students beyond mere information accumulation. These skills will enable them to acquire knowledge, learn lifelong, and develop as members of society.

Currently, the problem encountered by citizens does not lie in the quantity of information available but in the quality of this information and their difficulty in using it effectively in different contexts in which they must make decisions, considering personal, professional or social values and intentions (Pérez-Gómez, 2008). The ability to manage information critically and its use for reasoned decision-making is related to critical thinking (hereinafter, CT) and is considered a fundamental aspect of scientific competence (Blanco-López et al., 2015).

For this reason, curricula in different countries (Ministerio de Educación y Formación Profesional, 2022, in the case of Spain) include the development of CT skills in students, preparing them to become responsible and autonomous citizens. In this regard, teachers are crucial in forming autonomous and reflective individuals with CT. This involves the development of skills such as information management and communicative skills, which are essential for building more sustainable societies.

M. M. López-Fernández (✉)
Science Education, Universidad de Málaga, Málaga, Spain
e-mail: mmarlf@uma.es

A. J. Franco-Mariscal (ed.), *Critical Thinking in Science Education and Teacher Training*, Contemporary Trends and Issues in Science Education 64,
https://doi.org/10.1007/978-3-031-78578-8_13

13.2 Role-Playing as a Strategy for Developing Critical Information Analysis and Decision-Making Applied to the Socio-scientific Issue of Russian Gas

The games offer participants the opportunity to enter into new identities, explore different options and their respective consequences, foster social skills and cognitive competences, as well as record concrete and objective information in the framework of the experience (Perrotta et al., 2013; Zeng et al., 2020). Role-playing and simulation games constitute a methodological strategy in which an actual situation is represented, and each student assumes a previously assigned role (Cruz-Lorite et al., 2023; Franco-Mariscal et al., 2023).

Role-playing games are beneficial for developing CT in science education, particularly in critical analysis of information (López-Fernández et al., 2022) and decision-making (Simonneaux, 2001), among others. They also encourage problem-solving, teamwork, argumentation and communication (Franco-Mariscal et al., 2023). Furthermore, role-playing motivate students to explore and learn about topics related to the contexts addressed (Craciun, 2010; López-Fernández et al., 2021; Sagmeister et al., 2021).

The management and critical analysis of information is essential for learners in preparing for their assigned role and making decisions after the role-play. It involves handling information from different sources and being able to distinguish reliable sources from unreliable ones (Howell & Brossard, 2021), as well as reflecting on the topic addressed to make a decision or change their mind in situations that are part of their present and future personal and professional lives (Mao et al., 2022).

Choosing an appropriate problem for the role-play is essential because society faces complex challenges, such as climate change and pandemics, which require active citizen participation. It is, therefore, paramount that students are adequately prepared to engage in debates around socio-scientific issues, as they significantly impact the well-being of humanity.

Since the beginning of the conflict between Russia and Ukraine, Europe's dependence on Russian gas has become a critical issue. Despite the European Union's efforts to guarantee its energy security by diversifying its energy imports away from Russia, the fact is that many countries still depend on the purchase of this gas (European Commission, 2023). For instance, approximately 8% of Spain's gas imports come from Russia. This dependence of countries on Russian gas can be considered a socio-scientific issue since, in addition to the direct economic repercussions, it also involves social, scientific-technological, environmental, political and ethical issues.

The economic aspects of the issue, related to energy supply and demand, mean that the impact of this crisis extends to the global economy (Mbah & Wasun, 2022). Spain and other countries face economic risks due to their heavy dependence on Russian gas. Volatility in supply or changes in prices can negatively affect financial stability. Furthermore, exploring alternative energy sources, such as renewables or diversification of gas sources, involves assessing the short- and long-term economic costs and benefits.

The social aspects of the problem relate to the impact on the population and access to energy resources. Thus, changes in gas supply can affect citizens in terms of energy costs, job stability and quality of life. Society must be involved in decision-making to ensure that proposed solutions reflect the needs and concerns of the population.

Scientific and technological considerations are linked to managing gas dependency and decisions related to energy sources. Some alternatives are the search for sustainable solutions (Pirani et al., 2014; Thumann, 2006), the transition to renewable energy sources (De Vries et al., 2007) or energy efficiency (International Energy Agency, 2014). A possible sustainable alternative is to reduce motorised mobility linked to fossil fuels (Martínez, 2018). Potential alternative sources include purchasing liquefied gas from other regions, boosting domestic gas production, or investing in developing other energy sources. In the latter case, Spain already invests in solar, wind and hydroelectric energy. Finally, some energy efficiency proposals are related to their promotion in buildings or transport.

Environmental aspects are associated with the effects of the conflict, such as the destruction of ecosystems and air and soil pollution (Pereira et al., 2022). They are also closely related to scientific and technological aspects, so assessing how dependence on Russian gas affects the environment and how the proposed alternatives contribute to or alleviate existing environmental problems is necessary.

The political dimension is present in the interaction between countries and the implications of dependence on an external energy source (Dickel et al., 2014). International relations also test the ethical aspects of the issue when considering switching gas sources, especially if there are pre-existing trade agreements. The transition to alternatives must meet ethical standards, ensuring sustainability and minimising adverse environmental impact.

In this context, this chapter presents the findings of a role-playing activity implemented with Spanish eighth-grade students aimed at addressing critical analysis of information and decision-making regarding energy alternatives to purchasing Russian gas.

13.3 Research Questions

This study attempts to answer the following research questions (RQ):

- RQ1: How do eighth-grade students develop critical information analysis skills when evaluating information from various Internet sources to prepare characters for a role-playing about energy alternatives to buying Russian gas?
- RQ2: What decisions do students make regarding the energy alternative of Russian gas before and after participating in the role-play?

13.4 Method

13.4.1 Participants

The study's sample consisted of 62 eighth-grade students aged 13–14 taking the compulsory subjects of Physics and Chemistry in a secondary school in Málaga (Spain). The participants consisted of two class groups with students sharing similar characteristics, each comprising 31 students. In one group, 51.61% were boys and 48.39% were girls, while in the other group, 54.84% were girls and 45.16% were boys.

13.4.2 Experience Description

The activity consists of a role-playing that uses a parliamentary debate as a scenario on what energy alternative to the purchase of Russian gas by our country should be adopted. This role-play was an initial activity within a teaching–learning sequence on energy sources and production. The experience was carried out at the end of 2022, a few months after the start of the war in Ukraine, when winter was approaching and Russia was threatening to close the pipeline supplying gas to Europe. Figure 13.1 illustrates the tasks related to the role-playing.

- *Task 1. Presentation of the socio-scientific issue and initial decision-making*

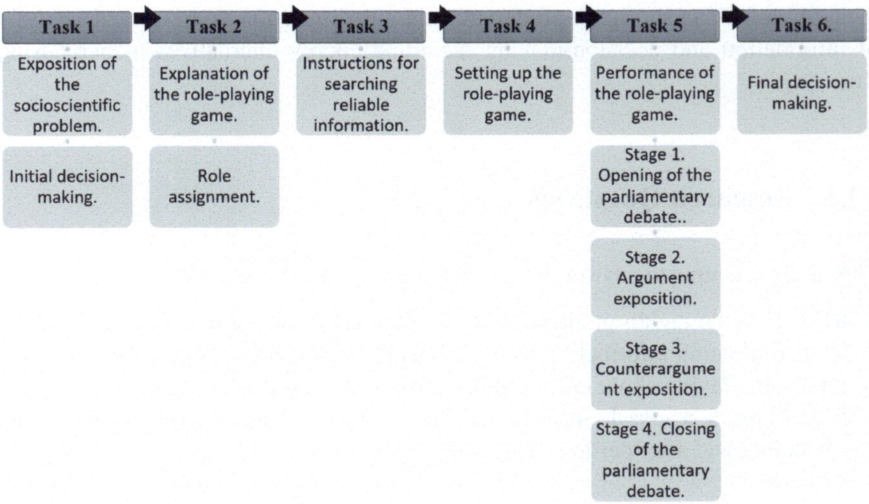

Fig. 13.1 Sequence of tasks in the role-playing game

Firstly, students were introduced to the socio-scientific issue arising from the conflict between Russia and Ukraine. Like other European Union countries, Spain supplies gas from Russia as part of its energy supply. After the outbreak of the conflict, the country is considering whether to seek alternative gas to Russian gas or continue to buy gas from other countries. The students were then asked an open-ended question, individually and in writing, in which they had to indicate which alternative they would prefer to see implemented.

- *Task 2. Explanation of the role-play and role allocation*

It was explained to the students that a role-play would take place in which the class would become the Congress of Deputies, and the students would be deputies from different parliamentary groups to address and decide on a response to the problem. The students were divided into seven groups, namely the government team whose role was to propose, moderate and conclude the game, and six parliamentary groups, each with different energy interests: in favour of buying gas from other countries (group 1), in favour of wind energy (group 2), solar and photovoltaic energy (group 3), hydropower (group 4), geothermal energy (group 5) or tidal energy (group 6). Each group consisted of four or five students, with one of them playing the role of spokesperson.

- *Task 3. Instruction in searching for reliable information on the Internet*

To foster the development of critical information analysis in students, it is essential to provide them with criteria that make it possible to evaluate the quality and reliability of sources, especially those available online. To this end, they received instruction based on the criteria proposed by Fornás (2003) for assessing the quality and reliability of Internet content. These criteria covered the following aspects: (a) the identification of the author, giving more excellent value to information coming from collective decisions rather than individuals, and within the latter group, to those with specialised expertise; (b) the clarity of the message, stressing the importance of precise wording and the clear distinction between factual information and the opinions expressed, (c) independence of the message, placing particular importance on the ability to separate information from possible advertising messages or bias, (d) use of bibliographical references, stressing the importance of supporting the information presented with documented and verifiable sources. Students were asked to ensure the information consulted could not have been published earlier than 2015.

- *Task 4. Preparation of the role-play*

The different groups of students searched for reliable information to construct their arguments. Specifically, they were asked to fill out a form in which they had to indicate at least five arguments to defend their position, the source consulted, and the date of publication of the information. This task was started in class so the teacher could accompany them in critically analysing the information and resolving any doubts. The students were given one week to complete the task.

- *Task 5. Staging the role-play*

The role-play was staged in four stages:

- *Stage 1* (*Opening of the parliamentary debate*). The president of the government came to the rostrum of the congress and started the parliamentary session by explaining the issue.
- *Stage 2* (*Presentation of arguments*). The president gives the floor to the spokespersons of the various parliamentary groups, who have three minutes in the gallery to present their arguments in favour of the alternative assigned to them. In the meantime, the other MEPs take notes on the arguments put forward by the different parliamentary groups. After the speeches, there is a five-minute parliamentary break. At this time, MEPs and spokespersons exchange notes to construct counter-arguments in response to the arguments put forward by the other spokespersons.
- *Stage 3* (*Presentation of counter-arguments*). In an orderly fashion and moderated by the government group, MEPs begin a debate from their seats to exchange counter-arguments.
- *Stage 4* (*Closing of the parliamentary debate*). The government team meets to reach a consensus on the best alternative to purchasing Russian gas based on the best arguments given during the parliamentary session. Once decided, it closes the debate and reports on the alternative adopted.

- *Task 6. Final decision making*

After the game, the students again answered the initial open-ended question, individually and in writing, on which energy alternative to Russian gas they would prefer to see implemented.

13.4.3 ENCIC-CT Model Application

The research is grounded in the ENCIC-CT model (Franco-Mariscal et al., 2025) to develop CT in eighth-grade students. This study explored a contemporary socio-scientific issue—the search for alternative energy sources to Russian gas amidst the Ukraine war (Layer 1)—with a focus on the skills domain, particularly the abilities of critical information analysis and decision-making (Layer 2). The process incorporated the scientific practice of argumentation (Layer 3), using parliamentary role-play as a key teaching strategy to strengthen CT (Layer 4).

13.4.4 Data Collection and Analysis

This study uses different data collection instruments depending on the CT dimension under investigation.

- *Critical Analysis of Information*

The instrument used was the debate preparation worksheet for each group of students, which includes the arguments constructed by the students to be used later in the debate, the source consulted and the date of publication of this information. In total, 12 worksheets were collected and analysed. Table 13.1 illustrates the record for the parliamentary group in favour of wind energy.

The quality of the information sources consulted was determined using the criteria established by Franco-Mariscal et al. (2014) in a previous study.

(a) Highly reliable source of information: A source that is considered reliable, accurate and credible. These sources are essential when seeking accurate and truthful information to make informed decisions, conduct research or support arguments.
(b) Medium-reliability information source: A source that is somewhere in between regarding reliability and credibility. This source type may have some positive aspects but some limitations or areas of concern.
(c) Low-reliability source of information: Information with significant deficiencies in credibility and reliability. These sources should be used cautiously, and their information should be verified with more reliable sources whenever possible.

Based on these criteria, the author developed a coding scheme to classify the information sources. Table 13.2 summarises the categories used in the study and their reliability. Clear definitions were provided for each category to ensure consistency in the classification process, enabling a systematic and rigorous assessment of the quality of the sources used in the research.

The author and an external researcher independently coded the sources to ensure reliability in the analysis. They discussed any ambiguous cases until reaching a consensus. Subsequently, the responses within each category were tallied, and percentages were calculated for each type of source. This approach not only aids in identifying the most trustworthy sources but also guarantees that the data collected is accurate and validated, which is essential for ensuring the reliability of the study's findings.

The data collected in the preparation file determined the frequency of publications per year to assess the timeliness of the information source used.

- *Decision-Making*

The instrument used to study this dimension was the open-ended, individual, written question asked to the students before and after the role-play about which alternative they preferred to be carried out. This question provides information about the position on the issue at both moments of the intervention, allowing us to obtain data about possible opinion changes.

The responses were categorized by the author and the external researcher based on the type of energy source mentioned. Each response was assigned to a specific energy source category, such as solar, wind, hydroelectric, etc., and frequencies and percentages for each category were calculated. Both the author and the researcher

Table 13.1 Example of a worksheet completed by a parliamentary group

Parliamentary group in favour of promoting wind energy as an alternative		
Complete the worksheet with five arguments to defend your role:		
Argument	Source/author	Publication date
1. Wind energy is renewable, clean and cheap. It adapts to any space and is quick and easy to install	https://www.antala.es/ventajas-y-disadvantages-wind-energy/ Antala (Industrial company)	13 September 2019
2. Wind energy is the most widely used renewable energy in most parts of Spain. In Andalusia, it accounts for 50%	https://institucional.us.es/catedratel/?p=2601 University of Seville	15 June 2015
3. Floating bases have already been created where wind turbines are placed in the ocean to take up less land. Also, because they are floating, they do not get in the way of the fish	https://elpais.com/clima-y-medio-ambiente/2021–12-10/espana-quiere-become-the-reference-of-energy-eolica-floating-in-europe.html El País (Spanish newspaper)	2021
4. Maintenance is almost zero, so only construction personnel are needed	https://montegar.es/ventajas-y-desventajas-of-wind-energy/ Montegar (Energy installation company)	2021
5. Bladeless wind turbines that are cheaper, quieter, and less dangerous for birds have been developed	https://www.elconfidencial.com/medioambiente/energia/2021–03–25/energy-wind-energy-environmental-impact-biodiversity_3000759/ El confidencial (Spanish newspaper)	2021

Table 13.2 Categories used and reliability

Source reliability	Categories
High	– National or international institution – Expert in the field
Medium	– Private entity – Spanish general press – Magazine/web of divulgation – Energy comparator
Low	– Personal opinion – Search engine

reviewed and resolved any ambiguous cases through discussion until a consensus was reached. Potential changes in decision-making were represented using a Sankey diagram.

13.5 Results

The findings obtained in the role-play activity for each CT dimension analysed are shown below.

13.5.1 Results of the Critical Analysis of the Information

Each group of students used a variety of sources to search for information on the Internet, indicating a balanced approach and a desire to obtain information from different resources (Table 13.3).

The most consulted sources for constructing arguments were the websites of private companies in the energy sector (21.8%). This is significant, as group of students must develop a critical vision to understand that an entity with economic interests may not publish reliable content or show only part of the information. The Spanish general press and national or international institutions are in second and third place with an outstanding percentage (18.3% each). The latter case suggests that students rely on academic sources for reliable information. Other sources used were popular magazines (15.0%) and consultation with experts in the field (8.3%). In addition, they show interest in energy comparators (8.3%). Although common search engines are used (8.3%), they have a relatively lower proportion than other sources. In addition, some group of students rely on their knowledge, experience, or personal opinion when searching for information online, which is reflected in 1.7%.

26.6% of the sources used were of high reliability, with national or international institutions standing out as reliable sources, especially if they are governmental, academic or internationally renowned organisations or experts in the field backed by a solid educational background or experience.

Table 13.3 Categorisation of sources consulted by students

Source type	Description	Examples	Reliability	Percentage (%)
Private entity	Organisation or company with economic interests in the energy sector	– Endesa – Montegar – Atlanta	Medium	21.8
National or international institution	Organisation or entity with functions intended for society	– University of Cantabria – Government of Mexico – Association of Promoters and Producers of Renewable Energies of Andalusia	High	18.3
Spanish general press	Media with coverage on various topics of interest	– El Mundo – BBC – El País	Medium	18.3
Magazine/ website of divulgation	Platform or channel for the dissemination of science-related content	– PQ Magazine – National Geographic	Medium	15.0
Energy comparator	Online tool for the comparison and evaluation of energy offers or companies	– PriceGas – Housage	Medium	8.3
Search engine	Online tool for searching for information on the Internet	– Google Scholar – Google	Low	8.3
Expert in the field	Expert in an area related to the energy or sustainability sector	– Florencia Quiroga (Sustainability expert) – Enrique Arriols (Ecology and Environment Specialist)	High	8.3
Other		– The student himself	Low	1.7

63.4% of the sources found were of medium reliability. Private entities and energy comparators stand out among these sources, as their commercial interests may influence objectivity. Also, the general press, magazines, and popular websites may be affected by time constraints, sensationalist approaches and biases, as trained professionals in the field must often produce the information.

In 10.0% of cases, low-reliability sources were used. These included a student's personal opinion, which, in terms of reliability, is not commonly perceived as an objective source backed by experience or expert knowledge. Search engine results,

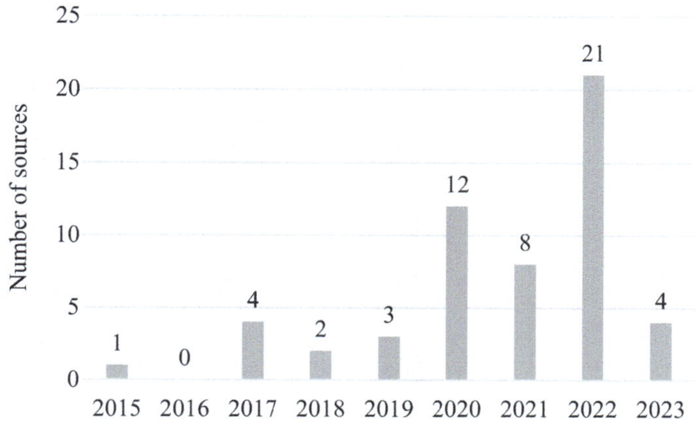

Fig. 13.2 Number of sources consulted by students by year

which are not sources of information but tools for accessing information, were also considered. The reliability of such results will depend on the sources to which they lead.

Regarding the timeliness of the sources used, it was observed that each group of students predominantly relied on very recent sources, with 45/55 (81.8%) being from 2020 to 2023, as illustrated in Fig. 13.2. However, it is relevant to note that the publication date was not provided in 5/60 searches performed (8%).

13.5.2 Decision-Making Results

Before the role-play, most students (59.68%) opted for solar energy instead of buying Russian gas. 24.19% preferred wind energy, and 11.29% favored purchasing gas from other countries. Students indicated no additional energy source as an alternative, nor the option for our country to produce more energy that would allow self-sufficiency without needing energy. 4.84% of the students gave inadequate answers, indicating they had never considered where the energy we consume at home or school comes from.

After the role-play, the three initial alternatives were maintained, albeit with some percentage variations and new energy options emerged. Decisions oriented towards solar energy (38.71%) and the purchase of gas (6.45%) experienced notable reductions, while those focused on wind energy showed a slight increase (27.42%). In addition, decisions based on other alternatives, such as tidal (9.68%), hydro (6.13%) or geothermal (1.61%), were observed in a smaller minority. Inadequate responses were not found after the role-play.

The Sankey diagram (Fig. 13.3) shows the percentages of decisions made before and after the role-play and the observed opinion changes.

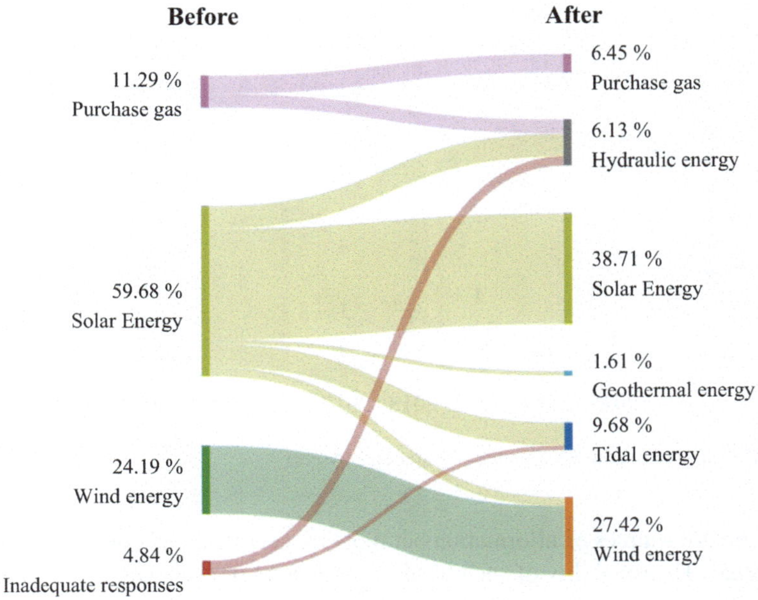

Fig. 13.3 Sankey diagram with decision-making before and after the role play

13.6 Discussion and Educational Implications

Several reasons regarding the energy problem show that the students critically analysed the information. On the one hand, the diversity of sources used reflects their research skills and their willingness to obtain information from different perspectives. On the other hand, students preferred medium-quality sources, especially those from private entities. Although they also made use, to a lesser extent, of high-quality sources, such as national and international institutions, there needed to be more consultation of experts in the field. Finally, the preference for topical sources over the last four years reflects their recognition of the importance of up-to-date data in their work. These findings align with research such as Lin and Tsai (2008) or Kim and Sin (2011).

From the ENCIC-CT model, these results highlight the importance of fostering critical information analysis skills among eighth-grade secondary school students (Layer 2). They must understand the importance of valuing information from collective decisions rather than from individuals, prioritising those with specialised expertise. Furthermore, the importance of recognising the independence of the message and paying special attention to the ability to distinguish between the information itself and potential advertising messages or biases is emphasized (Fornás, 2003), using information from various sources and the ability to discern between reliable and unreliable sources (Howell & Brossard, 2021). On the other hand, some searches identify a minor challenge in the need for publication dates, highlighting the need to emphasise the importance of verifying and citing publication dates in future research.

Regarding the decisions adopted in the problem, although most students preferred wind energy both before and after the activity, it is perceived that the role-play allowed them to reconsider the issue (López-Fernández et al., 2021). Some evidence was the emergence of new energy alternatives (geothermal, hydro or tidal) after the role-play, as well as the emergence of changes of opinion coinciding with Simonneaux (2001). These findings reveal that the critical analysis of information conducted during the experience was valuable for developing the scientific practice of argumentation (Layer 3 of the ENCIC-CT model), using appropriate arguments and counterarguments in the role-playing activity (Layer 4). This approach enabled students to become familiar with new energy sources and their potential. This finding aligns with the observations of López-Fernández et al. (2021) and Simonneaux (2001), who emphasize that role-playing can engage students in exploring and learning about issues relevant to the contexts discussed, while also encouraging them to reassess their decisions based on critical information. The observed changes reflect a more significant consideration of sustainability (Pirani et al., 2014; Thumann, 2006) of the transition to renewable energy sources (De Vries et al., 2007) or energy efficiency (International Energy Agency, 2014).

Regarding educational implications, the proposals put forward in the classroom must encourage secondary school students to recognise the importance of verifying the credibility of their sources of information. In addition, diversification in the consultation of sources should be promoted to obtain a complete view of the alternative perspectives that may arise in the face of complex problems, considering the context of a world saturated by Internet content (Kilag et al., 2024; Kim & Sin, 2011; Lin & Tsai, 2008).

On the other hand, some students' reliance on their own experience, knowledge or personal opinions when seeking information can be an opportunity to foster autonomy and informed decision-making. Educators can guide these students to use their expertise effectively while reminding them of the importance of updating and verifying information (Kim & Sin, 2011).

Future work aims to analyse how students argue their decision-making, what quality the different elements of their arguments have during the role-play, what knowledge they acquire after the activity or how they communicate their positions.

Acknowledgements This work is part of the Spanish National Plan R+D+i Project entitled "Citizens with critical thinking: A challenge for teachers in science education", reference PID2019-105765GA-I00, funded by MCIN/AEI/10.13039/501100011033.

References

Blanco-López, A., España-Ramos, E., González-García, F. J., & Franco-Mariscal, A. J. (2015). Key aspects of scientific competence for citizenship: A Delphi study of the expert community in Spain. *Journal of Research in Science Teaching, 52*(2), 164–198. https://doi.org/10.1002/tea.21188

Craciun, D. (2010). Role-playing as a creative method in science education. *Journal Science and Arts, 10*(1), 175–182.

Cruz-Lorite, I. M., Cebrián-Robles, D., Acebal, M. C., & Blanco-López, Á. (2023). Preservice primary teachers' positions on nuclear power before and after a role play. *Journal of Turkish Science Education, 20*(3), 379–398. https://doi.org/10.36681/tused.2023.022

De Vries, B. J., Van Vuuren, D. P., & Hoogwijk, M. M. (2007). Renewable energy sources: Their global potential for the first-half of the 21st century at a global level: An integrated approach. *Energy Policy, 35*(4), 2590–2610. https://doi.org/10.1016/j.enpol.2006.09.002

Dickel, R., Hassanzadeh, E., Henderson, J., Honoré, A., El-Katiri, L., Pirani, S., Rogers, H., Stern, J., & Yafimav, K. (2014). *Reducing European dependence on Russian gas: Distinguishing natural gas security from geopolitics*. Oxford Institute for Energy Studies.

European Commission. (2023). Report from the Commission to the European Parliament, the Council, the European Economic and Social Committee and the Committee of the regions. State of the Energy Union Report 2023. Available in: https://energy.ec.europa.eu/topics/energy-strategy/energy-union/eighth-report-state-energy-union_en

Fornás, C. (2003). Criterios para evaluar la calidad y fiabilidad de los contenidos en Internet [Criteria for assessing the quality and reliability of online content]. *Revista Española de Documentación Científica, 26*(1), 75–80.

Franco-Mariscal, A. J., Cano-Iglesias, M. J., España-Ramos, E., & Blanco-López, A. (2025). The ENCIC-CT model for the development of critical thinking. In A. J. Franco-Mariscal (Ed.), *Critical thinking in science education and teacher training* (Chap. 1) (pp. 3–42). Springer. https://doi.org/10.1007/978-3-031-78578-8_1

Franco-Mariscal, A. J., Hierrezuelo-Osorio, J. M., Cano-Iglesias, M. J., & Blanco-López, A. (2023). (Coords.). *El juego de rol como estrategia para desarrollar habilidades de pensamiento crítico. Aplicado al aula de las ciencias* [Role-playing game as a strategy to develop critical thinking skills. Applied in the science classroom]. Pirámide.

Franco-Mariscal, A. J., España-Ramos, E., & Blanco-López, A. (2014). Uso de Internet para analizar las relaciones entre drogas y salud bucodental. Una experiencia en 4º de ESO [Using the Internet to analyze the relationships between drugs and oral health: An experience in 4th year of secondary education]. In E. Soriano, A. J. González, & V. C. Cala (Eds.), *Retos actuales de educación y salud transcultural* [1] [Current challenges in transcultural education and health] (pp. 54.1–54.8). University of Almería.

Howell, E. L., & Brossard, D. (2021). (Mis)informed about what? What it means to be a science-literate citizen in a digital world. *Proceedings of the National Academy of Sciences, 118*(15), e1912436117. https://doi.org/10.1073/pnas.1912436117

International Energy Agency (IEA). (2014). *Energy efficiency indicators: Fundamentals on statistics*. OECD/IEA.

Kilag, O. K., Lisao, C., Lastimoso, J., Villa, F. L., & Miñoza, C. A. (2024). Bildung-oriented science education: A critical review of different visions of scientific literacy. *International Multidisciplinary Journal of Research for Innovation, Sustainability and Excellence, 1*(2), 313–320. https://risejournals.org/index.php/imjrise/article/view/349

Kim, K. S., & Sin, S. C. J. (2011). Selecting quality sources: Bridging the gap between the perception and use of information sources. *Journal of Information Science, 37*(2), 178–188. https://doi.org/10.1177/0165551511400958

Lin, C. C., & Tsai, C. C. (2008). Exploring the structural relationships between high school students' scientific epistemological views and their utilization of information commitments toward online science information. *International Journal of Science Education, 30*(15), 2001–2022. https://doi.org/10.1080/09500690701613733

López-Fernández, M. M., González-García, F., & Franco-Mariscal, A. J. (2022). How can socio-scientific issues help develop critical thinking in chemistry education? A reflection on the problem of plastics. *Journal of Chemical Education, 99*(10), 3435–3442. https://doi.org/10.1021/acs.jchemed.2c00223

López-Fernández, M. M., González-García, F., & Franco-Mariscal, A. J. (2021). Should we ban single-use plastics? A role-playing game to argue and make decisions in a Grade-8 school chemistry class. *Journal of Chemical Education, 98*(12), 3947–3956. https://doi.org/10.1021/acs.jchemed.1c00580

Martínez, J. D. (2018). Motorized mobility, environmental impact, alternatives and future prospects: Considerations for the Área Metropolitana del Valle de Aburrá. *Revista de Salud Pública, 20*(1), 126. https://doi.org/10.15446/rsap.v20n1.57038

Mao, W., Cui, Y., Chiu, M. M., & Lei, H. (2022). Effects of game-based learning on students' critical thinking: A meta-analysis. *Journal of Educational Computing Research, 59*(8), 1682–1708. https://doi.org/10.1177/07356331211007098

Mbah, R. E., & Wasum, D. F. (2022). Russian-Ukraine 2022 war: A review of the economic impact of Russian-Ukraine crisis on the USA, UK, Canada, and Europe. *Advances in Social Sciences Research Journal, 9*(3), 144–153. https://doi.org/10.14738/assrj.93.12005

Ministerio de Educación y Formación Profesional. (2022). Real Decreto 217/2022, de 29 de marzo, por el que se establece la ordenación y las enseñanzas mínimas en la Educación Secundaria Obligatoria [Royal Decree 217/2022, of March 29, establishing the organization and minimum teachings for Compulsory Secondary Education]. *Boletín Oficial del Estado*, núm. 76, 30 de marzo.

Pereira, P., Bašić, F., Bogunovic, I., & Barcelo, D. (2022). Russian-Ukrainian war impacts the total environment. *Science of the Total Environment, 837*, 155865. https://doi.org/10.1016/j.scitotenv.2022.155865

Pérez-Gómez, A. I. (2008). ¿Competencias o pensamiento práctico? La construcción de los significados de representación y de acción [Competencies or practical thinking? The construction of the meanings of representation and action]. In J. Gimeno (Ed.), *Educar por competencias, ¿qué hay de nuevo?* [Educating through competencies, what's new?]. Morata.

Perrotta, C., Featherstone, G., Aston, H., & Houghton, E. (2013). *Game-based learning: Latest evidence and future directions* (NFER Research Programme: Innovation in Education). NFER.

Pirani, S., Henderson, J., Honoré, A., Rogers, H., & Yafimava, K. (2014). *What the Ukrainian crisis means for gas markets.* Oxford Institute for Energy Studies.

Sagmeister, K. J., Schinagl, C. W., Kapelari, S., & Vrabl, P. (2021). Students' experiences of working with a socio-scientific issues-based curriculum unit using role-playing to negotiate antibiotic resistance. *Frontiers in Microbiology, 11*, 577501. https://doi.org/10.3389/fmicb.2020.577501

Simonneaux, L. (2001). Role-play or debate to promote students' argumentation and justification on an issue in animal transgenesis. *International Journal of Science Education, 23*(9), 903–927. https://doi.org/10.1080/09500690010016076

Thumann, M. (2006). Diversification des sources—la meilleure stratégie pour les relations énergétiques UE-Russie. *Russie Nei Visions* (ifri), *10*.

Zeng, J., Parks, S., & Shang, J. (2020). To learn scientifically, effectively, and enjoyably: A review of educational games. *Human Behavior and Emerging Technologies, 2*(2), 186–195. https://doi.org/10.1002/hbe2.188

Chapter 14
Promoting Argumentation with the Zapworks Augmented Reality Tool. Designs of Educational Proposals by Preservice Science Teachers

Noelia Margarita Moreno-Martínez and Antonio Joaquín Franco-Mariscal

14.1 Introduction

It is essential to rethink science education and adopt innovative methodologies to address the current and future demands of a constantly evolving knowledge society and adapt to new learning styles in the digital age. The inclusion of emerging technologies such as augmented reality can foster effective teaching–learning scenarios oriented, on the one hand, towards the development of favourable student attitudes towards science from a more interactive and playful approach (Chen & Liu, 2020; Jiménez, 2019; Martínez et al., 2017; Urzúa, 2021), and, on the other hand, towards educational paradigms based on a connectivist networked learning approach (Downes, 2008; Siemens, 2004) and from the methods of discovery learning and project-based learning (Moursund, 1999; Thomas, 2000), problem-based learning (Graaff & Kolmos, 2003) and game-based learning based on a gamified methodological procedure (Kapp, 2012; Lee & Hammer, 2011; Li et al., 2013; Sailer et al., 2013). These approaches will enable the construction of knowledge from the perspective of critical thinking (hereinafter, CT) (Faridi et al., 2021; Syawaludin et al., 2019).

Augmented reality technology allows the direct or indirect visualisation of real-world elements combined (or augmented) with computer-generated virtual elements, the fusion of which gives rise to a mixed reality (Cobo & Moravec, 2011). Azuma (1997) conceives it as a technology that combines real and virtual elements to create

N. M. Moreno-Martínez (✉)
Personality, Psychological Assessment and Treatment, Universidad de Málaga, Málaga, Spain
e-mail: nmarg@uma.es

A. J. Franco-Mariscal
Science Education, Universidad de Málaga, Málaga, Spain
e-mail: anjoa@uma.es

A. J. Franco-Mariscal (ed.), *Critical Thinking in Science Education and Teacher Training*, Contemporary Trends and Issues in Science Education 64,
https://doi.org/10.1007/978-3-031-78578-8_14

real-time interactive scenarios and record them in 3D. De Pedro and Méndez (2012) understand augmented reality as a technology capable of complementing perception and interaction with the real world, providing the user with a real scenario augmented with additional computer-generated information. This way, physical reality is combined with virtual elements, providing a mixed reality in real-time. Kato (2010) indicates that augmented reality can be defined as virtual objects or annotations that can be superimposed on the real world as if they exist, e.g. it maintains the real world by supplementing it with virtual information. In short, augmented reality is defined as an environment where virtual and real integration occurs (Cabero et al., 2016; Moreno-Martínez, 2020). Therefore, this technology transferred to science education makes possible an innovative methodology with a great didactic, formative and inclusive potential with a versatile, open, interactive and dynamic character to favour the teaching–learning processes of science contents (Arici et al., 2019; Cheng & Tsai, 2013; Irwanto et al., 2022; Nielsen et al., 2016; Sahin & Yilmaz, 2020).

As outstanding properties, augmented reality tools applied to science education scenarios allow for complementing, amplifying, reinforcing and enriching the information provided by the real physical space of the classroom, making scientific content more understandable and accessible and with which the student can interact (Abdinejad et al., 2021; Eriksen et al., 2020; Probst et al., 2021; Schmid et al., 2020; Wong et al., 2021). Furthermore, through the use of apps or mobile devices, augmented reality allows for overcoming spatio-temporal limitations in the science classroom, facilitating the construction of knowledge individually and collectively, and promoting the development of CT and reflective thinking.

In order to ensure effective integration of this technology in the classroom, it is important to provide preservice science teachers with comprehensive training ranging from the design of proposals to the practical application of augmented reality in science education. It will allow them to experience and fully understand the educational opportunities offered by augmented reality, facilitating its implementation in their future teaching practice. This content creation can be promoted through collaborative approaches based on the Makerspaces philosophy (Moreno-Martínez & Morales, 2021; Vuorikari et al., 2019).

14.2 Augmented Reality with Zapworks

In recent years, there has been a proliferation of augmented reality tools useful as teaching strategies, a recent tool being Zapworks (https://zap.works). The Zapworks web platform is a tool based on augmented reality technology that allows the creation of amplified scenarios from images that act as markers to generate layers of virtual multimedia information consisting of images, texts, videos, link buttons between augmented reality scenes created on the platform, direct access buttons to external websites, etc. All this information can be visualised after scanning the marker with the Zappar mobile application. The Zapworks web platform allows, for instance, to amplify the information of images related to science textbooks or any science-related image from royalty-free repositories.

Zapworks offers a wide array of educational benefits for science education. Firstly, this tool enhances interactive and dynamic learning by allowing students to engage with various types of multimedia content, including 3D elements and simulations. This variety facilitates a deeper understanding of concepts, particularly with 3D elements that help visualize abstract ideas such as molecular structures, human organs, or physical phenomena. Additionally, simulations provide opportunities to experience scenarios that would be challenging or impossible to replicate in a traditional classroom setting.

Furthermore, the platform enables the creation of gamified experiences, making learning more engaging and motivating. A standout feature of Zapworks is its ability to inspire creativity among preservice teachers by allowing them to design personalized content for their students. This transforms each activity into a unique project that can be tailored to the interests and knowledge levels of the students, with the added advantage of easily updating the content. In terms of accessibility, Zapworks excels due to its user-friendly interface, which requires no advanced technical or programming skills. It is also compatible with multiple platforms, enabling users to access their projects from any device. Furthermore, Zapworks fosters collaboration and teamwork by facilitating the creation of joint projects and allowing for easy sharing once the projects are completed.

In summary, augmented reality tools such as Zapworks allow the design of scenarios in which the student can develop some of the dimensions of CT described in the ENCIC-CT model (Chap. 1) (Franco-Mariscal et al., 2025). The development of CT in science education is a key factor in the formation of citizens and should be understood as a useful tool to understand and explain what is happening around us (Albertos & De la Herrán, 2018; Moreno-Martínez & Franco-Mariscal, 2022; Rumpagaporn et al., 2007).

Moreover, augmented reality is an innovative methodological alternative with great potential in science education and, therefore, deserves to be explored in science teacher training as a preliminary step to its implementation in classrooms at various educational levels. In this context, this chapter provides instructions for preservice teachers to demonstrate the potential of the augmented reality tool ZapWorks and to promote the design of proposals focused on developing argumentation as a CT skill.

14.3 Research Questions

This study poses the following research questions (RQ):

- RQ1. How do preservice science teachers address the technical aspects of augmented reality scenarios when designing educational proposals with Zapworks, particularly concerning the elements they consider most appropriate?

- RQ2. To what extent can preservice science teachers design educational proposals for secondary school students by applying Zapworks that include activities to develop argumentation?

14.4 Method

14.4.1 Participants and Context

The sample of this study consisted of 57 preservice science teachers of the subject Teaching Innovation and Initiation to Educational Research of the Master's Degree in Secondary Education Teaching of Physics and Chemistry (38.60%; N = 22) and Biology and Geology (61.40%; N = 35) of the University of Malaga (Spain) of the academic year 2021–2022. 64.90% (N = 37) of the participants were women and 35.10% (N = 20) were men, all of them aged between 21 and 47 years, where the highest frequencies were found in 22 years (19.30%; N = 11), 24 years (19.30%; N = 11) and 23 years (15.80%; N = 9).

Most preservice teachers (89.50%; N = 51) were unaware of and had not used augmented reality technology before this experience. Of the 15.90% (N = 6) who knew about it, only 5.30% (N = 3) had used it, but most could not indicate specific tools. 89.50% (N = 51) considered teacher training in the didactic use of augmented reality tools in science education entirely or significant. Only 10.60% (N = 6) rated it as somewhat or not very important (Moreno-Martínez & Franco-Mariscal, 2023).

The two authors of this chapter carried out the design and classroom implementation of the experience. The first author is an expert in emerging technologies, and the second author is a professor and researcher in science education with 20 years of experience.

14.4.2 Experience Description

The experience was developed in two sessions of two hours each. In the theoretical-practical session, the educational possibilities of augmented reality technology and the Zapworks tool were presented through examples of the creation of augmented reality scenarios with the Zapworks web platform and the Zappar mobile app. The preservice teachers had the opportunity to carry out a practical activity in the classroom simultaneously with the professor (first author).

After the session, the preservice teachers, in pairs, had to design and develop a proposal with Zapworks for secondary school students. This proposal had to focus on one of the following subjects: physics, chemistry, biology or geology, and include activities that encouraged argumentation as a CT skill based on everyday situations or problems. The design had to be submitted in a written report, which they had one week to complete. The second session was devoted to the oral presentation of the designs, focusing on their contributions to the argumentation development.

Participants received instruction on argumentation prior to this experience to prepare them for designing argumentation activities. The professor conducted a training session that explained how various authors understand argumentation, highlighting its significance in everyday life, science education, and its contribution to CT. Toulmin's (1958) model of argumentation, as well as Osborne et al.'s (2016) model, were presented. The first task involved identifying the essential elements of an argument (evidence, warrant, and conclusion) through a simple example: observing animal tracks in the snow and using them as evidence to answer the question, *Who is walking in the snow* (Cebrián-Robles et al., 2021). The second task required participants to argue whether a young redhead with fair skin and freckles, shown in an image, was likely to get sunburned (Cebrián-Robles et al., 2021). This exercise aimed to underscore the importance of refuting arguments and understanding scientific preconceptions. Additionally, various dimensions of CT were introduced, with a particular focus on the ENCIC-CT model (Franco-Mariscal et al., 2025).

14.4.3 ENCIC-CT Model Application

In the context of the ENCIC-CT model, Zapworks as a teaching strategy (Layer 4) enables preservice teachers to design scenarios in any scientific discipline focused on socioscientific or everyday problems (Layer 1).

Additionally, it facilitates the development of CT dimensions across the three domains outlined in the model (Layer 2). The knowledge domain of CT is present in the scientific knowledge that is integrated into the designs and is associated with everyday problems, as well as in the vision of science that is presented. Likewise, the designs must offer opportunities for students to develop skills such as critical analysis of information, argumentation, decision-making or comprehensive analysis of the problem. These skills are promoted thanks to augmented reality, which allows the integration of layers of digital information and three-dimensional models relevant to science education. These layers can contain texts, images, videos, 3D models or simulations that question different types of information, thus promoting the students' critical analysis of the data. Finally, within the domain of dispositions, tasks utilizing augmented reality should promote metacognition. This research will focus exclusively on the skill of argumentation.

The integration of scientific argumentation practice (Layer 3) into the design is essential for developing CT and can be included in various ways. One option is to raise questions around the issues presented to encourage the development of sound arguments and evidence-based decision-making. Another alternative is to provide interactive simulations or 3D models that allow students to explore science experientially. By interacting with these simulations or models, students can collect data or formulate hypotheses, which can then support their arguments about the scientific phenomena studied. In all cases, when a real-life problem is addressed through augmented reality, the materials included in the proposal should aim to present the issue comprehensively from all perspectives.

14.4.4 Data Collection and Analysis

The instrument used for data collection was the submitted report, which had to include an explanation of the scenarios developed with Zapworks and screenshots of them, the science content addressed, the proposed activities and how they contribute to scientific argumentation, and a critical reflection on the educational possibilities of Zapworks in relation to CT in science education. The designs were created by pairs of preservice teachers and, in some cases, individually, resulting in 30 reports.

The elements included in the designed scenarios were analyzed to address RQ1. For each design, the presence or absence of texts, images, videos, 3D elements, or simulations was examined.

In order to address RQ2, given that the preservice teachers' proposals have not been implemented, the analysis examines how suitable the activities are for promoting argumentation. The researchers reviewed each activity in the designs, assigning one of the verbs from the revised Bloom's taxonomy (Anderson & Krathwohl, 2001) to each. These verbs represent specific actions that describe the cognitive skills and mental processes students need to develop to master a particular area of knowledge. Each researcher carried out this analysis individually, and any disagreements were resolved through consensus. The analysis identified 13 actions: identify, describe, summarize, observe/visualize, list/complete, draw, search for information, construct models, debate, explain, make a judgment, interpret/justify, and compare. These actions were then categorized into two groups based on their relation to argumentation. This classification was based on the understanding that actions related to argumentation involve more complex cognitive processes, corresponding to the "evaluating" level in the revised Bloom's taxonomy. Consequently, argumentation activities were categorized under actions such as debating, explaining, making a judgment, interpreting/justifying, and comparing. It was found that actions not related to argumentation corresponded to the initial levels of Bloom's revised taxonomy, specifically remembering, understanding, or applying. Frequencies and percentages were calculated for each action category, as well as the average number of total activities, the average number of argumentation activities, and the percentage of argumentation activities included in each design.

On the other hand, a thorough analysis was conducted to determine whether the critical reflection on the educational possibilities of Zapworks in relation to CT explicitly referenced argumentation in the proposed activities. This analysis involved calculating the frequencies and percentages of reports that mentioned argumentation in their reflections. Additionally, it examined how these statements correlated with the actual integration of CT in the designed activities.

14.5 Results

14.5.1 Technical Aspects of Augmented Reality Scenarios with Zapworks

The designs proposed by the preservice teachers were very varied and covered all four disciplines. Similarly, the elements used to create the scenarios included texts, images, videos, 3D elements, and simulations, though in varying proportions. While all the designs incorporated texts, images, and videos, only 26.6% of them included 3D elements (6/30 reports) and simulations (2/30). The most frequently used 3D elements depicted parts of the human body, such as the brain, digestive system, or reproductive system. The simulations employed were from the PhET Colorado educational project, developed by the University of Colorado Boulder (2024), and were used to model atoms or molecules.

A representative proposal from each subject has been chosen to illustrate the technical aspects of the developed scenarios and some activities that promote argumentation.

14.5.1.1 Zapworks and Physics

This design aims for students to be able to argue about different everyday situations that occur through state changes. To this end, the proposal provides information on the different state changes through examples in videos or images.

The left image in Fig. 14.1 illustrates the scenario background, showing the six changes in the state of matter from which the rest of the scenes are developed. This background image acts as a marker to generate the augmented reality scenario. It comprises an additional digital information layer containing multimedia elements such as images, videos, and interactive buttons, which give access to resources with different activities and an evaluation questionnaire. Each scene associated with a change of state comprises an image and a video (Fig. 14.1, right image).

The knowledge acquired through Zapworks should enable students to argue about different questions related to everyday problems, such as what happens when you leave a bottle of water in the freezer for a long time and then take it out, why is it important in some regions of the world to boil water before drinking it, or why does dew form in the morning, etc. The goal is not only for students to provide the name of the change of state involved when answering questions, but also to develop more elaborative and argumentative responses.

14.5.1.2 Zapworks and Chemistry

This design incorporates chemical elements found in common objects and materials and explores their applications in daily life (Franco-Mariscal et al., 2018). The

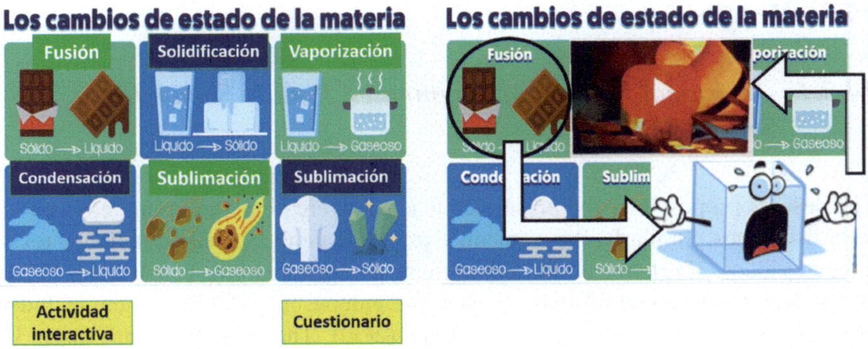

Fig. 14.1 Scenario background of the proposal on state changes with the Zapworks application (left image) and amplified scenario (right image) [*Key of terms* Melting (fusión), solidification (solidificación), vaporisation (vaporización), condensation (condensación), sublimation (sublimación)]

augmented reality scenario consists of a periodic table image as the background (Fig. 14.2). Through the Zapworks platform, this image is enriched with multimedia content that includes representative images of everyday objects associated with the different chemical elements, videos illustrating the composition of these objects, text identifying the objects and their correspondence with the elements of the periodic table, as well as direct access buttons to 3D models representing these elements. These 3D models are available in the 3D object repository of the Sketchfab platform (https://sketchfab.com). Figure 14.2 shows the visualisation of the amplified augmented reality scenario, which combines the multimedia above elements to provide additional information once the background image of the periodic table is scanned with the Zappar mobile app.

As a proposed activity, using objects that illustrate the chemical element in everyday life, different questions are posed in which students must make a reasoned

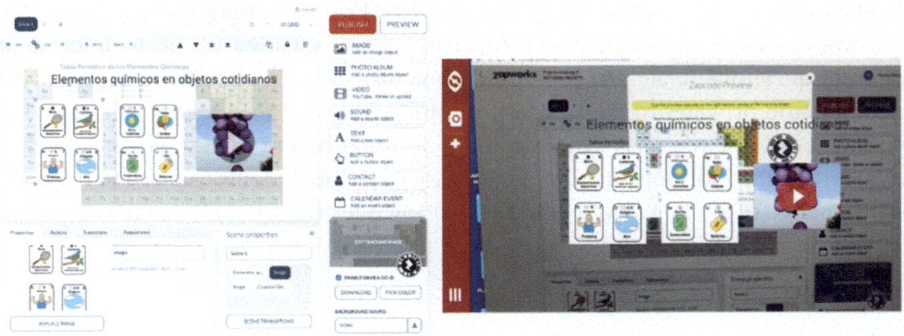

Fig. 14.2 Zapworks web platform (left image) and projection of the amplified scenario (right image)

decision. Thus, based on the image of a bicycle made of magnesium alloy, it is proposed to argue whether this material is the most suitable for its manufacture or whether steel or aluminium would offer more significant advantages. The students' arguments should provide evidence based on various aspects: physical, chemical, technological, environmental, economic, social, etc. (Bernal-Herrera et al., 2023). Another example was to argue about how excessive consumption of potassium-rich foods such as bananas affects health.

14.5.1.3 Zapworks and Biology

This proposal uses the promotion of healthy eating among young people as a daily life issue, as obesity rates are increasing in this age group (Chai et al., 2022). The scenario is based on Harvard T.H. Chan's proposal[1] for a healthy eating plate to guide the creation of healthy, balanced meals, whether served at the table or packed in a lunch box, which should include vegetables, whole grains, proteins, fruits, healthy fats and water. Each ingredient gives access to other scenarios with information in text format (questions), images or videos.

The main activity guiding the proposal is to elaborate on a healthy dish based on the student's diet and argue how this model fits into a healthy diet. To do this, they must use the information in the proposal and other sources regarding the characteristics of a healthy diet and the biomolecules present in each food. Although this is the main activity, the different scenarios also present other problems related to health and nutrition that the student must solve in a reasoned way, such as which is better to consume bottled or tap water (Fig. 14.3).

14.5.1.4 Zapworks and Geology

This design aims to raise awareness of the importance of rock formation in many aspects of daily life, such as its use as a raw material for construction (kitchen worktops, floors, etc.), minerals or resources such as oil extracted from rocks, etc.

The augmented reality scenario consists of a background image that includes the names of the different rocks (Fig. 14.4, top left image) and a button that gives access to a YouTube video explaining the rock types according to their origin. A click on each rock type (igneous, metamorphic or sedimentary) unlocks new scenarios presenting specific images and details about each rock variety (Fig. 14.4, top right image), as well as practical examples of their application in everyday objects (Fig. 14.4, bottom images). Different buttons allow you to switch from one scenario to another.

After working through all the information in the proposal, students should argue the most suitable materials for constructing the various components of a house, aiming to ensure its energy efficiency. Thus, they could argue the use of granite for the floor, being a strong and durable rock with a variety of colours or textures; slate

[1] https://www.hsph.harvard.edu/nutritionsource/healthy-eating-plate/.

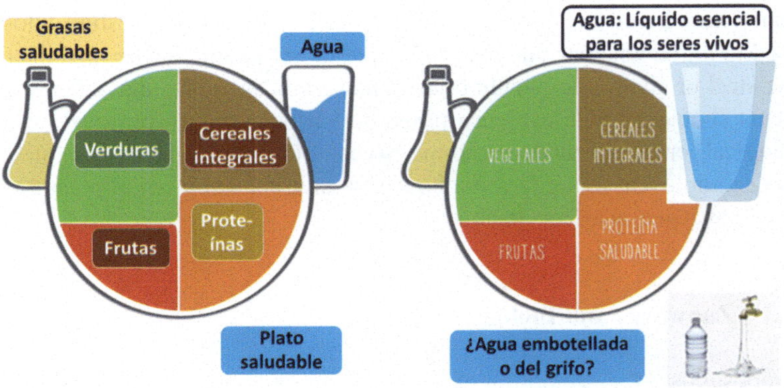

Fig. 14.3 Scenario background of the Zapworks healthy eating proposal (left image) and amplified scenario (right image) [*Key of terms* Left image: vegetables (verduras), whole grains (cereales integrales), fruits (frutas), proteins (proteínas), healthy fats (grasas saludables), water (agua), healthy dish (plato saludable). Right image: water: an essential liquid for living things (Agua: Líquido esencial para los seres vivos), bottled or tap water (¿agua embotellada o del grifo?)]

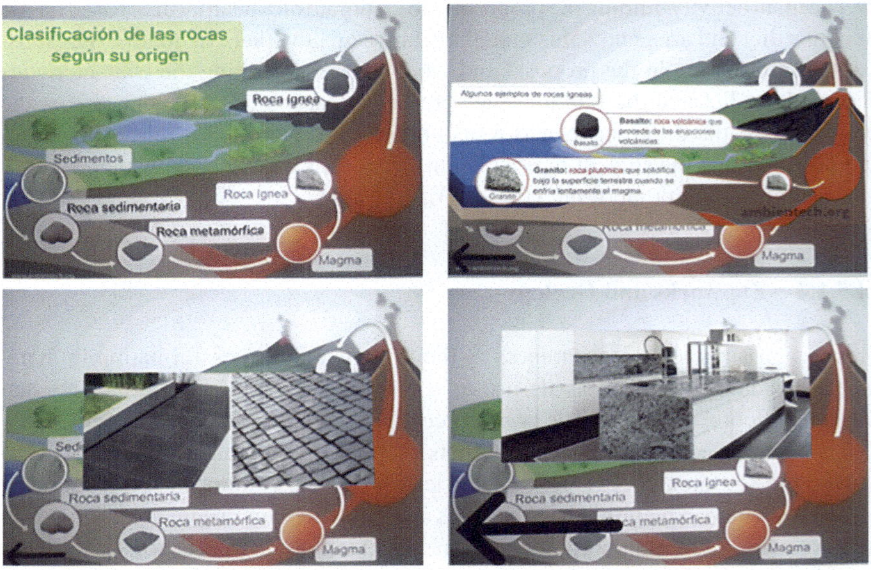

Fig. 14.4 Scenario in Zapworks on rock types [*Key of terms* Sedimentary rock (roca sedimentaria), igneous rock (roca ígnea), metamorphic rock (roca metamórfica), magma (magma)]

for the roof due to its durability and weather resistance; or limestone for the chimney, being an excellent thermal conductor that can absorb and retain heat efficiently, even after the fire has been extinguished, which can be beneficial for energy efficiency in the home.

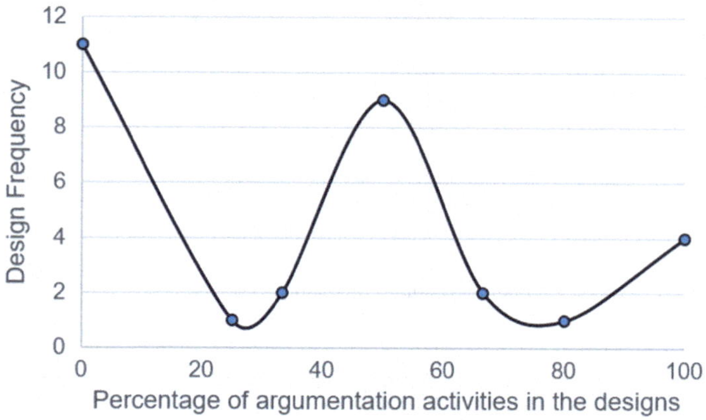

Fig. 14.5 Percentage of argumentation activities in the designs

14.5.2 Analysis of the Argumentation Activities in Designs

The designs featured an average of 2.4 activities, including 0.9 argumentation activities per design on average. Figure 14.5 illustrates the percentage distribution of the presence of argumentation activities across the designs. As shown, none of the proposed activities involved argumentation in 11/30 designs; half of the activities required argumentation in 9/30 designs; and in only 4/30 designs, all activities were related to argumentation.

Table 14.1 displays the frequency and percentage of each type of proposed activity, distinguishing between those related to argumentation and those that are not. Explanation and identification activities are notable as the most prevalent, respectively, for activities associated with and not associated with argumentation.

On the other hand, only 46.6% of the reports (14/30) explicitly stated that their designs included argumentation activities, even though this was a requirement of the task. The comparison between the reports that explicitly mentioned argumentation activities and the analysis of these activities is detailed in Table 14.2.

As seen, the vast majority of reports that claim to include argumentation activities do so only partially. Similarly, a significant number of reports that do not explicitly state this still include some argumentation activities.

14.6 Discussion and Educational Implications

Zapworks stands out as a highly valuable tool for enriching the teaching–learning process in science for several reasons. Firstly, this platform is versatile and adaptable, which makes it well-suited for designing educational experiences across the four scientific disciplines studied: physics, chemistry, biology and geology. However,

Table 14.1 Frequency and percentage of each type of activity in the designs

Cognitive skill	Activity example	Frequency	Percentage (%)
Argumentation activities			
Explain	Find a rock and explain its origin	11	15.1
Interpret/justify	Justify the state of the endometrium during ovulation using a menstrual cycle graph that represents phases versus days	7	9.6
Debate	Defend one cellular organelle against another in a debate	5	6.8
Compare	Compare different atomic models, outlining their differences and similarities	4	5.5
Make judgments	Identify the learning outcomes from the experience, their usefulness, and their relevance to daily life	1	1.4
Subtotal		28	38.4
No argumentation activities			
Identify	Identify the stages of mitosis in a diagram	14	19.2
Observe/visualize	Watch a video showing the different planets in the solar system	7	9.6
Describe	Describe the path of food through our body	6	8.2
Build models	Build molecular structures with clay	6	8.2
Search for information	Research information about different drugs and their effects on the brain	4	5.5
Summarize	Summarize the content of a video	3	4.1
List/complete	List the parts of the digestive system	3	4.1
Draw	Draw a eukaryotic cell	2	2.7
Subtotal		45	61.6
Total		73	100.0

while preservice teachers successfully created effective science teaching designs incorporating various elements such as texts, images, videos, 3D models, and simulations (RQ1), it is noteworthy that the limited use of 3D elements and simulations indicates a potential gap in understanding or familiarity with these technological tools.

These proposals encourage students to interact actively with the learning materials and contribute significantly to improving their understanding and application of scientific knowledge. Moreover, by connecting science topics to everyday life

Table 14.2 Comparison between the declaration of argumentation activities and their analysis

	Frequency	Percentage (%)
Designs that claim to include argumentation activities...		
But they actually aren't	4	13.3
But only 25% involve argumentation	1	3.3
But only 50–66.6% involve argumentation	8	26.7
And 100% involve argumentation	1	3.3
Designs that do not claim to include argumentation activities...		
And none involve argumentation	7	23.3
But 33.3% involve argumentation	2	6.7
But 50–80% involve argumentation	4	13.3
But 100% involve argumentation	3	10.1

situations or problems within the ENCIC-CT model, they provide valuable opportunities to develop scientific argumentation skills, which further enriches the learning experience (Abdinejad et al., 2021; Cortés et al., 2021; Schmid et al., 2020; Wong et al., 2021).

However, despite the potential of Zapworks, there are several aspects that require attention. The first concerns the difficulties encountered by preservice science teachers in integrating argumentation activities into their designs. Only 38.4% of the proposed activities required skills related to argumentation, such as explaining, interpreting/ justifying, debating, comparing, or making judgments, for their resolution.

Another issue to consider is that a significant number of these preservice teachers explicitly stated in their reports that the proposed activities involved argumentation. However, in reality, this was often not the case or occurred only partially, with activities more commonly involving actions such as identifying, remembering, or applying, according to the revised Bloom's taxonomy (Anderson & Krathwohl, 2001).

These findings are concerning, given that the participants in this study had received prior instruction in scientific argumentation, highlighting the challenge of designing activities that enable students to interpret data or integrate evidence into justifications. These are two well-documented obstacles in the literature on argumentation (Bernal-Herrera et al., 2023; Bravo & Jiménez-Aleixandre, 2013; Kanari & Millar, 2004). Although the instruction provided only included activities related to identifying the elements of an argument rather than designing such activities, the findings suggest a need to incorporate this aspect into argumentation training. Additionally, analyzing activities in terms of the cognitive actions they require from students should be included as a measure to verify the effective use of argumentation. In this regard, the verbs from the revised Bloom's taxonomy could assist preservice teachers in both analyzing and designing future activities by helping to identify the cognitive skills and mental processes involved in argumentation. Furthermore, developing a list of potential evidence and warrants that students might use in their arguments could also help ensure that activities effectively involve argumentation.

Hence, the findings from this experience allow us to address RQ2, indicating that preservice science teachers can effectively design educational proposals for secondary school students using Zapworks. However, they need support to incorporate activities that enhance scientific argumentation.

On the other hand, it is important to highlight that for augmented reality technology to be effective in schools, it is necessary to raise awareness and train teachers in its proper use. It implies providing technical and practical training on these tools and applications and offering a didactic perspective so that teachers understand their educational potential and can effectively integrate them into their practices. Prior training of preservice teachers in digital competences, which is increasing, favours the use of these technologies. Moreover, the effective integration of technologies in the classroom should not be limited to providing technological equipment in schools but should be seen as a fundamental educational innovation. This innovation should offer new methodological opportunities that facilitate the task of the agents involved in the educational process (Cebrián, 2011).

Ultimately, in the same discursive line as Ibáñez et al. (2011), we consider that to appreciate the potential of disruptive technologies for learning thoroughly, we must go beyond the individual use of the devices and evaluate their integrated use in practice or the learning experience. Hence, it is important to raise awareness and train teachers to implement innovative methodologies to adjust content and resources to develop the competences required in the digital era in which students become active participants in the construction of knowledge at the service of their integral development.

Acknowledgements This work is part of the Spanish National Plan R+D+i Project entitled "Citizens with critical thinking: A challenge for teachers in science education", reference PID2019-105765GA-I00, funded by MCIN/AEI/10.13039/501100011033.

References

Abdinejad, M., Ferrag, C. Qorbani, H. S., & Dalili, Sh. (2021). Developing a simple and cost-effective markerless augmented reality tool for chemistry education. *Journal of Chemical Education, 98*(5), 1783–1788. https://doi.org/10.1021/acs.jchemed.1c00173

Albertos, D., & De la Herrán, A. (2018). Desarrollo del pensamiento crítico en estudiantes de Educación Secundaria: Diseño, aplicación y evaluación de un programa educativo [Development of critical thinking in secondary education students: Design, implementation, and evaluation of an educational program]. *Profesorado, Revista de Currículum y Formación del Profesorado, 22*(4), 269–285. https://doi.org/10.30827/profesorado.v22i4.8416

Anderson, L. W., & Krathwohl, D. R. (Eds.). (2001). *A taxonomy for learning, teaching, and assessing: A revision of a bloom's taxonomy of educational objectives.* Longman.

Arici, F., Yildirim, P., Caliklar, S., & Yilmaz, R. M. (2019). Research trends in the use of augmented reality in science education: Content and bibliometric mapping analysis. *Computers and Education, 142*, 103647. https://doi.org/10.1016/j.compedu.2019.103647

Azuma, R. (1997). A survey of augmented reality. *Presence: Teleoperators and Virtual Environments, 6*(4), 355–385.

Bernal-Herrera, P., Cano-Iglesias, M. J., Franco-Mariscal, A. J., & Blanco-López, A. (2023). Impacto de un debate sociocientífico en las habilidades argumentativas y en la toma de decisiones del profesorado de secundaria en formación inicial [Impact of a socio-scientific debate on the argumentative and decision-making skills of preservice secondary teachers.] *Enseñanza de las Ciencias, 41*(3), 113–132. https://doi.org/10.5565/rev/ensciencias.5861

Bravo, B., & Jiménez-Aleixandre, M. P. (2013). ¿Criaríamos leones en granjas? Uso de pruebas y conocimiento conceptual en un problema de acuicultura [Would we raise lions in farms? Use of evidence and conceptual knowledge in a problem about aquaculture.] *Revista Eureka sobre Enseñanza y Divulgación de las Ciencias, 10*(2), 145–158. https://doi.org/10.25267/Rev_Eureka_ensen_divulg_cienc.2013.v10.i2.01

Cabero, J., Leiva, J. J., Moreno-Martínez, N. M., Barroso, J., & López, E. (2016). *Realidad Aumentada y Educación. Innovación en contextos formativos* [Augmented reality and education: Innovation in educational contexts]. Octaedro.

Cebrián, M. (2011). Los centros educativos en la sociedad de la información y el conocimiento [The educational centers in the society of information and knowledge]. In M. Cebrián de la Serna, & M. J. Gallego (Coords.), *Procesos educativos con TIC en la sociedad del conocimiento* [Educational Processes with ICT in the Knowledge Society] (pp. 23–31). Pirámide.

Cebrián-Robles, D., Franco-Mariscal, A. J., & Blanco-López, A. (2021). Secuencia de tareas para enseñar argumentación en ciencias a profesorado en formación inicial a través de CoRubric. Ejemplificación en una actividad sobre una central salina [*Task sequence for teaching argumentation in science to pre-service teachers through CoRubric. Exemplification in an activity about a saline power plant*]. *Didáctica de las Ciencias Experimentales y Sociales, 40*, 149–168. https://doi.org/10.7203/dces.40.18178

Cobo, C., & Moravec, J. W. (2011). *Aprendizaje invisible. Hacia una nueva ecología de la educación* [Invisible learning: Towards a new ecology of education]. Col.lecció Transmedia XXI. Laboratori de Mitjans Interactius. Univesity of Barcelona.

Cortés, F., Frattini, G., Krapp, L. F., Martínez-Hung, H., Moreno, D. M., Roldán, M., Salomón, J., Stemkoski, L., Traeger, S., Dal Peraro, M., & Abriata, L. A. (2021). MoleculARweb: A web site for chemistry and structural biology education through interactive augmented reality out of the box in commodity devices. *Journal of Chemical Education, 98*(7), 2243–2255. https://doi.org/10.1021/acs.jchemed.1c00179

Chai, J. J. K., O'Sullivan, C., Gowen, A. A., Rooney, B., & Xu, J. L. (2022). Augmented/mixted reality technologies for food: A review. *Trends in Food Science and Technology, 124*, 182–194. https://doi.org/10.1016/j.tifs.2022.04.021

Chen, S. Y., & Liu, S. Y. (2020). Using augmented reality to experiment with elements in a chemistry course. *Computers in Human Behavior, 111*, 106418. https://doi.org/10.1016/j.chb.2020.106418

Cheng, K. H., & Tsai, C. C. (2013). Affordances of augmented reality in science learning: Suggestions for future research. *Journal of Science Education and Technology, 22*, 449–462. https://doi.org/10.1007/s10956-012-9405-9

De Pedro, J., & Méndez, C. L. M. (2012). Realidad Aumentada. Una Alternativa metodológica en la educación primaria nicaragüense [Augmented Reality: A Methodological Alternative in Nicaraguan Primary Education]. *IEEE-RITA, 7*, 102–108.

Downes, S. (2008). An introduction to connective knowledge. In T. Hug (Ed.), *Media, knowledge and education—Exploring new spaces, relations and dynamics in digital media ecologies*. Innsbruck University Press.

Eriksen, K., Nielsen, B. E., & Pittelkow, M. (2020). Visualizing 3D molecular structures using an augmented reality app. *Journal of Chemical Education, 97*(5), 1487–1490. https://doi.org/10.1021/acs.jchemed.9b01033

Faridi, H., Tuli, N., Mantri, A., Singh, G., & Gargrish, S. (2021). A framework utilizing augmented reality to improve critical thinking ability and learning gain of the students in physics. *Computer Applications Engineering Education, 29*, 258–273. https://doi.org/10.1002/cae.22342

Franco-Mariscal, A. J., Cano-Iglesias, M. J., España-Ramos, E., & Blanco-López, A. (2025). The ENCIC-CT model for the development of critical thinking. In A. J. Franco-Mariscal (Ed.), *Critical thinking in science education and teacher training* (Chap. 1) (pp. 3–42). Springer. https://doi.org/10.1007/978-3-031-78578-8_1

Franco-Mariscal, A. J., España-Ramos, E., & Blanco-López, A. (2018). Teaching students about chemical elements using daily-life contexts. In O. E. Finlayson, E. McLoughlin, S. Erduran, & P. Childs (Eds.), *Electronic proceedings of the ESERA 2017 conference. Research, practice and collaboration in science education* (pp.710–718). Dublin City University.

Graaff, E., & Kolmos, A. (2003). Characteristics of problem based learning. *International Journal of Engineering Education, 19*(5), 657–662.

Ibáñez, A., Asensio, M., & Correa, J. (2011). Mobile learning y patrimonio: Aprendiendo historia con mi teléfono, mi GPS y mi PDA [Mobile learning and heritag: Learning history with my phone, my GPS, and my PDA]. In A. Ibánez (Coord.), *Museos, redes sociales y tecnología 2.0* [Museums, social media and 2.0 technology] (pp. 59–88). University of the Basque Country.

Irwanto, I., Dianawati, R., & Lukman, I. (2022). Trends of augmented reality applications in science education: A systematic review from 2007 to 2022. *International Journal of Emerging Technology in Learning, 17*(13), 157–175. https://doi.org/10.3991/ijet.v17i13.30587

Jiménez, Z. A. (2019). Teaching and learning chemistry via augmented and immersive virtual reality. In T. Gupta, & R. E. Belford (Eds.), *Technology integration in chemistry education and research (TICER)* (vol. 1318, pp. 31–52). American Chemical Society Symposium Series. https://doi.org/10.1021/bk-2019-1318.ch003

Kanari, Z., & Millar, R. (2004). Reasoning from data: How students collect and interpret data in science investigations. *Journal of Research in Science Teaching, 41*(7), 748–769. https://doi.org/10.1002/tea.20020

Kapp, K. (2012). Games, gamification, and the quest for learner engagement. *Training and Development, 66*(6), 64–68.

Kato, H. (2010). *Return to the origin of augmented reality.* Presentation at IEEE International Symposium on Mixed and Augmented Reality 2010 (Seoul, Korea). Panel discussion: «The Future of ISMAR: Converging Science, Business, and Art» (organized by Henry Fuchs and Christian Sandor). http://www.youtube.com/watch?v=b33eqcVz7X8

Lee, J., & Hammer, J. (2011). Gamification in education: What, how, why bother? *Academic Exchange Quarterly, 15*(2), 146–151.

Li, C., Dong, Z., Untch, R. H., & Chasteen, M. (2013). Engaging computer science students through gamification in an online social network based collaborative learning environment. *International Journal of Information and Education Technology, 3*(1), 72–77. https://doi.org/10.7763/IJIET.2013.V3.237

Martínez, H., García, A., & Escalona, J. (2017). Augmented reality models applied to chemistry education on college. *Revista Cubana De Química, 29*(1), 13–25.

Moreno-Martínez, N. M. (2020). Realidad aumentada para la construcción del conocimiento en contextos amplificados educativos. [Augmented reality for knowledge construction in amplified educational contexts]. In M. Raposo & M. Cebrián (Coords.), *Tecnologías para la formación de educadores en la sociedad del conocimiento* [Technologies for the training of educators in the knowledge society] (pp. 243–254). Pirámide.

Moreno-Martínez, N. M., & Morales, M. B. (2021). COVID-19 from a techno-educational perspective through makerspaces. *International Journal of Educational Research and Innovation, 15*, 57–72. https://doi.org/10.46661/ijeri.4898

Moreno-Martínez, N. M., & Franco-Mariscal, A.J. (2022). La herramienta de realidad aumentada Zapworks para desarrollar habilidades de pensamiento crítico. Un estudio de caso con estudiantes de Logopedia [The augmented reality tool Zapworks for developing critical thinking skills. A case study with speech therapy students]. In D. Cobos, E. López, A. H. Martín, L. Molina, & A. Jaén (Coords.), *Educar para transformar: Innovación pedagógica, calidad y TIC en contextos formativos* [Educating to transform: Pedagogical innovation, quality, and ICT in formative contexts] (pp. 2253–2262). Dykinson.

Moreno-Martínez, N. M., & Franco-Mariscal, A. J. (2023). Posibilidades didácticas de la herramienta de realidad aumentada ZapWorks en la enseñanza de las ciencias. Una experiencia con estudiantes de un Máster en Profesorado [Didactic possibilities of the augmented reality tool ZapWorks in science education. An experience with students of a Master's in Teaching program]. *Revista Tecnología, Ciencia y Educación, 24*, 91–118.

Moursund, D. (1999). *Project-based learning: Using information technology*. Society for Technology in Education.

Nielsen, B. L., Brandt, H., & Swensen, H. (2016). Augmented reality in science education—Affordances for student learning. *NorDiNa, 12*(2), 157–174.

Osborne, J., Henderson, J. B., MacPherson, A., Szu, E., Wild, A., & Yao, S. (2016). The development and validation of a learning progression for argumentation in science. *Journal of Research in Science Teaching, 53*(6), 821–846. https://doi.org/10.1002/tea.21316

Probst, Ch., Fetzer, D., Lukas, S., & Huwer, J. (2021). Effekte von Augmented Reality (AR) zur visualisierung eines dynamischen Teilchenmodells virtuelle Modelle zum Anfassen. *Chemkon, 28*, 1–7. https://doi.org/10.1002/ckon.202000046

Rumpagaporn, M. W., & Darmawan, I. G. N. (2007). Students' critical thinking skills in a Thai ICT schools pilot project. *International Education Journal, 8*(2), 125–132.

Sahin, D., & Yilmaz, R. M. (2020). The effect of Augmented Reality Technology on middle school students' achievements and attitudes towards science education. *Computers and Education, 144*, 103710. https://doi.org/10.1016/j.compedu.2019.103710

Sailer, M., Hense, J. Mandl, H., & Klevers, M. (2013). Psychological perspectives on motivation through gamification. *Interaction Design and Architecture Journal, 19*, 28–37. https://ediatum.ub.tum.de/doc/1222424/file.pdf

Schmid, J. R., Ernst, M. J., & Thiele, G. (2020). Structural chemistry 2.0: Combining augmented reality and 3D online models. *Journal of Chemical Education, 97*(12), 4515–4519. https://doi.org/10.1021/acs.jchemed.0c00823

Siemens, G. (2004). Connectivism: A learning theory for the digital age. *International Journal of Instructional Technology and Distance Learning, 1*(2). https://jotamac.typepad.com/jotamacs_weblog/files/Connectivism.pdf

Syawaludin, A., Gunarhadi, & Rintayati, P. (2019). Development of augmented reality-based interactive multimedia to improve critical thinking skills in science learning. *International Journal of Instruction, 12*(4), 331–344. https://doi.org/10.29333/iji.2019.12421a

Thomas, J. W. (2000). *A review of research on project-based learning*. The Autodesk Foundation.

Toulmin, S. E. (1958). *The uses of argument* (2003, 3rd ed.). Cambridge University Press.

University of Colorado Boulder. (2024). *PhET interactive simulations*. https://phet.colorado.edu/

Urzúa, M. D. (2021). *Augmented reality for learning chemistry*. Institute for the Future of Education. https://observatory.tec.mx/edu-bits-2/augmented-reality-for-learning-chemistry

Vuorikari, R., Ferrari, A., & Punie, Y. (2019). Makerspaces for education and training: Exploring future implications for Europe. *Publications Office of the European Union*. https://doi.org/10.2760/946996

Wong, C. H. S., Tsang, K. C. K., & Chiu, W. K. (2021). Using Augmented Reality as a Powerful and Innovative Technology to Increase Enthusiasm and Enhance Student Learning in Higher Education Chemistry Courses. *Journal of Chemical Education, 98*(11), 3476–3485. https://doi.org/10.1021/acs.jchemed.0c01029

Chapter 15
Digital Storytelling as a Strategy to Develop Critical Thinking in Preservice Preschool Teachers

Isabel María Cruz-Lorite, Juan Zagalaz, and Antonio Joaquín Franco-Mariscal

15.1 Introduction

The audiobook format, or "sound recording of a book" (Magadán & Rivas, 2020, p. 6), gained worldwide popularity in the early twenty-first century, influencing reading habits and increasing interest in literature. In Spain, the use of audiobooks has grown significantly in recent years, with 55% of the population consuming them, generating annual revenues of 10 million euros in 2020 (Rivas-García, & Magadán-Díaz, 2022). The COVID-19 pandemic is likely to have contributed to the rise of this trend as emerging technologies become more relevant.

The educational audiobook has enormous potential (Kim, 2021; Singh & Alexander, 2022). In addition, storytelling and dramatisation of stories are helpful at different academic levels, especially valuable in preschool (Wright et al., 2008). According to Alterio and McDrury (2003), "storytelling fosters reflective practice and provides an excellent framework, given its potential as a personalised application (p. 20)".

Moreover, this concept has grown over time (Rahiem, 2021), and emerging educational technologies have transformed the complexity of storytelling by enabling collaborative approaches (O'Byrne et al., 2018). Combined with the rise of

I. M. Cruz-Lorite (✉)
Department of Education, University of Nicosia, Nicosia, Cyprus
e-mail: imclorite@gmail.com

J. Zagalaz
Didactics of Musical Expression, Universidad de Málaga, Málaga, Spain
e-mail: juanzagalaz@uma.es

A. J. Franco-Mariscal
Science Education, Universidad de Málaga, Málaga, Spain
e-mail: anjoa@uma.es

A. J. Franco-Mariscal (ed.), *Critical Thinking in Science Education and Teacher Training*, Contemporary Trends and Issues in Science Education 64,
https://doi.org/10.1007/978-3-031-78578-8_15

podcasting and audiobooks, they have posed the transformation from the classical educational storytelling concept to digital storytelling, presenting different educational functionalities (Thooft, 2011).

The term digital storytelling used in this study is defined as an educational resource inspired by the audiobook concept. Its distinctive features include dramatisation, the integration of storytelling, a clear motivation conceived as a pedagogical tool, and its connection to educational innovation and technologies, as it is designed to be presented in audio format. According to Robin (2016), "digital storytelling combines the art of telling stories with a mixture of digital media, including text, pictures, recorded audio narration, music and video (p. 17)".

The educational effectiveness of digital storytelling depends on its key elements: theme, audience, purpose, narrator and characters (Engel et al., 2018). In addition, it must be well constructed, have an appropriate sequence of events, have the element of surprise (Banister & Ryan, 2001) and involve an immersive experience (Van Laer et al., 2014).

Several authors have linked the use of digital storytelling to increased motivation in their students (Peleg et al., 2017; Ya-Ting & Wan-Chi, 2012), regardless of the level of education. Ya-Ting and Wan-Chi (2012) point out that digital storytelling is emerging as a transformative technology to improve teaching, addressing aspects such as content acquisition, development of critical thinking (hereinafter, CT), motivation and literal sense of information. Also, including stories in the classroom enhances student engagement and participation (Goldbaum, 2000).

The narrative format is uncommon in science education, except preschool education (Hadzigeorgiou, 2016). Despite this, it can promote understanding of abstract concepts (Hu et al., 2021) and be more motivating than other formats, such as logical-scientific communication (Green, 2006). As Walan and Enochsson (2019) point out, a dramatisation of content to improve science education can enhance a wide range of skills, as well as promote cognitive, social and emotional development in students (Bravo et al., 2021) and, therefore, the development of CT skills (Celume et al., 2019). One example is the study by Åkerblom et al. (2019), who tested how preschoolers's conceptions of water, molecules, and chemistry improved after participating in a playfully dramatised preschool education activity.

However, the development of stories applicable to preschool science education warrants further exploration (Erickson & Ward, 2015; Hu et al., 2021), particularly in terms of adapting language and the way science is communicated (Fiske & Dupree, 2014) to avoid pseudoscientific notions (Hernández & Solorzano, 2017). This is linked to pedagogical content knowledge (PCK), a specialized type of knowledge that allows teachers to present subject matter in a manner that is understandable and accessible to their students (Akerson & Volrich, 2006; Shulman, 1987). Therefore, incorporating content drama in the classroom is a challenge that requires adequate teacher training.

In summary, the pedagogical function of storytelling in science education is widely recognised and increases the effectiveness of improving children's understanding of scientific concepts and nurturing their positive attitudes toward science (Ayra & Maul, 2012; Erten et al., 2013). However, its application could be improved because writing stories is complex (Hu et al., 2021). These ideas lead to the need for action in the initial training of preschool teachers. This chapter presents a study with students of the Degree in Preschool Education at the University of Malaga (Spain) whose main objective is to show the effectiveness of digital storytelling for developing CT in science education.

15.2 Digital Storytelling and Critical Thinking

Digital storytelling offers opportunities to CT development in the three domains proposed in the ENCIC-CT model through everyday life problems (Franco-Mariscal et al., 2025).

In the story's development, students' understanding and application of scientific and technological knowledge, along with their vision on science (Hadzigeorgiou, 2016), significantly contribute to the knowledge domain of CT. The research on storytelling in science education shows that this knowledge is approached in two ways. Some storytelling approaches are based on historical stories, such as semibiographical narratives of scientists and their work, aiming to inspire children's motivation and engagement in learning activities (Hadzigeorgiou et al., 2012). Others employ imagined stories illustrating sequences of events to directly address scientific concepts, promoting children's understanding (Engel et al., 2018; Morais, 2015).

The contribution of this chapter lies in the use of problems and situations from everyday life to develop scientific knowledge to be integrated into digital storytelling. These problems should be relevant for preschoolers and, in most cases, are related to their curiosity about the world around them. This leads them to ask questions about how things work, how plants grow (Rodríguez-Melero et al., 2021), why some objects float, how rainbows are formed, etc., problems very suitable for stories in digital storytelling for children. In addition, this approach of teaching in context in everyday situations and problems is recognised in the literature, where it is noted that it motivates students and promotes positive attitudes towards science (De Jong, 2008; King & Ritchie, 2012), allowing them to develop CT.

Critical analysis of information features prominently in the skills domain since to create a coherent and convincing science narrative, students must analyse elements related to the plot, characters, conflict and scientific knowledge associated with the story. They also need to evaluate the quality and reliability of information sources when seeking information about science for the story.

Dispositions are also present in creating and recording digital storytelling, particularly in communicating science and emotions (Sinatra et al., 2014). When developing digital storytelling on a scientific topic, students must communicate their ideas through narration and audio production, which involves selecting appropriate scientific terminology and considering the intended audience. Creativity is very much at the forefront at this stage.

Emotions manifest themselves in different ways. On the one hand, during design and recording, students may experience excitement, nervousness, satisfaction, frustration or joy. These emotions may arise because of the creativity involved in the storytelling or the challenge of playing the characters. On the other hand, digital storytelling is designed to evoke emotions in listeners, so during recording, students must strive to convey these emotions effectively through their narration, intonation and vocal expression to elicit an emotional response from the listeners (Sinatra et al., 2014).

As a final disposition, reflection and self-evaluation should be highlighted. Thus, after carrying out the digital storytelling, students can reflect on their creation process, identify areas for improvement and evaluate the effectiveness of their work in terms of how they communicate science and how it has impacted their peers.

15.3 Research Questions

This study poses the following research questions (RQ):

- RQ1: What impact does the design and development of digital storytelling for preschoolers have on preservice preschool teachers (PPTs) in the knowledge domain?
- RQ2: What is the impact on PPTs' critical analysis of information related to scientific knowledge?
- RQ3: How the PPTs address the emotions in their digital storytelling and what emotions do PPTs experience when designing and developing digital storytelling for preschoolers?

15.4 Method

15.4.1 Participants and Context

The study sample consisted of 119 PPTs taking the subject Didactics of Nature Sciences in the third year of the Preschool Education Degree at the University of Malaga (Spain) during the academic years 2019–2020 (N = 55) and 2020–2021 (N = 64), with a gender distribution of 117 women and 2 men.

During the COVID-19 pandemic, there were two implementations, one per school year, carried out by this chapter's first and third authors, in which a large part of the sessions were delivered virtually. Although the first implementation was conducted entirely in a virtual environment due to movement restrictions, the relaxation of health measures allowed PPTs to meet physically during the second implementation.

The activity required the PPTs to submit both group materials (worksheets and digital storytelling) and individual materials (questionnaires), leading to varying sample sizes for the different analyses conducted. For the worksheets and digital storytelling, all 119 PPTs participated, while there was an attrition rate of 29 PPTs for the questionnaires, resulting in a final sample size of 90 cases.

15.4.2 Experience Description

The activity involved creating digital storytelling for science education for preschoolers. Participants were organised into four to five PPTs, with 10 groups in the first implementation and 14 in the second. The digital storytelling had to be approximately eight minutes long, and each PPT had to represent at least one character, and there also had to be a narrator. The story of the digital storytelling had to be original and based on an everyday problem related to science education. The story could focus on one or more daily life problems related to science topics from a list created by the teachers (properties of matter, energy, living and inert beings, animals and plants, the human body, natural phenomena, illness and health, etc.). The recommended outline for the story was introduction, plot and denouement (Mateos, 2020).

The development of digital storytelling was carried out in three phases:

- *Phase 1: Design of the idea and development of the characters.* The PPTs were provided with a worksheet in which they had to include general information about the initial idea, the problem, and the characters of the audio story (Table 15.1). In addition, the cast of characters to be portrayed in the recording had to be specified. The teachers reviewed this first worksheet to check the appropriateness of the problem and the idea, making suggestions for improvement. After approval, the PPTs were recommended to prepare a script for the story, which the teachers also reviewed in cases where some groups requested.
- *Phase 2: Record and edit the digital storytelling.* Each group was free to choose the media they considered most appropriate for the audio recording. However, various support resources were provided, such as examples of scientific digital storytelling, a bibliography on the use of stories and digital storytelling in education, information on recording and editing software, and freely accessible repositories of sound effects and music. In addition, they were given the teacher's rubric for evaluating digital storytelling, which covered narrative, oral, content-related, technical and value-inclusive aspects. In this way, the aim was to assess some dimensions of CT of the ENCIC-CT model in the domains of knowledge, skills and dispositions.

The QR code in Fig. 15.1 shows digital storytelling about plants and another about the scientific method, both in Spanish.

- *Phase 3: Dramatisation.* A dramatisation was proposed to share the digital storytelling with the PPTs more attractively. Each group performed a theatrical representation, choosing the appropriate format and means. Most groups decided

Table 15.1 Example of a phase 1 worksheet completed by a group

<table>
<tr><td colspan="2">Everyday problem: how does a plant grow?</td></tr>
<tr><td colspan="2">Selected topic: parts of a plant</td></tr>
<tr><td colspan="2">Digital storytelling title: lending a hand</td></tr>
<tr><td colspan="2">Initial idea</td></tr>
<tr><td colspan="2">Conchi was a seed on the legs of a bird flying over a small garden. She threw herself directly into the ground, feeling she needed a change in her life
Conchi questions her existence. Suddenly, Water appears, trying to find out why she is so sad, finding out that she needs help. Water begins to fall on her, and she explains that, with her help, she will be able to grow her roots little by little
After a few days, Conchi started taking root, so she wondered how to grow her stem from the ground
At that moment, Gas appears and explains that, with his help, he can obtain oxygen and expel carbon dioxide, and a small stalk begins to grow out of the earth
After a few days, the stem has grown, and Conchi realises that the seed lacks the nutrients it needs to grow stronger. Light approaches her and explains that with her help, she can transform the substances into nutrients. Conchi then realises that she is growing healthy and strong
Water, Gas and Light appear again to tell Conchi that not only one element has an influence but that all of them together can contribute to her growth</td></tr>
<tr><td>Character</td><td>Description</td></tr>
<tr><td>Narrator</td><td>It tells the story</td></tr>
<tr><td>Conchi Seed</td><td>The changes in it allow us to know the parts of a plant, helped by the characters that intervene in its growth (Water, Gas and Light)</td></tr>
<tr><td>Water</td><td>Factor influencing the germination of a seed and the growth of a plant</td></tr>
<tr><td>Light</td><td>Factor influencing the germination of a seed and growth of a plant</td></tr>
<tr><td>Gas</td><td>Gases involve photosynthesis and plant respiration (oxygen and carbon dioxide)</td></tr>
</table>

Fig. 15.1 QR code showing examples of digital storytelling by PPTs

to represent the characters by making costumes and props and accompanying them with a digital presentation. As a novelty, one group used a puppet theatre (Fig. 15.2). This phase could only be developed in the second implementation due to health restrictions caused by the pandemic.

Fig. 15.2 Dramatisation of two digital storytelling during the second implementation

15.4.3 ENCIC-CT Model Application

The experience is framed within the ENCIC-CT model (Franco-Mariscal et al., 2025) and addresses its four application layers. The PPTs' creations should focus on daily life problems related to science or technology, or socio-scientific issues (Layer 1 of the model). While the experience enables PPTs to develop various dimensions of CT (Layer 2), this study specifically investigates the dimension of knowledge application within the knowledge domain; the dimension of critical analysis of information within the skills domain; and the dimension of emotional engagement within the dispositions domain. Layer 3 of the model incorporates the scientific practice of argumentation, requiring PPTs to craft well-supported arguments that are critically analyzed in order to construct the narrative structure of the digital storytelling and develop scientific ideas clearly and persuasively. This involves presenting evidence, reasoning, and examples that support the main message, along with critically reflecting on their arguments. Finally, digital storytelling serves as the teaching strategy (Layer 4) employed in this context.

15.4.4 Data Collection and Analysis

The data collection instruments included worksheets with the initial versions of the stories, digital stories produced by PPTs (audio files), and an activity assessment questionnaire. The latter consisted of four questions that the PPTs answered individually online (see Table 15.2). Each question aimed to collect information on a specific domain of CT in the ENCIC-CT model and some of its dimensions (Franco-Mariscal et al., 2025). Specifically, question Q1 aimed to obtain information on knowledge, Q2 on skills, focusing on critical analysis of information, and Q3 on emotional engagement as dispositions. Finally, Q4 sought an overall numerical assessment of the activity. Other authors have already used the emotions questionnaire provided in Q3 (Gil & Martínez, 2015).

A mixed analysis combining qualitative and quantitative methods was conducted. The worksheets included open-ended questions about the plot and characters of the digital storytelling, but they were not systematically analyzed as they contained only

Table 15.2 Assessment questions on digital storytelling activity

Q1. (a) Evaluate what you have learned as a future teacher using digital storytelling (b) Evaluate the usefulness of digital storytelling for preschoolers learning science content (c) Evaluate the best and worst aspects of the activity
Q2. Did you have to look for information or research to prepare for digital storytelling? If yes, please specify which science content you researched and which sources you consulted
Q3. How did you feel doing the digital storytelling? For each of the pairs of emotions, choose where you feel by marking one of the seven points on the Likert scale (very/somewhat/fairly/indifferent/fairly/somewhat/very): scared/calm, stressed/relaxed, concerned/unconcerned, insecure/secure, disgusted/pleased, distrustful/confident, unhappy/happy, bored/engaged, dissatisfied/satisfied, disinterested/interested Example: (1) very scared; (2) somewhat scared, (3) fairly scared, (4) indifferent, (5) fairly calm, (6) somewhat calm, (7) very calm
Q4. Rate the digital storytelling activity from 0 to 10 points

brief information from the early stages of the activity. However, a targeted analysis was conducted on these documents to extract examples that illustrate the findings of this study. The audio files, on the other hand, were systematically analyzed and evaluated using the rubric mentioned before (Sect. 15.4.2.) that assessed 16 aspects grouped into five dimensions (narrative, oral, content-related, technical, and values inclusion) across three levels of proficiency. Due to space limitations, the rubric is not included, but examples from this analysis are provided.

Regarding the questionnaire, the three sections of Q1 were analysed together using an open coding process, resulting in a series of response categories. Each category was subdivided into learning-related, positive and negative aspects. Q2 was analysed independently also through an open coding process extracting different categories. Q3 and Q4 were analysed quantitatively by counting the ratings made. The qualitative analyses were performed using Atlas.ti v.9 software, and the quantitative analyses were completed by calculating percentages for each category.

15.5 Results

15.5.1 Results on Knowledge

This section provides findings from the knowledge domain of CT in the PPTs' activity of designing digital storytelling. First, two examples of how PPTs integrate scientific knowledge about everyday problems when designing digital storytelling are shown, followed by the assessments in this domain.

15.5.1.1 Integration of Scientific Knowledge into Everyday Life Problems

- Digital storytelling: In search of the terrestrial "chusi".

This digital storytelling approached the scientific method through the problem of everyday life: what would happen if food that is essential for our lives were no longer available? The story is told through the dialogue between two girls and two aliens who arrive on Earth in search of food. The different stages of an investigation developed in this way:

- The research problem is posed through the search for food necessary for the life of the aliens on their planet, called "chusi", where it has disappeared.
- Observation is shown through the information the aliens provide to the girls gathered through their five senses (size, colour, shape, etc., of the missing food).
- Hypotheses are formulated by the girls based on the observations given by the aliens, e.g. when told that the food was red, the girls hypothesised that it might be strawberry jelly.
- Data collection is shown by the girls collecting all the information given by the aliens in a notebook.
- The experiments were exemplified by tasting an apple to see if its taste was similar to that of the target food, as the characteristics suggested that it might be this food.
- The girls concluded after tasting different foods.

As can be seen, digital storytelling presents a model of the scientific method adapted to the age of children.

- Digital storytelling: A long wait.

This digital storytelling addressed the problem of everyday life: Why did my snowman disappear? By exploring the disappointment of two brothers when they realised that the snowman they had built in their garden had turned into a puddle of water the next day. From this situation, their mother explained to them the changes in the state of water with household utensils (small pot or freezer).

15.5.1.2 Knowledge Assessment

From the analysis of Q1, 541 contributions emerged, distributed across several categories. Most were related to general aspects of digital storytelling as an educational resource (usefulness, motivation, versatility of topics, etc.). Approximately one-fifth (102 contributions) were related to science education and could be associated with the knowledge domain of CT. These contributions belonged to two categories: (a) importance of science education through problems and adaptation of scientific knowledge, and (b) meaningful learning of scientific knowledge. The frequencies shown in Table 15.3, distinguishing between contributions related to learning, positive and negative aspects. As can be seen, the positive aspects and learning predominated over the negative aspects.

- *Importance of science education through problems and adaptation of scientific knowledge*

Aware of the fact that science is less often taught than other areas in preschool education, the PPTs highlighted these ideas:

The importance of creating digital storytelling about science for children due to the scarcity of these resources for science education:

> *There are few resources on scientific topics for children. In addition, these subjects are worked on less at this stage and should be promoted* (PPT32).
>
> *It is a way of introducing new knowledge through an entertaining and understandable resource for them* (PPT81).
>
> *Digital storytelling is a tool that makes it easier for children to understand biology, history, mathematics, etc.* (PPT54).

The diversity of science problems and topics highlights the importance of selecting those that are relevant to children's experience and understanding:

> *I have learnt that with any subject or problem, digital science storytelling can be developed for infants* (PPT32).
>
> *Digital storytelling can attract our students' attention if we know how to choose the subject that interests them* [...] (PPT86).

In some cases, the creation of the story was shown as a negative aspect, mentioning difficulties in the choice and focus of the topic, the decision between several possibilities for the plot and its development, the balance of the characters and the organisation of knowledge.

> *The worst aspect was coming up with the story's main idea, as we had quite a few ideas, but they needed to be more convincing* (PPT84).
>
> *The worst thing has been dealing with all the possibilities and managing our doubts while planning the story* (PPT31).

It is important to include content appropriate to the age of the pupils, as scientific knowledge is often complex for this stage.

Table 15.3 Frequency of categories found in the answers to Q1

	Learning	Positive Aspects	Negative Aspects	Total
Importance of science education through problems and adaptation of scientific knowledge	33	27	9	69
Meaningful learning of scientific knowledge	9	24	0	33
Total	42	51	9	102

As long as the content of digital storytelling is age-appropriate, the explanation is clear and child-friendly language is used, digital storytelling is a handy resource that offers many learning opportunities. It is often easier to explain scientific content through a story than in other ways (PPT57).

There is a need to adapt more advanced or abstract scientific knowledge, summarising information or using vocabulary and expressions that are more accessible and understandable to learners.

Digital storytelling can attract our students' attention as long as we know how to [...] *adapt the scientific language to a language they can understand* (PPT86).

I have learnt to explain, facilitating children's understanding, scientific concepts complicated for this educational stage, such as "observation" or "experiment" (PPT54).

It is new and made specifically for them, with vocabulary and expressions appropriate to their age (PPT72).

I have learned to adapt scientific terms into a more accessible vocabulary for children (PPT58).

In addition to adapting the vocabulary to children's audiences, some digital storytelling chooses to incorporate the main scientific ideas as the main characters of the plot to simplify the understanding of scientific concepts. One example was the digital storytelling "Lending a hand" (Table 15.2) addressing the problem of how a plant grows, which used the factors responsible for growth, such as seed, water, and light, as central characters while giving a prominent role to the gaseous processes involved for a complete understanding of the phenomenon.

Some PPTs (9) highlighted this adaptation of scientific content as a negative aspect because of its difficulty and the fear of introducing misconceptions:

The worst part was synthesising and adapting scientific content for infant pupils. We found it very difficult to find synonyms to explain these terms more straightforwardly and child-friendly (PPT88).

- *Meaningful learning of scientific knowledge*

The PPT's put the value of learning scientific knowledge in a meaningful way by showing these benefits:

It promotes understanding of scientific knowledge:

Digital storytelling is very useful, making it easier to understand knowledge as it is told as a story. Science seems complicated to transmit to children. However, by adapting the contents and transmitting them orally, they acquire the knowledge in a fun way. All this allows for meaningful learning (PPT79).

Children acquire knowledge faster and in a meaningful way (PPT38).

It contributes to the strengthening of memory through meaningful learning:

Digital storytelling can help them exercise memory and relate information to other knowledge (PPT79).

The need to integrate digital storytelling with other activities to promote meaningful learning.

There needs to be more than just a single hearing of digital storytelling for the children to acquire the knowledge. That is why they are very useful if they are worked on with subsequent activities (PPT71).

15.5.2 Results on the Skill to Critically Analyse Information

This section presents the findings of the critical analysis of information conducted during the digital storytelling creation process. It begins with two examples demonstrating how PPTs critically evaluated the information, followed by the assessments within this domain.

15.5.2.1 Critical Analysis of Information in Story Creation

The feedback provided by the teachers on the worksheet submitted in Phase 1 helped the PPTs to conduct a critical analysis of the scientific information used. An example of this process can be seen in the digital storytelling titled "Ramón's Mission," which addressed the everyday problem of how climatic conditions can affect ecosystems through a story set in ancient Egypt. In this narrative, the protagonist, Zacarías, tries to solve a locust plague caused by a Pharaoh.

In its initial version, the PPTs proposed the following solution to the problem: "Zacarías calls his cousin Ramón to bring as much rain as possible to end the locust plague." In her feedback, the teacher recommended that they investigate the conditions required to effectively eliminate a locust plague and assess whether the central idea of the story—that rain could resolve the plague—was supported by accurate scientific information.

As a result of this critical analysis of scientific information, the PPTs significantly revised this part of the story in its final version: "Zacarías calls his friend Ramón to ask for help, as a plague of locusts has appeared in Egypt after the recent storms and is damaging the crops. Ramón travels to Egypt to try to 'take away' the rain." This change is based on the understanding that rain, rather than being a solution, actually promotes the proliferation of locusts by increasing the available vegetation for their feeding. In fact, many plagues in African countries have occurred due to the mass migration of locusts to more humid areas following periods of drought.

Another example is found in the Phase 1 of the digital storytelling 'Lending a Hand' mentioned above (Table 15.2), where the introduction of the character 'Gas' to explain the processes of photosynthesis and respiration could confuse children, leading them to believe that it is an additional factor necessary for the germination and growth of a plant, alongside the seed, water, or the presence of light (Rodríguez-Melero et al., 2021). In the initial version, the teacher suggested that the PPTs reconsider whether 'Gas' was the most accurate term to describe the medium containing the gases essential for photosynthesis and plant respiration—oxygen and carbon dioxide. In the final version of the digital story, the PPTs chose to rename the character from 'Gas' to 'Air,' clarifying that both gases are found in the air.

These examples illustrate how the process of critical analysis and the pursuit of precise scientific information allowed the PPTs to revise their narratives, resulting in stories that are more scientifically accurate. However, despite the revisions, some digital stories, such as the first example, still present challenges that need to be addressed, such as the idea of 'bringing or taking away rain' from a place. This critical analysis underscores the importance of grounding educational narratives in solid scientific principles, which enhances both student understanding and the overall quality of the stories.

15.5.2.2 Assessment of Critical Information Analysis

Regarding Q2, 68/119 PPTs indicated that they had searched for information related to the topic of their digital storytelling for its elaboration, while 21 PPTs stated that no research was required. In one case, the answer needed to be clarified. 36 PPTs who searched for information did so on general websites, 15 PPTs consulted science subject notes, and 10 mentioned academic sources such as Google Scholar and scientific articles. At the same time, 9 used books from non-university educational stages. Other PPTs used websites of newspapers, organisations, science blogs and forums, and Wikipedia or YouTube.

As can be seen, most of the information sources used come from non-academic sources or lack sufficient verification, which could have resulted in conceptual errors in the creation of the digital storytelling if the teachers hadn't provided feedback on the stories during Phase 1, as illustrated in the previous section.

15.5.3 Results on Emotions

This section examines both how the PPTs addressed emotions in their digital storytelling and how they felt while engaging in the activity.

15.5.3.1 Emotions in Digital Storytelling

Regarding the treatment of emotions in the digital storytelling, the activity required PPTs to incorporate affective vocabulary into their stories, which was an evaluation criterion under the aspect "Cognitive, Affective, Volitional, and Perceptual Vocabulary" of the rubric. Since each story had specific characteristics, not all groups included affective vocabulary in their digital stories, as they focused on developing other types of vocabulary more aligned with their narrative. Nevertheless, 16 out of the 24 analyzed digital storytelling (8 from the first year and 8 from the second) did incorporate affective vocabulary.

The groups that included affective vocabulary were explicit in addressing emotions, clearly stating which emotion was felt at specific points in the narration and complementing it with a tone of voice that matched the sentiment. Below is an example excerpt from the dialogue in the digital storytelling "Lending a Hand" (Table 15.2):

> Water: *Hi! I'm Water, what's your name?* [cheerful tone of voice].
> Conchi: *My name is Conchi* [sad tone].
> Water: *What's wrong? You seem sad* [worried tone].
> Conchi: *Uh... [sigh] I don't know what's wrong, but I feel like something inside me needs to change. I'm tired of being just another seed* [sad tone].

The characters in this digital storytelling explore various emotions and states of mind, such as boredom, confusion, insecurity, and happiness, as well as values like friendship and teamwork.

15.5.3.2 Emotions Assessment

The analysis of Q1 and Q3 provided qualitative and quantitative information, respectively, on the emotions that surfaced in the activity. The analysis of Q1 revealed that the emotions experienced by the PPTs in each phase were influenced by their ability to use recording and editing software and by their previous experiences in dramatisation. Thus, if these experiences were positive, emotions such as interest, engagement and confidence emerged, while if they were limited, emotions such as insecurity and frustration predominated.

During the recording and editing phase of the digital storytelling, a good number of PPTs perceived positively the use of the tools necessary for the creation of digital storytelling (75 ratings), but there was also a significant number of negative perceptions associated (41 ratings), due to the insecurity generated by the lack of knowledge of the tools.

> *I find it an exciting activity that allows them* [the children] *to acquire and develop many skills such as language, communication, forms of expression, respect for turns* [of speech], *listening skills, expanding vocabulary, etc.* (PPT06). (Example of interest as emotion).

> *The audio story recording was the best, as we had much fun and laughed a lot, changing the voices to interpret the different planets* (PPT85) (Example of engagement as emotion).

> *The worst thing* [was] *the need for more knowledge about the programs to record, edit and manipulate the audio* (PPT79). (Example of insecurity as emotion).

For the dramatisation phase, 55 contributions were found, of which 21 aspects were related to learning, 21 to negative and 13 to negative aspects.

> *The best thing was to act out the story. We made a puppet theatre, and I loved creating the puppets* (PPT84) (Example of engagement as emotion).

> *The worst thing was that we could not practice* [theatre performance] *because COVID gave us insecurity* (PPT52). (Example of insecurity as emotion).
>
> *What I liked least* [...] *was the performance of the digital storytelling in class. It did not go well; nerves played tricks on us* (PPT85) (Example of frustration as emotion).

In addition, the PPTs were aware that digital storytelling should also evoke emotions in children:

> *I have learnt the importance of intonation and voice to give more emotion to the audio story and make the content more engaging* (PPT58).
>
> [I have learned] [...] *the importance of emotions when we tell something* (PPT37).

Despite this variety of emotions experienced during the activity, the overall assessment of the whole experience mainly showed positive emotions (Fig. 15.3). Thus, the PPTs felt mostly relatively calm (35.6%), somewhat stressed (36.7%), somewhat concerned (53.3%), somewhat secure (32.2%), very pleased (40.0%), somewhat confident (37.8%), somewhat happy (45.6%), somewhat engaged (51.1%), very satisfied (48.9%) and somewhat or very interested in the same percentage (48.9%).

Finally, the PPTs' average rating of the activity (Q4) was 9.3 out of 10 points, which is considered very satisfactory.

15.6 Discussion and Educational Implications

Overall, the experience with PPTs regarding digital storytelling design was considered satisfactory. Despite this, some PPTs highlighted some negative aspects related to the creation of the story, such as the difficulty in choosing the scientific topic, developing the plot or organising the knowledge, aspects already partially reflected in the study by Zagalaz et al. (2020) carried out with PPTs.

The results also revealed some difficulties in being confident in creating proper digital storytelling for science education. These include the difficulty in explaining more abstract concepts through digital storytelling (Kokkotas et al., 2010), the adaptation of scientific knowledge to children's language (Fiske & Dupree, 2014) or the fear of introducing erroneous or pseudo-scientific ideas (Hernández & Solorzano, 2017), which led them to seek information that did not always come from contrasted sources. The significant proportion of PPTs (68 out of 119) who had to search for and analyze information to address scientific knowledge is also noteworthy.

All these difficulties could justify the scarcity of digital storytelling in science education and could be related to the poor attitude towards science of future preschool teachers (Pendergast et al., 2017; Riegle-Crumb et al., 2015) and to the appearance of these professionals' emotions such as insecurity or stress when developing science activities in the classroom (Smit et al., 2021). In fact, self-confidence is a key factor that limits and mediates teachers' educational practices, as demonstrated by Abd-El-Khalick et al. (1998) in their study on the nature of science with secondary school teachers.

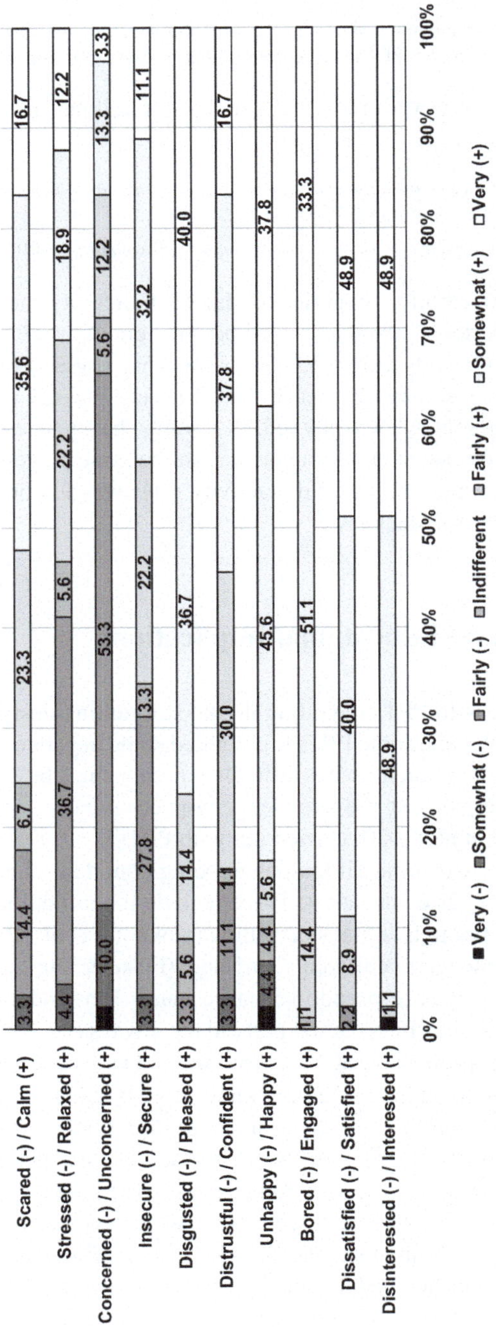

Fig. 15.3 Emotions felt during the realisation of digital storytelling

Notably, the PPTs were able to create varied digital storytelling addressing different topics from everyday problems that a child may face. In most cases, they focused on imagined stories illustrating the sequence of events to address the scientific notions directly for promoting children's understanding (Engel et al., 2018; Kokkotas et al., 2010; Morais, 2015) and in no case focused on the life of scientists and their work (Hadzigeorgiou et al., 2012).

Some PPTs indicated that digital storytelling should be a variety of activities. Still, some authors advocate creating complete experiences by integrating storytelling with other activities, such as outings in the natural environment (Caduto & Bruchac, 1997), thus promoting meaningful learning.

The relevance of emotions and their impact on science teaching–learning processes has been widely demonstrated in various studies (Sinatra et al., 2014). Our experience has allowed them to be considered by PPTs from a dual perspective. On the one hand, this is evidenced by the interest and confidence they have gained from previous experience in dramatisation and the use of recording and editing programmes. On the other hand, it highlights the importance of the construction of emotions in the audience when narrating stories, a factor already pointed out in the study by Rahiem et al. (2020).

This study highlights the difficulty of PPTs in composing stories to explain scientific concepts, making learning experiences memorable and exciting for children, which has some educational implications. The first implication relates to the need to include a subsequent implementation phase of the designed digital storytelling with preschoolers in this experience. This implementation could be carried out during external practices in preschools, which would allow PPTs to learn about the weaknesses and strengths of their digital storytelling as shown by the children themselves and to make proposals for improvement regarding the scientific knowledge addressed the need to adapt them, if necessary, the characters, or the script.

In relation to the ENCIC-CT model, the experience has shown that digital storytelling (Layer 4) is particularly effective for addressing daily life problems relevant to preschoolers (Layer 1). This experience allows for the exploration of the CT domains of knowledge, skills, and dispositions. Specifically, it involves: knowledge application, where PPTs applied their scientific knowledge to develop their stories; critical analysis of information, where they conducted research and evaluated scientific content to include in their digital storytelling; and emotional engagement, where they integrated emotional aspects to engage preschoolers with the stories (Layer 2). As a result of the study, the development of communicative skills is also evident, as PPTs adapted their stories to the educational level, which required effective communication, such as the adaptation of scientific vocabulary. For Layer 3, the development of digital storytelling was supported by the need for strong arguments. Additionally, some digital stories even included the scientific practice of inquiry to construct the narrative (e.g., digital storytelling "In Search of the Terrestrial 'Chusi'"—see Sect. 15.5.1.1.).

As a second implication, digital storytelling should not only be a powerful technology tool for schools to promote CT. Its capacity should be explored as a resource for science outreach in other informal or non-formal education areas, such as museums, community organisations, healthcare centres, science fairs and other science outreach events for children, and promoting its dissemination through social networks. Nowadays, its extension to the whole population is easy through computers, tablets, smartphones and other digital devices.

In this line, a third educational implication could be to expand or adapt the experience to the creative podcasting format, which involves students creating their podcasts, which, it is hoped, will help them develop their understanding of a topic (Bartle et al., 2011), which is experiencing an important boom as a means of communication, and which shares a large number of elements with digital storytelling, such as planning, the pedagogical perspective, and the management of audio files. Moreover, as with digital storytelling, podcasting allows students to generate their knowledge about the topics they are working on and fits in with project-based learning (Brooks-Kirkland, 2006). Finally, it should be noted that the implementation of podcasts in education requires, in addition to the relevant planning, a process of reflection, which should end with a planned evaluation that will positively impact future designs.

Acknowledgements This work is part of the Spanish National Plan R+D+i Project entitled "Citizens with critical thinking: A challenge for teachers in science education", reference PID2019-105765GA-I00, funded by MCIN/AEI/10.13039/501100011033.

References

Abd-El-Khalick, F., Bell, R. L., & Lederman, N. G. (1998). The nature of science andinstructional practice: Making the unnatural natural. *Science Education, 82*(4), 417–436.

Åkerblom, A., Součková, D., & Pramling, N. (2019). Preschool children's conceptions of water, molecule, and chemistry before and after participating in a playfully dramatized early childhood education activity. *Cultural Studies of Science Education, 14*, 879–895. https://doi.org/10.1007/s11422-018-9894-9

Akerson, V. L., & Volrich, M. L. (2006). Teaching nature of science explicitly in a first-grade internship setting. *Journal of Research in Science Teaching, 43*(4), 377–394.

Alterio, M., & McDrury, J. (2003). *Learning through storytelling in higher education*. Routledge. https://doi.org/10.4324/9780203416655

Ayra, D. J., & Maul, A. (2012). The role of the scientific discovery narrative in middle school science education: An experimental study. *Journal of Educational Psychology, 104*(4), 1022–1032. https://doi.org/10.1037/a0028108

Banister, F., & Ryan, C. (2001). Developing science concepts through story-telling. *School Science Review, 83*, 75–84.

Bartle, E. K., Longnecker, N., & Pegrum, M. (2011). Collaboration, contextualisation and communication using new media: Introducing podcasting into an undergraduate chemistry class. *International Journal of Innovation in Science and Mathematics Education, 19*(1), 16–28.

Bravo, F. A., Hurtado, J. A., & González, E. (2021). Using robots with storytelling and drama activities in science education. *Education Sciences, 11*, 329. https://doi.org/10.3390/educsci11070329

Brooks-Kirkland, A. (2006). Podcasting for learning. *School Libraries in Canada, 25*(4), 44–48.

Caduto, M. J., & Bruchac, J. (1997). *Keepers of the earth: Native American stories and environmental activities for children*. Fulcrum Publishing.

Celume, M. P., Besançon, M., & Zenasni, F. (2019). Fostering children and adolescents' creative thinking in education. Theoretical model of drama pedagogy training. *Frontiers in Psychology, 9*, 2611. https://doi.org/10.3389/fpsyg.2018.02611

De Jong, O. (2008). Context-based chemical education: How to improve it? *Chemical Education International, 8*(1), 1–7.

Engel, A., Lucido, K., & Cook, K. (2018). Rethinking narrative: Leveraging storytelling for science learning. *Childhood Education, 94*(6), 4–12. https://doi.org/10.1080/00094056.2018.1540189

Erickson, T., & Ward, T. (2015). *The master communicator's hand-book.* Changemakers Books.

Erten, S., Kiray, S. A., & Sen-Gumus, B. (2013). Influence of scientific stories on students' ideas about science and scientists. *International Journal of Education in Mathematics Science and Technology, 1*(2), 122–137. https://doi.org/10.4628/ijemst.v1i2.30

Fiske, S. T., & Dupree, C. H. (2014). Gaining trust as well as respect in communicating to motivated audiences about science topics. *Proceedings of the National Academy of Sciences, 11*(4), 13593–13597. https://doi.org/10.1073/pnas.1317505111

Franco-Mariscal, A. J., Cano-Iglesias, M. J., España-Ramos, E., & Blanco-López, A. (2025). The ENCIC-CT model for the development of critical thinking. In A. J. Franco-Mariscal (Ed.), *Critical thinking in science education and teacher training* (Chap. 1) (pp. 3–42). Springer. https://doi.org/10.1007/978-3-031-78578-8_1

Gil, P., & Martínez, M. (2015). Emociones auto-percibidas en las clases de educación física en primaria [Self-perceived emotions in primary physical education classes]. *Universitas Psychologica, 14*(3), 923–935. https://doi.org/10.11144/Javeriana.upsy14-3.eapc

Goldbaum, E. (2000). Grant to advance case study approach. *University of Buffalo Reporter, 31*(21), 1.

Green, M. C. (2006). Narratives and cancer communication. *Journal of Communication, 56*(1), 163–183. https://doi.org/10.1111/j.1460-2466.2006.00288.x

Hadzigeorgiou, Y. (2016). Narrative thinking and storytelling in science education. In Y. Hadzigeorgiou (Ed.), *Imaginative science education. The central role of imagination in science education* (pp. 83–119). Springer. https://doi.org/10.1007/978-3-319-29526-8_4

Hadzigeorgiou, Y., Klassen, S., & Klassen, C. F. (2012). Encouraging a "romantic understanding" of science: The effect of the Nikola Tesla story. *Science & Education, 21*(8), 1111–1138. https://doi.org/10.1007/s11191-011-9417-5

Hernández, M. M., & Solorzano, N. M. V. (2017). Vulgarización científica en cuentos ecológicos: problemas y peligros [Scientific popularization in ecological tales: issues and dangers]. *Revista Ciencia UNEMI, 10*(23), 90–103.

Hu, J., Gordon, C., Yang, N., & Ren, Y. (2021). "Once upon a star": A science education program based on personification storytelling in promoting preschool children's understanding of astronomy concepts. *Early Education and Development, 32*(1), 7–25. https://doi.org/10.1080/10409289.2020.1759011

Kim, N. Y. (2021). E-book, audio-book, or E-audio-book: The effects of multiple modalities on EFL comprehension. *English Teaching, 76*(4), 33–52. https://doi.org/10.15858/engtea.76.4.202112.33

King, D., & Ritchie, S. (2012). Learning science through real world contexts. In B. Fraser, K. Tobin, & C. McRobbie (Eds.), *Second international handbook of science education* (pp. 69–77). Springer.

Kokkotas, P., Rizaki, A., & Malamitsa, K. (2010). Storytelling as a strategy for understanding concepts of electricity and electromagnetism. *Interchange, 41*(4), 379–405. https://doi.org/10.1007/s10780-010-9137-9

Magadán, M., & Rivas, J. I. (2020). El impacto del libro electrónico sobre la cadena de valor editorial española: Un estudio de casos [The impact of the e-book on the Spanish publishing value chain: A case study]. *Revista Española de Documentación Científica, 43*(1), e258. https://doi.org/10.3989/redc.2020.1.1650

Mateos, A. (2020). La literatura infantil como medio para educar en valores. Análisis de cuentos [Children's literature as a means to educate in values. Analysis of tales]. *Papeles Salmantinos de Educación, 24*, 31–53.

Morais, C. (2015). Storytelling with chemistry and related hands-on activities: Informal learning experiences to prevent "chemophobia" and promote young children's scientific literacy. *Journal of Chemical Education, 92*(1), 58–65. https://doi.org/10.1021/ed5002416

O'Byrne, W. I., Houser, K., Stone, R., & White, M. (2018). Digital storytelling in early childhood: Student illustrations shaping social interactions. *Frontiers in Psychology, 9*, 1800. https://doi.org/10.3389/fpsyg.2018.01800

Peleg, R., Yayon, M., Katchevich, D., Mamlok-Naaman, R., Fortus, D., Eilks, I., & Hofstein, A. (2017). Teachers' views on implementing storytelling as a way to motivate inquiry learning in high-school chemistry teaching. *Chemistry Education Research and Practice, 18*(2), 304–309. https://doi.org/10.1039/C6RP00215C

Pendergast, E., Lieberman-Betz, R. G., & Vail, C. O. (2017). Attitudes and beliefs of Prekindergarten teachers toward teaching science to young children. *Early Childhood Education Journal, 45*(1), 43–52. https://doi.org/10.1007/s10643-015-0761-y

Rahiem, M. D. H. (2021). Storytelling in early childhood education: Time to go digital. *ICEP, 15*(4), 1–20. https://doi.org/10.1186/s40723-021-00081-x

Rahiem, M. D. H., Surayyah, N., Krauss, S. E., & Rahim, H. (2020). Moral education through Dramatización storytelling: Insights and observation from Indonesia Kindergarten teachers. *International Journal of Learning, Teaching and Educational Research, 19*(3), 475–490.

Riegle-Crumb, C., Morton, K., Moore, C., Chimonidou, A., Labrake, C., & Kopp, S. (2015). Do inquiring minds have positive attitudes? The science education of preservice elementary teachers. *Science Education, 99*(5), 819–836. https://doi.org/10.1002/sce.21177

Rivas-García, J. I., & Magadán-Díaz, M. (2022). An overview of the audiobook marketplace in Spain. *Publishing Research Quarterly, 38*, 168–179. https://doi.org/10.1007/s12109-022-09862-9

Robin, B. R. (2016). The power of digital storytelling to support teaching and learning. *Digital Education, 30*, 17–29.

Rodríguez-Melero, A. M., Cáceres, M. J., & Franco-Mariscal, A. J. (2021). ¿Cómo hacemos crecer una planta? Una indagación con niños de 3 años de educación infantil [How do we grow a plant? An inquiry with 3-year-olds]. *Enseñanza de las Ciencias, 39*(3), 231–253. https://doi.org/10.5565/rev/ensciencias.334

Sinatra, G. M., Broughton, S. H., & Lombardi, D. (2014). Emotions in science education. In R. Pekrun & L. Linnenbrink-Garcia (Eds.), *International handbook of emotions in education* (pp. 415–436). Routledge/Taylor & Francis Group.

Singh, A., & Alexander, P. A. (2022). Audiobooks, print and comprehension: What we know and what we need to know. *Educational Psychology Review, 34*, 677–715. https://doi.org/10.1007/s10648-021-09653-2

Smit, R., Robin, N., & Rietz, F. (2021). Emotional experiences of secondary pre-service teachers conducting practical work in a science lab course: Individual differences and prediction of teacher efficacy. *Disciplinary and Interdisciplinary Science Education Research, 3*(5), 1–17. https://doi.org/10.1186/s43031-021-00034-x

Shulman, L. (1987). Knowledge and teaching: Foundations of the new reform. *Harvard Educational Review, 57*, 1–22.

Thooft, N. A. (2011). *The effect of audio books on reading comprehension and motivation*. The College of St. Scholastica ProQuest Dissertations Publishing.

Van Laer, T., de Ruyter, K., Visconti, L. M., & Wetzels, M. (2014). The extended transportation-imagery model: A meta-analysis of the antecedents and consequences of consumers' narrative transportation. *Journal of Consumer Research, 40*(5), 797–817. https://doi.org/10.1086/673383

Walan, S., & Enochsson, A. B. (2019). The potential of using a combination of storytelling and drama, when teaching young children science. *European Early Childhood Education Research Journal, 27*(6), 821–836. https://doi.org/10.1080/1350293X.2019.1678923

Wright, C., Bacigalupa, C., Black, T., & Burton, M. (2008). Windows into children's thinking: A guide to storytelling and dramatization. *Early Childhood Education Journal, 35*(4), 363–369.
Ya-Ting, C. Y., & Wan-Chi, I. W. (2012). Digital storytelling for enhancing student academic achievement, critical thinking, and learning motivation: A year-ling experimental study. *Computer and Education, 59*(2), 339–352. https://doi.org/10.1016/j.compedu.2011.12.012
Zagalaz, J., Muñoz, R., & Porras, V. (2020). Diseño y desarrollo de audiocuentos como recurso educativo: una experiencia interdisciplinar en las aulas de primaria y de Grado en Educación Primaria [Design and development of audiobooks as an educational resource: an interdisciplinary experience in primary school classrooms and primary education degree]. In E. Encabo (Ed.), *Música y pantallas. Cultura, sociedad y educación* [Music and screens. Culture, society, and education] (pp. 229–250). Ediciones Difácil.

Part VI
Research Ideas and Concluding Remarks

Chapter 16
Research Ideas for Developing Critical Thinking

Antonio Joaquín Franco-Mariscal

16.1 Introduction

This chapter presents three proposals for applying the ENCIC-CT model, offering readers new ideas for further research on the development of students' critical thinking (hereinafter, CT). The first proposal focuses on using a mobile app for online role-playing games to enhance argumentation and decision-making, illustrated by the socio-scientific issue of building a sports complex in a natural setting. The second proposal explores the CT dimensions related to knowledge application and vision of science through movies and TV series, addressing socio-scientific issues such as cloning and the genetic inheritance, as well as everyday problems like the safety and stability of structures. The third and final proposal emphasizes the development of personal autonomy, metacognition, and reflection as key dispositions of CT through Self-Reflection activities. Students engage in synthesis, self-assessment, extension, and discussion of a CT-related activity conducted in class, exemplified by a role-playing exercise.

16.2 Research Idea 1: Enhancing Critical Thinking Through Role-Playing Mobile Applications

Antonio Joaquín Franco-Mariscal, María del Mar López-Fernández, and María José Cano-Iglesias

---

A. J. Franco-Mariscal (✉)
Science Education, Universidad de Málaga, Málaga, Spain
e-mail: anjoa@uma.es

A. J. Franco-Mariscal (ed.), *Critical Thinking in Science Education and Teacher Training*, Contemporary Trends and Issues in Science Education 64,
https://doi.org/10.1007/978-3-031-78578-8_16

16.2.1 Mobile Applications and Critical Thinking

Despite the potentialities described in this book, some challenges still need to be addressed to develop CT through science education (Santos, 2017). One of them is the need to create an additional motivation in the proposals brought to the classroom, which generates an intrinsic motivation for learning. Although the scientific practices and teaching strategies used in the ENCIC-CT model are recognized in the literature as effective and motivating for science education, as highlighted in Chap. 1 (Franco-Mariscal et al., 2025), it is important to acknowledge that some students may require additional scaffolding to enhance their learning process.

One of these elements could be the use of mobile applications (apps), which offer opportunities to develop CT. There is no doubt about the exponential growth in digital technologies and the availability of mobile software applications applied to education (Hinze et al., 2023). Specialists worldwide have highlighted mobile learning since the ninth edition of the Horizon Report (2021) as an emerging and disruptive technology impacting teaching, learning, and research (Michael & Chen, 2006). In addition, the COVID-19 pandemic contributed to strengthening the use of digital technologies in education (Fasae et al., 2022).

On the other hand, the literature on the integration of mobile technology into higher education reveals an increasing focus on how mobile devices are used within the classroom environment, both physical and online, rather than on how mobile applications may be used for either teaching or the research process (Hinze et al., 2023; Lai & Smith, 2018). This last idea is relevant since, in the educational domain, the use of mobile applications is an innovation for teachers and students and for improving the quality of education as they attempt to transform the curriculum, methodologies and ways of transmitting knowledge, offering new alternatives for teaching–learning and, therefore, for the development of CT.

Although some authors encourage the development of CT through mobile applications (Lin et al., 2014), there are still few experiences in science education. Since 2023, the ENCIC research group has been working in this line on a research project, designing some mobile applications that promote the development of CT using the ENCIC-CT model. Specifically, the Junta de Andalucía project entitled "Mobile applications for scientific and technological argumentation on climate, environmental and resource efficient actions" (ArgumAPP) (21ProyExcel_00176) (https://argumapp.encic.es/) promotes CT through argumentation.

One of the challenges lies in finding visually appealing formats to present proposals for developing CT in digital environments by contextualising everyday problems and avoiding reliance on lengthy texts or quiz-like questions. In their review study, Cheng et al. (2015) found that adventure or role-playing was the most common genre, followed by simulation and puzzle games. These approaches allow for interactive immersion that promotes student participation and engagement, facilitating a deep understanding of scientific knowledge while developing critical problem-solving skills in a dynamic and stimulating environment.

The first research idea presents a mobile application developed for this purpose. Specifically, the *App for Role-Play Debates* combines the educational benefits of role-playing games in developing CT (Franco-Mariscal et al., 2023) with the added advantages offered by digital technologies.

16.2.2 Method

16.2.2.1 Proposal Description

The *App for Role-Play Debates*, developed by the ENCIC group, is an interactive platform where users can participate in various online role-playing games. The application, built in Unity, uses C# programming language and PHP for database connectivity. Its interface is designed to be clear and accessible, making it easy to review and manage debate participation. The app includes seven role-playing games, all of which have been previously tested in face-to-face settings (Franco-Mariscal et al., 2023). These scenarios cover topics such as nuclear energy, genetically modified foods, urban space for caregiving, river pollution, the destruction of a natural setting, the ban on plastic use, and breastfeeding versus formula feeding.

Users must register to access the app, ensuring personalized access. The application offers profiles for both teachers and students. With the teacher profile (Fig. 16.1), users can configure a new role-play from a list of predefined debates (via the Create debate [Crear debate] button); access previously created debates, making it easier to track and manage their activities (View my debates [Ver mis debates] button); join a debate created by another teacher using a code provided by its creator (Join a debate [Unirse a un debate] button); or generate a key for their students to join the debate ("Teacher key" [Clave de profesor] button).

When creating a debate (Fig. 16.1, right), the teacher assigns a name to the debate to identify the student group, selects a theme from the available games, decides whether roles will be randomly assigned to students, sets the maximum number of participants for each role, specifies the stage of the role-playing game, and provides the URL for the rubric used to evaluate student participation. Additionally, the teacher has the option to change the game's stage once student progress has been verified.

Students access the designated role-playing through a code provided by the teacher. The app presents an overview of the socio-scientific issue, supplemented with various links for deeper exploration. Students can select a role within the game, create arguments with supporting sources, respond to other participants, receive notifications of replies, and develop a final synthesis on the issue. Interactions are displayed in a chat-like format, allowing students to present their arguments during different stages of the role-playing, fostering a continuous flow of debate and discussion.

Fig. 16.1 Teacher profile screens: accessing a debate (left image) and creating a debate (right image). *Source* Own elaboration

16.2.2.2 ENCIC-CT Model Application

The mobile application is designed with role-playing as a teaching strategy (Layer 4 of the ENCIC-CT model) (Franco-Mariscal et al., 2025). Through this approach, the app presents users with a variety of socio-scientific issues (Layer 1) that are specifically designed for debate in a collaborative setting with other students.

This methodology not only facilitates the exploration and discussion of significant issues from both scientific and social perspectives but also fosters the development of CT dimensions focused on argumentation and decision-making within the skills domain (Layer 2). Furthermore, the app integrates elements designed to develop the scientific practice of argumentation (Layer 3), which is essential for constructing well-supported and coherent arguments. Argumentation has a relevant consideration in the application as it is present in all its phases, so students are required to use in their arguments, at least evidence, warrant and conclusion as essential elements, desirable the use of other auxiliary components such as backing, modal qualifiers and rebuttal (Erduran & Jiménez-Aleixandre, 2007).

Additionally, this application allows the development of other dimensions included in the ENCIC-CT model. During their argumentation, students must deal with the socio-scientific issue comprehensively, considering the different aspects that make it up (scientific, technological, economic, social, health, legal, etc.). The vision of science can be considered implicit in the integral approach to the issue since one of the aspects of this dimension is scientific, a relevant aspect of the problem, although not the only one. Likewise, in developing their arguments, students apply and improve their scientific knowledge of the issue, broadening it with ideas different from those they already possess. The preparation of the activity also involves a critical analysis of the information to construct arguments to support the positions defended in the role-play. The online role-play ends as an informed decision on the issue, thus addressing this dimension of CT. Personal autonomy/metacognition/reflection are also present in the different stages of the role play. Their development helps to create new ideas and the formation of independent opinions on the issue. Personal autonomy involves preparing for the debate, either individually or in small groups, managing information about the issue and using the arguments they consider most important depending on the role assigned and the course of the debate. Finally, role-playing can evoke different emotions depending on the arguments presented in the debate and foster social engagement in some students.

16.2.3 Application Example

As an example, the app offers a role-playing game that explores the issue of whether or not to build a sports complex in a natural setting. This scenario considers the environmental values of the area, its significance for biodiversity and ecosystems, and the diverse interests and perspectives of the stakeholders involved (environmental, political, economic, social, etc.) (Valencia et al., 2023).

The issue dates back to 2008 when plans were made to build a sports complex with luxury residences, hotels, and a commercial area in the natural zone of Arraijanal, Málaga (Spain). The project was abandoned due to environmental concerns and the possible presence of archaeological remains. A year later, a proposal was introduced to create a large coastal park, and part of the land was donated by the city council for the construction of a sports complex for Málaga's football team. Construction began in 2018 but was halted by environmental activists protesting tree cutting and monitoring for potential archaeological finds. The camp was cleared by police several times.

The app features roles supporting the issue, such as defense attorney, business consultant, city council representative, politician, sports club member, and a representative from a sports tourism company. Opposing roles include prosecuting attorney, scientist, environmental activist, politician, local resident, and a representative from a nature tourism company. Neutral roles include a judge, jury members, and journalists. Each role is assigned to a group of students, with each student having individual autonomy within the app. The characters interact in the setting of a trial.

The role-playing takes place over three stages, each spaced one week apart. In Stage 1 (argument development), each student must research and submit arguments related to the issue in the chat to participate in the game. Students can present their arguments but are not allowed to exchange arguments or respond to other roles.

In Stage 2 (counterargument development), each student must respond to the arguments made by other characters, engaging in message exchanges. A minimum of five interactions is required, and all arguments directed at their role must be addressed. In Stage 3 (synthesis development), all participants representing the same role must collaborate to create a unified synthesis, and the judge will then deliver a final verdict on the issue.

Figure 16.2 provides examples of arguments given by preservice teachers for this role-play. The left image illustrates an argument in the Stage 1 in which the role-playing politician defended the economic aspects of the construction of the sports complex. The image on the right shows the conclusion of the scientists' role.

Fig. 16.2 Examples of arguments from Stage 1 (left image) and Stage 3 (right image) of the role-play on the construction of a sports complex in a natural setting. *Left image* Politician. Argument: the construction of the sports complex will generate a significant boost in employment both during its construction and throughout its operation, in the short, medium, and long term. This development is expected to substantially stimulate the city's economy. Given the existing metro infrastructure and bus lines connecting to the area near the construction site, the development of this project will significantly enhance the local public transportation system, benefiting both metro and bus services. Date: October 31, 2023. *Right image* Scientists: as scientists, we conclude that the construction of the sports complex in Arraijanal not only jeopardizes the integrity of a unique ecosystem in our province but also undermines established protection categories, including protected natural areas, zones of special conservation, and special protection areas for birds. These categories are intended to preserve biodiversity and natural habitats for the benefit of both the community and the environment. The lack of thorough assessments and the disregard for the biological wealth present in Arraijanal raise serious concerns about the sustainability of this project, posing a threat to the numerous endangered species, both flora and fauna, in the area, as well as to the coastal path and the safety zone of the Guadalhorce natural park. Date: November 15, 2023

Some research ideas for exploring the development of various CT dimensions are:

- Argumentation: The arguments created by students at different stages of the role-playing are stored in the database associated with the app. This data enables analysis of the quality of the arguments, the types of evidence and warrants used, and their frequency. The Toulmin model (2003) or the adaptation by Jiménez-Aleixandre (2010) can be useful for this analysis.
- Decision-making on the issue: This CT skill, along with any potential shifts in students' opinions, can be examined through a questionnaire administered before and after using the app. The questionnaire should ask students to provide a reasoned explanation for their decisions. The analysis focuses on identifying each participant's decision. Then, responses are categorized as "in favor" for those supporting the sports complex, "against" for those opposed, or "undecided" for those who did not clearly choose a side. After categorizing, frequencies and percentages are calculated. Changes in stance can be visualized using a Sankey diagram, and McNemar's statistical test can help identify significant differences between pre- and post-app usage.
- Comprehensive analysis of the problem : The evidence used in arguments and decisions made before and after using the app allows for investigation into the various aspects of the issue discussed in the debate (e.g., environmental, social, economic, legal). Comparing these aspects at both points in time helps assess the impact of the debate on a holistic understanding of the problem. This analysis involves creating categories based on students' outputs. Once categories are established, frequencies and percentages are calculated.

16.2.4 Concluding Remarks

This research idea is noteworthy because it combines the benefits of traditional role-playing games—such as argumentation and decision-making (Franco-Mariscal et al., 2023)—with the advantages of technology, specifically through an online mobile app that facilitates collaboration among a large number of students.

Unlike conventional role-playing games, which are typically conducted in a single session, this app allows students to engage with and reflect on the issue over a period of three weeks, thanks to remote access and flexible timing. This approach facilitates immediate interaction while also enabling students to reflect over time on the arguments provided by different characters, after conducting additional research to develop new counterarguments. Although the app currently does not support it, future updates could include features such as image, video, or audio uploads. Additionally, the app caters to various learning styles without requiring immediate response times. Finally, this approach effectively integrates the layers of the ENCIC-CT model, offering a comprehensive and powerful educational tool for enhancing CT in collaborative learning environments and providing new avenues for research in the field of CT.

16.3 Research Idea 2: Enhancing Critical Thinking Through Movies and Television Series

María del Mar López-Fernández, María José Cano-Iglesias, and Antonio Joaquín Franco-Mariscal

16.3.1 Movies and TV Series and Critical Thinking

Student interest in science remains a significant concern (Dávila et al., 2015) due to its perception as a complicated and boring subject, far removed from their daily lives (Rutjens et al., 2018). These attitudes also appear in preservice science teachers, particularly at the lower levels, who show distrust and rejection towards science, finding it boring and difficult to learn and making them feel uncomfortable teaching it (Sánchez et al., 2018). Therefore, it is important to develop actions involving all students, not only those more motivated towards science (Membiela et al., 2023).

To overcome this obstacle, different studies have proposed strategies to motivate and improve students' learning and attitudes towards science (Newhouse, 2016). Various authors have investigated the benefits of using fiction resources such as films, TV shows, series, stories, and novels to teach science (Hamalosmanoglu et al., 2020; Koutnikova, 2017). These forms of entertainment are not only widely consumed but also integral to our culture landscape (Ulu-Aslan & Bas, 2023). Proof of this is that since the COVID-19 pandemic, the consumption of films and TV series through paid platforms such as Netflix, Amazon Prime or Disney Plus has increased enormously (Saha & Srivastava, 2023).

The use of movies and TV series in science education offers multiple advantages. On the one hand, they are an effective resource for motivating students and making learning more relevant and meaningful by relating it to real-life experiences (Hamalosmanoglu et al., 2020). In a society increasingly dominated by science and technology, fostering a reflective understanding of the world around us is essential. Moreover, these resources facilitate understanding abstract concepts by creating mental images correlated with underlying scientific theories (Barnett & Kafka, 2007) while being visually appealing, encouraging students' engagement with science learning. Hence, movies and TV series can be useful tools for transmitting scientific knowledge (Quirantes, 2011).

16.3.2 Method

16.3.2.1 Proposal Description

Students watch a movie or TV series clip, lasting under five minutes, as a large group. In some cases, this clip consists of the full movie trailer. The audiovisual material is selected to provide an overview of the plot and characters, which helps in identifying everyday situations and problems depicted in the clip.

Working in groups of three or four, the students must identify sequences that present everyday or socio-scientific issues, often of interest within the fictional universe of the film. Finally, they must formulate and provide solutions to the proposed problem.

For preservice teachers, these sequences can be used to develop science teaching issues across various disciplines, such as physics, chemistry, biology, or geology. Based on the identified problems, they should design educational activities tailored to the level of students they are preparing to teach. This approach helps future educators integrate real-world problems and CT into their teaching practices, enhancing the relevance and engagement of their science lessons.

16.3.2.2 ENCIC-CT Model Application

The interaction between entertainment and science education contributes to developing CT within the ENCIC-CT model (Franco-Mariscal et al., 2025). Movies and TV series as teaching strategy (Layer 4) allow viewers to immerse themselves in plots that address everyday life situations and/or socio-scientific issues (Layer 1), offering thought-provoking scenarios on scientific, technological, ethical, and social issues (Evrekly et al., 2011). This allows participants to identify and explore these problems while encouraging them to solve them using their CT. It should be noted that, in most cases, these issues are not posed straightforwardly, but the viewer must be able to identify them.

Regarding the CT dimensions that can be developed (Layer 2), the application of knowledge to everyday problems and the vision of science are particularly notable within the knowledge domain. These dimensions help foster a deeper understanding of science, positioning movies and TV series as powerful tools for meaningful learning.

Another aspect to consider is that these resources offer a partial view of science. Not only do they sometimes present questionable ideas from a scientific point of view, but they also address controversial issues for which society does not yet have an answer (Rose, 2003). These questions can also be useful for science teaching and learning, as they are prone to discussion in the classroom and allow for the development of CT (Evrekly et al., 2011).

Thus, as Rodríguez (2010) suggests, film fosters the development of reflection, critique, and debate, which are essential for understanding the world, human nature, and the context of the issues portrayed in movies. Therefore, the scientific practice of argumentation (Layer 3) should play a crucial role in comprehending and addressing the issue presented in the resource.

Movies and TV series also offer opportunities for preservice teachers to design activities oriented toward their future students. These experiences will allow us to identify the weaknesses and strengths of the preservice teachers in identifying and applying scientific knowledge in other contexts and get to know their vision of science. In addition, these activities can help prevent the transmission of possible misconceptions to their future students.

16.3.3 Application Examples

The potential of some movies and TV series is shown below.

16.3.3.1 Jurassic Park Saga

The *Jurassic Park* saga (and later *Jurassic World*), both in the novels written by Michael Crichton and in the films directed by Steven Spielberg and other directors, revolves around the idea that humanity has managed to revive dinosaurs by cloning DNA contained in the blood of fossilised mosquitoes. Set on different fictional islands, the film focuses on a group of people trying to escape from a theme park full of genetically modified dinosaurs.

An example application for *Jurassic World* is detailed below, illustrating the layers of the ENCIC-CT model.

- Socio-scientific issue (Layer 1): Cloning.

Question: Is it possible to create a dinosaur, or is it fiction? Can we design a custom dinosaur?

- Movie sequence (Layer 4): Scientists demonstrating cloning techniques for the creation of dinosaurs.
- CT dimension (Layer 2): Application of knowledge about cloning and genetics.
- Argumentation process (Layer 3): Students should argue how the cloning process works, including the extraction and manipulation of genetic material, the transfer of DNA into recipient cells and the development of a cloned organism. In addition, they can discuss the ethical, social and environmental implications of cloning, including debates about genetic manipulation, conservation of endangered species and the impact on ecosystems. These discussions should be supported by evidence, such as scientific studies or historical examples, to ensure a well-rounded and informed perspective on the issue. It also allows them to explore genetic diversity, evolution and biotechnology. Students can analyse how cloning affects genetic

variability within a species and discuss how it can influence adaptation and long-term survival.

Thus, other scenes highlight physics concepts related to movement (e.g., Can a velociraptor catch up with a motorcyclist?) or forces (e.g., Can a velociraptor open a door?). Furthermore, these movies offer scenarios to explore geology, from forming volcanic islands to coastal erosion and plant diversity influenced by soil. Scenes of extreme natural phenomena and fossil excavations provide opportunities to discuss climate, the formation of mineral deposits or the preservation of fossil remains. A potential issue to address could be: How do human activities, such as urbanization and tourism, influence the formation and erosion of volcanic islands?

16.3.3.2 Harry Potter Saga

The Harry Potter series of books and films, written by J. K. Rowling, follows the adventures of young wizard Harry Potter and his friends at Hogwarts School of Witchcraft and Wizardry. Set in a fantasy world, the saga features various magical creatures, which opens up the possibility of exploring biological questions such as: How can we classify a magical creature based on its characteristics? How would these magical traits adapt to a specific environment? And how would the creature interact with other inhabitants of that environment?

The Quidditch match on broomsticks poses problems such as how changes in the speed and trajectory of the balls used might affect the strategies and performance of the players. Students should consider factors such as the impact of these changes on player positioning, scoring opportunities, and the overall dynamics of the game. This problem encourages an exploration of the physics and strategy involved in Quidditch, combining elements of motion and team dynamics to analyze how variations in the game's elements might influence play.

Professor Snape's lessons on potions allow chemistry problems to be explored. Students could argue about the different substances used in potions and how they can be combined to achieve specific effects and compare the magic in potions with the scientific principles of chemistry.

In addition, the multitude of characters and their relationships allow for the exploration of possible magical genetics and the inheritance of magical powers. An example can be found in *Harry Potter and the Philosopher's Stone*:

- Socioscientific issue in the Harry Potter universe (Layer 1): Genetic inheritance of magic.

Question: How is magic transmitted, and how might it impact a society with individuals who possess magical abilities—whether they are pure-bloods or half-bloods—as well as those without magical abilities (muggles)?

- Movie sequence (Layer 4): Hagrid reveals to Harry Potter his true identity as a wizard and gives him some details about his parents' life as wizards.

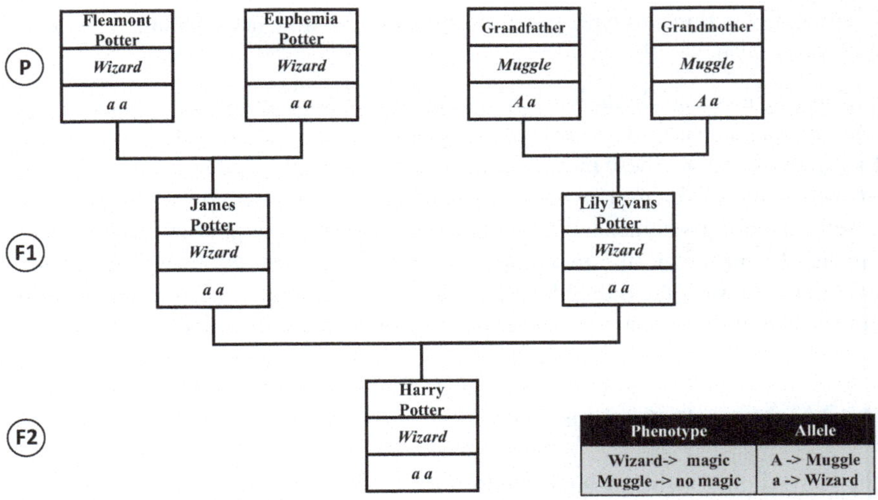

Fig. 16.3 Proposed family tree for Harry Potter's family to explain magical inheritance

- CT dimension (Layer 2): Application of knowledge about genetic inheritance, phenotype, genotype and allele.
- Argumentation process (Layer 3): Students can understand genetic inheritance by analysing the Potter family tree (Fig. 16.3). First, they should identify that inheritance in the Potter family can be neither intermediate nor co-dominant due to the existence of only two phenotypes (wizard and muggle). Next, they could explore the possibility of sex-linked inheritance, but in attempting to make the family tree, they must realise that this explanation is not feasible. Finally, they should consider that a suitable model must be based on two autosomal alleles: a dominant allele (A) that determines being a muggle and a recessive allele (a) that determines being a wizard.

Once the Potter family tree is revealed, magical genetic inheritance must be examined in terms of its social and ethical implications. Key arguments may address the social status of individuals with magical abilities and their opportunities within the magical world. Additionally, other arguments should provide evidence of the ethical dilemmas related to equality and discrimination, particularly in relation to muggles and half-blood wizards.

16.3.3.3 Indiana Jones Saga

This film saga, created by George Lucas, follows the adventures of archaeologist Indiana Jones, who frequently collaborates with the government to search for objects of great historical significance. Physics problems, particularly those related to kinematics and dynamics, play a prominent role in many of the saga's most iconic scenes.

For instance, in *Raiders of the Lost Ark*, Indiana Jones tries to escape from a giant boulder. A possible problem to explore is how the mass of the boulder, the incline of the passage, and the friction between the boulder and the ground affect the speed and behavior of the boulder during the chase. Another example is found in *The Temple of Doom*, where Jones must survive a thrilling chase through train cars inside a mine.

The movies also present situations involving geological obstacles, such as the formation of caves, canyons, and rivers, allowing for the exploration of geomorphology and hydrology. Finally, the scene in *The Kingdom of the Crystal Skull*, where Indiana Jones discovers an underground city inhabited by extraterrestrial beings, provides an opportunity to explore concepts of evolutionary biology and scientific speculation about extraterrestrial life. This leads to the question: How might living beings adapt to extreme environments on other planets?

As an example of application, the escape across the collapsing suspension bridge in *The Temple of Doom* is detailed.

- Everyday problem (Layer 1): Safety and stability of structures.

Questions: How do the properties of construction materials, the design of the bridge, and the load it supports influence the stability and safety of a suspension bridge? What measures can be implemented to ensure that a suspension bridge remains safe for daily use?

- Movie sequence (Layer 4): Indiana Jones and his companions find themselves on a rope and wooden suspension bridge over a river, fleeing from pursuing enemies. As they cross, the bridge starts to collapse and breaks in several places.
- CT dimensions (Layer 2): Application of knowledge related to materials, structures, and forces.
- Argumentation process (Layer 3): This scenario provides a basis for arguing various factors. It allows for evaluating how different materials (wood and rope) impact the bridge's load-bearing capacity and durability. It also facilitates an examination of how the bridge's design (length, width, and angle of inclination) affects its stability and how weight distribution and applied forces (e.g., all enemies advancing simultaneously on the same section of the bridge) contribute to the risk of structural failure. Additionally, arguments should address maintenance strategies to ensure the bridge remains in optimal condition. All these points should be supported by scientific evidence to substantiate the arguments presented.

16.3.3.4 Wall-E

This animated film is set in a dystopian future where the Earth has become desolate and covered in rubbish due to the depletion of natural resources. In this scenario, Wall-E, a small cleaning robot, is the planet's only inhabitant and spends days compacting rubbish and collecting objects. However, his routine is interrupted by the arrival of Eve, a robot explorer sent from a spaceship in search of signs of life on Earth. The film allows us to reflect on the importance of environmental conservation, biodiversity preservation, pollution's impact on ecosystems, and the need to adopt sustainable practices to protect the planet. In addition, concepts about the evolution of species and the adaptation of living beings to different environments can be explored. The plot also reflects on the relationship between technology and nature and how our actions can affect the Earth's ecological balance. An application example is provided below.

- Socio-scientific issue (Layer 1): Sustainable use of natural resources.

  Question: What is the future of planet Earth?

- Movie sequence (Layer 4): Wall-E shows Eve a plant he found in the rubbish.
- CT dimension (Layer 2): Application of knowledge about biodiversity, ecosystems, and degradation.
- Argumentation process (Layer 3): Students should reflect on the significance of the presence of a plant in a desolate environment, emphasizing the role that vegetation plays in maintaining life on Earth, including its contributions to oxygen production, carbon dioxide absorption, and the overall health of ecosystems. They should also discuss the negative impact of pollution and wasteful consumption on the environment and how these actions can destroy ecosystems. Additionally, they can argue that the plant serves as a critical indicator of the environment's capacity to recover from degradation, highlighting the importance of preserving biodiversity and protecting natural habitats from further destruction. Moreover, they can contend that the Earth's future is in jeopardy if current behaviors continue unchecked, leading to the collapse of ecosystems and the loss of vital resources needed for survival. The argumentation can extend to proposing solutions and strategies for mitigating environmental damage, such as adopting sustainable consumption practices, reducing waste, and promoting conservation efforts. Finally, students should be encouraged to use scientific evidence to support their arguments, drawing on examples from real-world environmental issues. This could include case studies of ecosystems that have been restored through conservation efforts or data on the effects of pollution on biodiversity. By grounding their arguments in evidence, students can better understand the urgency of addressing environmental challenges and the importance of taking action to ensure a sustainable future for the planet.

16.3.3.5 Peppa Pig

This animated children's series follows the adventures of a little pig named Peppa, along with her family and friends. Created by Astley and Baker, the series focuses on everyday situations in the lives of Peppa and her family, often involving outdoor activities and experiences in nature. The short episodes tackle problems about basic science concepts such as gravity, frictional forces, the cycle of animals, their habitats and the environment or the shape of the earth. Special mention should be made of the episode entitled *Mirrors*, where children can learn about the properties of light through examples from everyday life (Franco-Mariscal, 2016):

- Everyday problem (Layer 1): Effects of light.

Fig. 16.4 Peppa and George look at each other on a spoon. *Source* Own elaboration

Question: Why do Peppa and George appear normal when they look at each other on the outside of the spoon, but when they turn it upside down, they appear upside down?

- TV series sequence (Layer 4): Daddy Pig gives Peppa and George spoons to look at each other. Looking at each other, they look normal, but when they turn around, they look upside down (Fig. 16.4).
- CT dimension (Layer 2): Application of knowledge about optics, reflection in curved mirrors, and concave and convex surfaces.
- Argumentation process (Layer 3): Students should explain how an image is produced in a curved mirror and compare it to a plane mirror. Similarly, they should argue that when Peppa looks at her image on the outside of the spoon, she looks into a convex mirror, which always produces straight, virtual, smaller images no matter how far away the object is. However, when she looks inside, she uses a concave mirror, which can produce inverted or upright images depending on the distance from the object. As Peppa and George see themselves inverted at all times, they are far from the mirror's focus. Otherwise, they would see their right image. They can also use a ray diagram of a concave or convex mirror to explain the phenomenon.

16.3.4 Ideas for Data Collection and Analysis

Some of the instruments for data collection and analysis can include:

- Oral productions of students with the arguments used to solve the proposed problem. This analysis requires evaluating the various elements included in the arguments presented (Toulmin, 2003), as well as the adequacy of the application of scientific knowledge and the vision of science they present.

In the case of studies involving preservice teachers:

- Written productions of the designs concerning the resource used, which can be categorised according to the scientific disciplines addressed in the proposals and, within these, according to the issue tackled and the scientific knowledge involved. This analysis also enables the evaluation of the integration of different disciplines in the design of activities and their vision of science.
- Interviews with some preservice teachers who volunteered to evaluate the activity and the learning achieved. These interviews provided additional information on the impact of the activity on their training and learning, as well as their opinions and reflections on the process of designing activities with this resource. The interviews must be transcribed, and the responses coded to identify categories and themes in the data.

16.3.5 Concluding Remarks

Proposals based on movies and TV series emphasize that science is not an isolated body of abstract knowledge but a dynamic human activity intricately connected to society, culture, and various disciplines. This requires a solid knowledge of fundamental scientific concepts, which is an important dimension of CT. Such knowledge includes comprehending scientific principles and theories and recognizing their applications and relevance to the real world. The use of movies and TV series can effectively facilitate this understanding because when citizens grasp how science influences everyday life, they are better equipped to make informed political and ethical decisions. As an additional value of this proposal, it highlights that designing activities that address everyday problems or socio-scientific issues by applying knowledge from various disciplines strengthens their integration into the curriculum.

The vision of science as a dimension of CT is reflected in how students approached the issues, recognizing its relevance in everyday situations and connecting with students' familiar experiences and contexts. Additionally, these proposals can promote science for citizenship, for instance, by understanding fundamental aspects of genetic inheritance (López-Fernández & Franco-Mariscal, 2021), such as the transmission of hereditary traits, blood grouping, or knowledge about genetically inherited diseases, which are relevant and can also have a significant impact on personal life.

Furthermore, the role of argumentation in solving the proposed problems is fundamental because it helps students build and defend their viewpoints, using evidence to justify their interpretations and opinions. This process aids them in better understanding the complexities of the problem by applying scientific knowledge, developing skills for debating and exchanging ideas, which enriches collaborative learning and enhances students' ability to critically analyze media.

Lastly, knowing students' preferences regarding the movies and television they consume is important to achieve greater intrinsic motivation for learning. However, this should not be the only criterion for their selection as the materials should also be age-appropriate.

16.4 Research Idea 3: Enhancing Critical Thinking Through Self-Reflection Activity

Antonio Joaquín Franco-Mariscal, Elisabet Luque-Henares, and Ángel Blanco-López

16.4.1 Reflective Thinking and Critical Thinking

Initial training is a very fertile and important first period of training for learning the teaching profession and, therefore, very appropriate for providing future teachers with training proposals to strengthen their capacity for reflection, self-criticism and self-evaluation. Moreover, reflection on practice and scientific and didactic knowledge, as an essential professional competence, enables the development of didactic knowledge of content (Mellado, 2003).

The concept of reflection is associated with a problem that results in surprise or uncertainty (Hatton & Smith, 1995). This idea has been extended to education through the concept of the reflective teacher, which refers to a pedagogical approach in which educators continuously reflect on their teaching practice (Perrenoud, 1999). In this field, the experiential-intuitive model of Schön (1998) and the rationalist-technicist model of Dewey (1933) are considered references (Hébert, 2015).

Schön (1998) argues that teaching praxis is characterised by complexity, uncertainty, instability, singularity and conflicting values and that the technical perspective is inadequate for managing classroom problems. The teaching profession must be understood as a reflective activity which, in any case, includes some technical applications. In this context, Schön (1998) argues that teachers apply theoretical knowledge to practical situations and reflect on their experiences to improve their performance. A reflective teacher engages in a continuous cycle of observation, reflection and action. This involves analysing their actions and decisions in the classroom, considering how they influence student learning, and adjusting their approach if necessary.

According to Dewey (1933), "reflective thinking" is a thought process that involves "turning a subject over in the mind and giving it serious and consecutive consideration" (p. 3). According to this author the idea of reflection is based on four fundamental criteria: (1) it is a meaning-making process that moves the learner from one experience to the next with a greater understanding of its relationships and connections to other ideas and experiences; (2) it is a systematic, rigorous and disciplined way of thinking with its roots in scientific inquiry; (3) it has to take place in community; and (4) it requires attitudes that value personal and intellectual growth of self and others.

Dewey (1933) also distinguishes between reflective thinking and reflective action. For this author, reflective thinking constitutes a particular way of thinking, while reflective action refers to various actions to solve problems. Both processes are closely related (Schön, 1998). Thus, reflection is a systematic process derived from thinking phases that determine that and constitutes a means for learning, as it enables integrating a body of knowledge. For Perrenoud (1999), reflection can occur based on a problem linked to any phenomenon, event or object.

In short, reflective thinking is a cognitive process by which a person analyses, questions and evaluates their ideas, beliefs, experiences and actions. Reflective thinking is closely linked to CT. It constitutes one of its dimensions, recognised in some approaches, such as reflection, reflective thinking, metacognition or disposition. The ENCIC-CT model includes the personal autonomy/metacognition/reflection disposition (Franco-Mariscal et al., 2025).

Reflective teachers may use various techniques to reflect on their practice, such as self-assessment, observation of other teachers, analysis of classroom data and student feedback. These include the use of personal interviews (Smith, 1991) and written texts in different formats (Hatton & Smith, 1995; Mena et al., 2012), the analysis of preservice teachers' classroom performance through videos (Simon & Campbell, 2012) or group discussions on aspects related to teaching (Yoon & Kim, 2010). These strategies allow students to express their thoughts and actions, encouraging reflection (Hatton & Smith, 1995).

16.4.2 Method

16.4.2.1 Proposal Description

The *Self-Reflection* activity (Luque-Henares et al., 2022) is designed to develop preservice teachers' reflective skills related to CT. Its purpose is to help students improve their ability to produce independent, creative, and critical personal work, as well as to reflectively analyze their own work. This proposal was designed by the third author of this research idea.

The proposal is carried out individually and involves an oral presentation to classmates, reflecting on a previous activity completed in the course that was approached using the ENCIC-CT model. This activity could be any of the experiences described in the various chapters of this book. This presentation is supported by a digital slideshow organized into four sections: (1) a synthesis of two key ideas, from their perspective, addressed in the activity (two slides); (2) self-assessment, identifying two ideas from the activity that were not sufficiently clear (two slides); (3) extension, contributing two new ideas not covered (two slides); and (4) posing a question for discussion (a slide). The presentation time is seven minutes, with one minute per slide.

16.4.2.2 ENCIC-CT Model Application

The proposal addresses the ENCIC-CT model as follows: The activity on which the student reflects focuses on a socio-scientific or everyday life issue (Layer

1). Analyzing and reflecting on this problem fosters the development of personal autonomy, metacognition, and reflection (Layer 2), as the student is required to synthesize ideas, clarify points of confusion, expand on certain concepts, and propose a question for further discussion. The scientific practice of argumentation (Layer 3) is essential for constructing the arguments to support their discourse and persuade the audience. Lastly, *Self-Reflection* as a teaching strategy (Layer 4) is designed with the necessary elements to foster reflective thinking.

16.4.3 Application Example

This section presents a *Self-Reflection* by a preservice teacher on a classroom role-play activity addressing the issue of banning single-use plastics (López-Fernández et al., 2021).

The first key idea emphasized is the importance of clearly defining the socio-scientific issue, ensuring it is both relevant and current, and establishing the setting where the role-play will take place. Additionally, the discussion also highlights the necessity of having an appropriate number of characters, their alignment either for or against the issue, and the importance of students preparing their arguments in advance using a structured worksheet. The second key idea underscores the benefits of role-playing in tackling socio-scientific issues related to the environment, energy sources, food, or health, along with the skills it fosters, such as argumentation and decision-making, within the framework of developing CT. However, a drawback noted is the challenge of effectively staging the activity and utilizing non-verbal communication when conducted in a virtual format.

Regarding the ideas that were not sufficiently clear, the first aspect raised is determining the most suitable age for implementing role-playing in the classroom. This idea is argued through Piaget's theory of cognitive development (Piaget, 1950), which suggests that intelligence develops through a series of stages, each reflecting different levels of complexity and reasoning: the sensorimotor stage (0–2 years), where the world is explored through sensory experiences and motor actions; the preoperational stage (2–7 years), where children develop a symbolic system; the concrete operational stage (7–11 years), where thinking is applied to concrete objects and events using straightforward rules; and the formal operational stage (beginning around age 11), where abstract thinking and the use of more complex rules emerge. These considerations lead to the conclusion that the secondary school years are the most appropriate for role-playing activities, as students are capable of formal operations and can engage in role-playing with greater personal investment and intensity. The second unclear idea involves the challenges associated with implementing role-playing. These challenges include the difficulty of simulating every possible issue, ensuring that the activity goes beyond mere entertainment to foster meaningful learning, the need for well-designed group structures, effective classroom management, a high level of commitment from both teachers and students, the potential frustration some students may experience, and the complexity of evaluating such activities.

A new idea proposed is to teach students how to effectively search for information on the Internet. This proposal is supported by a study from the International Association for the Evaluation of Educational Achievement (2020), which found that only 2% of students can differentiate between relevant and irrelevant information online—a crucial skill for developing CT. The preservice teacher recommended using webquests to achieve this goal, as they are structured and guided online activities that promote educational use of the Internet. The webquest concludes with the creation of a website where the research findings are published, thereby fostering CT. As a second new idea, asynchronous virtual debates were proposed as another resource to enhance CT, involving the use of online forums for discussion and engagement in activities. The preservice teacher highlights the findings of Mendoza et al. (2016), which demonstrate that this strategy effectively promotes CT. The study observed improvements in students' grades and various CT indicators, such as deduction and explanation, following the implementation of asynchronous virtual debates.

Finally, the discussion questions were: Do you believe high school students will be able to carry out a role-play with well-supported arguments? Will there be notable differences in the execution of role-plays across different high school grade levels?

As observed, the four sections of the *Self-Reflection* provide evidence of the proposal's potential for developing this disposition of CT. Below are some useful ideas for analyzing students' reflective capacity. In this case, data collection tools should include the digital presentation and the oral recording of the presentation, which should be transcribed for analysis. In order to guide this analysis, the four levels of reflection of increasing complexity proposed by Hatton and Smith (1995), based on the categorization established by Van Manen (1977) and the ideas of Schön (1998), can be particularly helpful: (1) descriptive writing, where a situation is described (not considered reflection), (2) descriptive reflection, where reasons are provided about a situation, (3) dialogical reflection, where justified reasons are given about a situation, and (4) critical reflection, where justifications of a situation are extended to consider political or social aspects, among others. Other useful categories of analysis are those considered by Toom et al. (2014) derived from the phases or moments of reflective action defined as (1) introspection when they take back their thoughts about something related to teaching; (2) association, when these facts are linked to previous knowledge; (3) integration, when they relate their experience to a theoretical framework, (4) validation when they establish a balance between their thoughts and their actions and, (5) appropriation, understood as the moment in which they declare ownership of their learning.

16.4.4 Concluding Remarks

The example provided illustrates that the *Self-Reflection* activity has significant potential to enhance students' CT skills. By requiring students to synthesize key ideas, the activity helps them focus on essential elements from their perspective, ensuring clarity and depth in their analysis. The self-assessment component fosters reflective thinking by encouraging students to identify and address areas that need further clarification, thus deepening their understanding. The extension section, which introduces new ideas, enriches the discussion by bringing in fresh perspectives, thereby broadening the learning experience. Additionally, the inclusion of a discussion question at the end stimulates critical engagement and encourages further dialogue among peers.

A notable aspect of this proposal is the challenge of presenting and synthesizing information within a seven-minute timeframe using seven slides. This constraint demands thorough preparation and ensures that content is presented in a structured and efficient manner. Furthermore, the proposal not only improves the presenter's ability to effectively communicate their insights but also promotes active participation and reflection from the audience during the discussion phase.

Overall, by integrating synthesis, self-assessment, extension, and discussion, all grounded in a previous CT activity under the ENCIC-CT model, the *Self-Reflection* exemplifies a comprehensive approach to educational and professional development.

Acknowledgements This work is part of the Spanish National Plan R+D+i Project entitled "Citizens with critical thinking: A challenge for teachers in science education", reference PID2019-105765GA-I00, funded by MCIN/AEI/10.13039/501100011033. The research idea 1 is part of the R&D Project "Mobile applications for scientific and technological argumentation on climate, environmental, and resource-efficient actions" (21ProyExcel_00176), funded by the Andalusian Plan for Research, Development, and Innovation (PAIDI 2020) under the 2021 call from the Regional Government of Andalusia (Spain).

References

Barnett, M., & Kafka, A. (2007). Using science fiction movie scenes to support critical analysis of science. *Journal of College Science Teaching, 36*(4), 31–35.

Cheng, M. T., Chen, J. H., Chu, S. J., & Chen, S. Y. (2015). The use of serious games in science education: A review of selected empirical research from 2002 to 2013. *Journal of Computers in Education, 2*(3), 353–375. https://doi.org/10.1007/s40692-015-0039-9

Dávila, M. A., Borrachero, A. B., Mellado, V., & Bermejo, M. L. (2015). Emotions in students of that in learning physics and chemistry content, according to gender. *International Journal of Developmental and Educational Psychology (INFAD), 1*(1), 173–180. https://doi.org/10.17060/ijodaep.2015.n1.v1.26

Dewey, J. (1933). *How we think: A restatement of the relation of reflective thinking to the educative process.* Henry Regnery.

Erduran, S., & Jiménez-Aleixandre, M. P. (2007). *Argumentation in science education.* Springer. https://doi.org/10.1007/978-1-4020-6670-2

Evrekly, E., Inel, D., & Balim, A. G. (2011). Research on the effects of using concept cartoons and mind maps in science education. *Necatibey Eğitim Fakültesi Elektronik Fen ve Matematik Eğitimi Dergisi, 5*(2), 58–85.

Fasae, O., Alufa, F., Ayodele, V., Okoli, A., & Dele-Ajayi, O. (2022). Mobile technology for learning during covid-19: Opportunities, lessons, and challenges. In M. E. Auer & T. Tsiatsos (Eds.), *New realities, mobile systems and applications. Proceedings of the 14th IMCL Conference.* Lecture notes in networks and systems (vol. 411). Springer. https://doi.org/10.1007/978-3-030-96296-8_2

Franco-Mariscal, A. J. (2016). Óptica con Peppa Pig [Optics with Peppa Pig]. *Alambique, Didáctica de las Ciencias Experimentales, 83*, 57–63.

Franco-Mariscal, A. J., Cano-Iglesias, M. J., España-Ramos, E., & Blanco-López, A. (2025). The ENCIC-CT model for the development of critical thinking. In A. J. Franco-Mariscal (Ed.), *Critical thinking in science education and teacher training* (Chap. 1) (pp. 3–42). Springer. https://doi.org/10.1007/978-3-031-78578-8_1

Franco-Mariscal, A. J., Hierrezuelo-Osorio, J. M., Cano-Iglesias, M. J., & Blanco-López, A. (Coords.). (2023). *El juego de rol como estrategia para desarrollar habilidades de pensamiento crítico. Aplicado al aula de las ciencias* [*Role-playing game as a strategy to develop critical thinking skills. Applied in the science classroom*]. Pirámide.

Hamalosmanoglu, M., Kizilay, E., & Saylan Kirmizigül, A. (2020). The effects of using animated films in the environmental education course on prospective teachers' behavior towards environmental problems and their attitude towards solid waste and recycling. *International Online Journal of Education and Teaching, 7*(3), 1178–1187.

Hatton, N., & Smith, D. (1995). Reflection in teacher education: Towards definitions and implementation. *Teaching and Teacher Education, 11*, 33–49. https://doi.org/10.1016/0742-051X(94)00012-U

Hébert, C. (2015). Knowing and/or experiencing: A critical examination of the reflective models of John Dewey and Donald Schön. *Reflective Practice, 16*(3), 361–371. https://doi.org/10.1080/14623943.2015.1023281

Hinze, A., Vanderschantz, N., Timpany, C., Cunningham, S. J., Saravani, S. J., & Wilkinson, C. (2023). A study of mobile app use for teaching and research in higher education. *Technology, Knowledge and Learning, 28*, 1271–1299. https://doi.org/10.1007/s10758-022-09599-6

Horizon Report (2021). Educause horizon report 2021: Teaching and learning. Educause. https://library.educause.edu/-/media/files/library/2021/4/2021hrteachinglearning.pdf

International Association for the Evaluation of Educational Achievement, IEA. (2020). *Measuring global education goals: How TIMSS helps monitoring progress towar.ds sustainable development goal 4 using TIMSS (trends in international mathematics and science study).* https://www.iea.nl/fileadmin/user_upload/Studies/ICILS_2013/IEA_ICILS_Press_Release.pdf

Jiménez-Aleixandre, M. P. (2010). *10 ideas clave. Competencias en argumentación y uso de pruebas* [10 key ideas. competencies in argumentation and use of evidence]. Graó.

Koutnikova, M. (2017). The application of comics in science education. *Acta Educationis Generalis, 7*(3), 88–98. https://doi.org/10.1515/atd-2017-0026

Lai, K.- W., & Smith, L. (2018). Socio-demographic factors relating to perception and use of mobile technologies in tertiary teaching. *British Journal of Educational Technology, 49*, 492–504. https://doi.org/10.1111/bjet.12544

Lin, L., Widdall, C., & Ward, L. (2014). Improving critical thinking with interactive mobile tools and apps. *Social Studies and the Young Learner, 26*(4), 10–14.

López-Fernández, M. M., & Franco-Mariscal, A. J. (2021). Teaching genetics using The Simpsons. An innovative proposal with Spanish pre-service science teachers. In G. S. Carvalho, A. D. Afonsa, & Z. Anastácio (Eds.), *Fostering scientific citizenship in an uncertain world (proceedings of ESERA 2021)* (pp. 372–377). University of Minho.

López-Fernández, M. M., González-García, F., & Franco-Mariscal, A. J. (2021). Should we ban single-use plastics? A role-playing game to argue and make decisions in a Grade-8 school chemistry class. *Journal of Chemical Education, 98*(12), 3947–3956. https://doi.org/10.1021/acs.jchemed.1c00580

Luque-Henares, E., Franco-Mariscal, A. J., & Blanco-López, Á. (2022). El pensamiento crítico a través de la reflexión. Un estudio en el Grado en Educación Infantil [Critical thinking through reflection: A study in the degree in early childhood education]. In L. Villalustre & M. Cueli (Eds.), *Avances y desafíos para la transformación educativa* [Advances and challenges for educational transformation] (pp. 59–69). University of Oviedo.

Mellado, V. (2003). Cambio didáctico del profesorado de ciencias experimentales y filosofía de la ciencia [Didactic change of science education and philosophy of science teachers]. *Enseñanza de las Ciencias, 21*(3), 343–358. https://doi.org/10.5565/rev/ensciencias.3913

Membiela, P., Acosta, K., Yebra, M. A., & González, A. (2023). Motivation to learn science, emotions in science classes, and engagement towards science studies in Chilean and Spanish compulsory secondary education students. *Science Education, 107*(4), 939–963. https://doi.org/10.1002/sce.21793

Mena, M., García, M. L., & Tillema, H. (2012). Student teacher reflective writing: What does it reveal? *European Journal of Teacher Education, 1*, 1–17. https://doi.org/10.1080/02619768.2012.713933

Mendoza, E., Mariel, I., Gómez, M. G., & García, I. A. (2016). El desarrollo del pensamiento crítico mediante el debate asincrónico en foros virtuales en educación secundaria [The development of critical thinking through asynchronous debate in virtual forums in secondary education]. *Aletheia, 8*(1), 100–115.

Michael, D., & Chen, S. (2006). *Serious games. Games that educate, train and inform.* Course Technology, CENGAGE Learning.

Newhouse, C. P. (2016). STEM the boredom: Engage students in the Australian curriculum using ICT with problem-based learning and assessment. *Journal of Science Education and Technology, 26*(1), 44–57. https://doi.org/10.1007/s10956-016-9650-4

Perrenoud, P. (1999). *Dix nouvelles compétences pour enseigner Invitation au voyage* [*Ten new competences for teaching*]. ESF.

Piaget, J. (1950). *The Psychology of Intelligence.* Routledge, Taylor & Francis. https://doi.org/10.4324/9780203164730

Quirantes, A. (2011). Física de película: una herramienta docente para la enseñanza de Física universitaria usando fragmentos de películas [Film physics: A teaching tool for undergraduate physics education using film clips]. *Revista Eureka sobre Enseñanza y Divulgación de las Ciencias, 8*(3), 334–340. http://hdl.handle.net/10498/14397

Rodríguez, L. M. (2010). El cine, estrategia para el desarrollo del pensamiento [Film, a strategy for the development of thinking]. *Praxis and Saber, 1*(2), 87–110.

Rose, C. (2003). How to teach biology using the movie science of cloning people, resurrecting the dead, and combining flies and humans. *Public Understanding of Science, 12*(3), 289–296. https://doi.org/10.1177/0963662503123007

Rutjens, B. T., Heine, S. J., Sutton, R. M., & van Harreveld, F. (2018). Attitudes towards science. In J. M. Olson (Ed.), *Advances in experimental social psychology* (pp. 125–165). Elsevier Academic Press. https://doi.org/10.1016/bs.aesp

Saha, S., & Srivastava, A. (2023). Digitized entertainment—Factors influencing consumer adoption of OTT platforms. In *2022 OPJU international technology conference on emerging technologies for sustainable development* (*OTCON*) (pp. 1–6). IEEE. https://doi.org/10.1109/OTCON56053.2023.10113910

Sánchez, J., Cañada, F., & Dávila, M. A. (2018). Emotional responses to innovative science teaching methods: Acquiring emotional data in a general science teacher education class. *Journal of Technology and Science Education, 8*(4), 346–359. https://doi.org/10.3926/jotse.408

Santos, L. F. (2017). The role of critical thinking in science education. *Journal of Education and Practice, 8*(20), 159–173.

Schön, D. (1998). *The reflective practitioner*. Basic Books.

Simon, S., & Campbell, S. (2012). Teacher learning and professional development in science education. In B. Fraser, K. Tobin, & C. J. McRobbie (Eds.), *Second International Handbook of Science Education* (vol. 24, pp. 307–321). Springer. https://doi.org/10.1007/978-1-4020-9041-7_22

Smith, D. (1991). Educating the reflective practitioner in curriculum. *Curriculum, 12*, 115–124.

Toom, A., Husu, J., & Patrikainen, S. (2014). Student teachers' patterns of reflection in the context of teaching practice. *European Journal of Teacher Education, 38*(3), 1–21. https://doi.org/10.1080/02619768.2014.943731

Toulmin, S. E. (2003). *The uses of argument* (3rd ed.). Cambridge University Press.

Ulu-Aslan, E., & Baş, B. (2023). Popular culture texts in education: The effect of tales transformed into children's media on critical thinking and media literacy skills. *Thinking Skills and Creativity, 47*, 101202. https://doi.org/10.1016/j.tsc.2022.101202

Valencia, J., Acebal, M. C., & Franco-Mariscal, A. J. (2023). Juego de rol sobre la destrucción de un paraje natural. In A. J. Franco-Mariscal, J. M. Hierrezuelo-Osorio, M. J. Cano-Iglesias, & A. Blanco-López (Coords.), *El juego de rol como estrategia para desarrollar habilidades de pensamiento crítico. Aplicado al aula de las ciencias* [Role-playing as a strategy for developing critical thinking skills. Applied to the science classroom] (pp. 71–85). Pirámide.

Van Manen, M. (1977). Linking ways of knowing within ways of being practical. *Curriculum Inquiry, 6*, 205–228.

Yoon, H. G., & Kim, M. (2010). Collaborative reflection through dilemma cases of science practical work during practicum. *International Journal of Science Education, 32*(3), 283–301. https://doi.org/10.1080/09500690802516538

Chapter 17
Concluding Remarks

Antonio Joaquín Franco-Mariscal and Ángel Blanco-López

17.1 General Conclusions

This edited volume presents the ENCIC-CT model (Franco-Mariscal et al., 2025) for developing critical thinking (hereinafter, CT) within science education, illustrated through a wide range of experiences spanning different educational levels, from preschool to initial teacher training. A comprehensive analysis of the various experiences discussed allows us to draw the following general conclusions as a contribution to this topic.

17.1.1 First

The conducted experiences underscore that fostering CT in the science classroom necessitates the integration of elements inherent to science education, which have proven effective in our field. Key among these are socio-scientific issues (Sadler & Zeidler, 2005; Zeidler et al., 2019) and the scientific practices of argumentation, inquiry, and modeling (Next Generation Science Standards, 2013; National Research Council, 2012). Socio-scientific issues and everyday life problems are particularly valuable for enhancing CT as they establish connections between science and various social aspects, while also addressing multiple perspectives on these issues (Chen & Xiao, 2021; Simonneaux, 2001). These connections are crucial for making well-reasoned decisions based on a holistic understanding (Evagorou et al., 2012).

A. J. Franco-Mariscal (✉) · Á. Blanco-López
Science Education, Universidad de Málaga, Málaga, Spain
e-mail: anjoa@uma.es

Á. Blanco-López
e-mail: ablancol@uma.es

A. J. Franco-Mariscal (ed.), *Critical Thinking in Science Education and Teacher Training*, Contemporary Trends and Issues in Science Education 64,
https://doi.org/10.1007/978-3-031-78578-8_17

This approach not only aids in the comprehension of scientific knowledge but also empowers students to apply CT in real-world, meaningful contexts. Similarly, scientific practices are fundamental because they equip students with the tools and methodologies of science, fostering the development of strong CT skills that enable them to analyze, evaluate, and construct knowledge effectively and thoughtfully (Rocard et al., 2007).

17.1.2 Second

This book presents a comprehensive model of CT that, in addition to elements intrinsic to science education, incorporates other components from broader fields of education and philosophy, organized into three domains: knowledge, skills, and dispositions (Franco-Mariscal et al., 2025). These domains and their associated dimensions are supported by various authors (Bailin, 2002; Dwyer et al., 2014; Ennis, 1987; Facione, 1990; Halpern & Danna, 2023; Jiménez-Aleixandre & Puig, 2022; Paul & Elder, 2005; Vieira & Tenreiro, 2016; Vila et al., 2023). According to this model, important dimensions of CT include the application of scientific knowledge and knowledge about science, vision of science/epistemic knowledge about science, comprehensive analysis of problems, critical analysis of information, argumentation, decision-making, communicative skills, personal autonomy/metacognition/reflection, social engagement/activism, and emotional engagement.

17.1.3 Third

The implementation of the ENCIC-CT model has uncovered numerous opportunities for fostering CT. The various potential combinations among the elements within the model's four layers (socio-scientific issues—layer 1, CT domains and dimensions—layer 2, scientific practices—layer 3, and teaching strategies—layer 4) provide multiple ways to cultivating CT. This flexibility allows science teachers to transition from simple experiences that involve a single CT domain and/or dimension to more complex experiences that integrate the three core domains of knowledge, skills, and dispositions, along with their respective dimensions. Such adaptability enables the tailoring of experiences to different levels of complexity and student needs, offering clear pathways for advancing CT development. The annex highlights this versatility by detailing how CT can be nurtured through various socio-scientific issues or everyday life problems, scientific practices, and teaching strategies in each experience featured in this edited volume.

17.1.4 *Fourth*

While the experiences outlined in this book are specifically crafted to target particular domains or dimensions of CT, they also hold the potential to effectively cultivate other dimensions, given the interrelated nature of these CT components. This interconnectedness allows science teachers to harness these additional benefits, thereby maximizing the overall impact on the holistic development of CT (see Chap. 2).

17.2 Specific Considerations on the Dimensions of Critical Thinking in the ENCIC-CT Model

This section outlines specific considerations regarding the various dimensions of CT based on the results obtained from the different experiences, highlighting both advancements and areas that require further attention.

17.2.1 *Knowledge Domain*

17.2.1.1 Regarding the Application of Knowledge and the View of Science

Various experiences presented throughout this book highlight the critical importance of a solid foundation in scientific knowledge for effective argumentation. Without this knowledge, students struggle to make reasoned decisions regarding socio-scientific issues (Henderson et al., 2018). Moreover, it is essential that students not only grasp and apply the scientific concepts involved but that this knowledge is also suitable for their educational level. A strong understanding and appropriate application of scientific knowledge are essential for students to effectively tackle the issues at hand. If their understanding is insufficient or the knowledge poorly adapted, there is a risk that students will end up discussing irrelevant aspects or even fictitious problems, which would compromise the quality of their learning. This underscores the necessity of adapting scientific knowledge to the cognitive maturity of students, thereby fostering critical and well-informed reflection that enables active participation in discussions and decision-making on pertinent socio-scientific issues. The experiences explored in Chap. 4, regarding the naturalization of spaces, and Chap. 15, focusing on the creation of digital storytelling, exemplify this need.

Among the strategies that have significantly enhanced the dimensions of scientific knowledge application and vision of science, the use of movies and television series stands out. These mediums allow students to recognize everyday problems and propose solutions within familiar entertainment contexts (Chap. 16). In this way, students not only improve their understanding of scientific concepts but also learn to apply this knowledge critically in various situations, promoting a more integrated and practical view of science.

17.2.2 Skills Domain

17.2.2.1 Regarding Comprehensive Analysis of Problems

The comprehensive analysis of problems is a critical dimension of CT that must be addressed effectively to prevent a narrow focus on isolated dimensions. Experiences such as controversy mapping (Chap. 5) and short debates (micro-debates) (Chap. 8) have proven to be effective strategies, engaging students in research with a holistic approach. This ensures that all relevant dimensions of the problem are fully considered, which is essential for constructing well-informed and balanced arguments. By tackling problems comprehensively, students not only develop a deeper and more complete understanding of the topic (Evagorou et al., 2012) but also enhance their ability to analyze interrelationships among various factors, evaluate diverse perspectives, and make sound judgments. This broad, multifaceted view fosters more robust CT and enables students to address complex issues with greater effectiveness and creativity.

17.2.2.2 Regarding Argumentation

Argumentation as a dimension of CT is significantly enriched through the use of technology-focused teaching strategies and scientific practices like inquiry and modeling. Experiences utilizing energy simulations (Chap. 7), the collaborative platform Kialo (Chap. 9), the augmented reality tool Zapworks (Chap. 14), and a mobile role-playing game application (Chap. 16) have shown particular effectiveness in fostering students' argumentative skills. These technologies create interactive and dynamic environments where students can explore and debate complex problems, facilitating the development of well-founded and critical arguments. Inquiry and modeling further encourage the presentation of a broad range of evidence to support these arguments.

However, several challenges remain that need to be addressed. The first is the difficulty in adequately justifying evidence (Chap. 10), which often results in weak arguments (Kolstø, 2006). The second challenge arises from the struggles faced by preservice teachers in designing activities that effectively promote argumentation (Chap. 14). In order to overcome these challenges, it is crucial to provide specific training in argumentation for both teachers and students. This training should not only cover how to construct arguments but also involve a thorough analysis of all their components. Furthermore, preservice teachers must engage in deep reflection on the cognitive skills involved in the argumentative process, allowing them to design activities that genuinely stimulate and develop this dimension of CT. Addressing these challenges is vital, as argumentation is closely linked to decision-making and must be grounded in scientific evidence (Bravo & Jiménez-Aleixandre, 2018).

17.2.2.3 Regarding Critical Analysis of Information

The critical analysis of information is paramount, as it significantly impacts other dimensions of CT and is essential for developing sound argumentation and making informed decisions. Specifically, educating students on how to evaluate the quality of information disseminated through social media is crucial for empowering them to distinguish between reliable sources and false or misleading content (Lazer et al., 2018). This aspect has been addressed in several experiences throughout this book (Chaps. 3 and 7), underscoring its importance in modern education and within the ever-evolving landscape of digital information.

17.2.2.4 Regarding Decision-Making

Several experiences indicate that students can shift their positions on issues after engaging with the ENCIC-CT model. Strategies such as role-playing (Chap. 13) (Simonneaux, 2001) and micro-debates (Chap. 8) are particularly effective, as they enable a comprehensive analysis of the problem. Even when substantial changes in position are not observed, these experiences consistently lead to an improvement in the quality of arguments students use to articulate their views on socio-scientific issues or everyday life problems (Chap. 8).

17.2.2.5 Regarding Communicative Skills

While it is a key dimension for the development of CT, the experiences presented in this edited volume do not yield definitive conclusions, as it has not been subjected to specific analysis. Consequently, there is a need for more in-depth exploration of how to effectively address this dimension within the ENCIC-CT model.

17.2.3 Dispositions Domain

17.2.3.1 Regarding Personal Autonomy/Metacognition/Reflection

Various experiences highlighted in this book (Chaps. 4 and 7) demonstrate that students can develop significant personal autonomy when they effectively engage with dimensions of CT that are closely interconnected. Specifically, when involved in activities that foster critical analysis of information, students learn to evaluate sources and data more rigorously and objectively. This analytical process not only enhances their ability to question and validate information but also promotes greater independence in their learning and decision-making.

Moreover, honing argumentation skills empowers students to articulate and defend their viewpoints coherently and based on sound evidence, which further boosts their confidence in their capabilities and intellectual autonomy. These practices contribute to a greater ability for students to manage their own learning independently.

However, developing personal autonomy, metacognition, and reflection requires intentional training. Therefore, it is advisable to incorporate preliminary activities, such as Self-Reflection (Chap. 16), to support this development.

17.2.3.2 Regarding Social Engagement

The impact of social engagement and activism on the development of CT is particularly evident in experiences that present students with social issues that directly affect them, allowing for meaningful contributions. This is exemplified by the experience with Chilean students addressing an environmental problem in their region (Chap. 12). In contrast, for other significant societal issues, such as purchasing gas from Russia—issues perceived as more distant—students do not exhibit the same level of social engagement (Chap. 13).

17.2.3.3 Regarding Emotional Engagement

Emotional engagement has emerged as a crucial factor in promoting the development of various dimensions of CT. As highlighted in the experience of creating digital storytelling (Chap. 15), emotional involvement not only enhances the application of knowledge and critical analysis of information but also strengthens students' ability to articulate and justify their arguments in a more compelling and reflective manner. When students are enthusiastic about a topic, it can boost their motivation, interest, and effort, ultimately leading to a deeper development of CT.

However, it's important to acknowledge that emotional engagement can also lead to feelings of insecurity (Bellocchi, 2018), which may manifest as a lack of confidence in their ability to address complex issues or difficulty in making decisions. Therefore, while emotional engagement can serve as a powerful catalyst for the development of CT, it is essential to address and manage these potential insecurities by fostering a supportive learning environment where students feel secure in their knowledge and skills.

Acknowledgements This work is part of the Spanish National Plan R+D+i Project entitled "Citizens with critical thinking: A challenge for teachers in science education", reference PID2019-105765GA-I00, funded by MCIN/AEI/10.13039/501100011033.

Annex

Application of the ENCIC-CT model in the experiences presented in the edited volume

Chapter	Layer 1	Layer 2			Layer 3	Layer 4
	Daily life problem/SSI	Knowledge	Skills	Dispositions	Scientific practice	Teaching strategy
2	Energy problem (an artificial moon as a source of electrical energy in a city)	KA VS	CAP CAI A DM CS	PA	Argumentation	Media sources
3	Health problem (COVID-19 vacinnation)	KA	A DM	PA	Argumentation	
4	Educational problem (naturalization of spaces)		DM	PA	Argumentation modeling	Think-pair-share technique
5	Social problem (urban spaces for care)		CAP		Modeling	Controversy mapping
6	Health problem (pasteurisation)		CAI	PA	Argumentation	Technology (mobile devices)
7	Energy problem (energy production and consumption)	KA	CAI A CS	PA	Inquiry argumentation	Technology (simulation)
8	Different socio-scientific issues		A DM		Argumentation	Micro-debate
9	Health and social problem (breastfeeding in public places)		A		Argumentation	Technology (Kialo)
10	Health problem (yoghurt making)		A DM		Inquiry argumentation	
11	Health problem (yoghurt making, preservice teachers as learners) Different daily life problems (preservice teachers as trainee teachers)	KA	A DM		Inquiry	Different teaching strategies for designing their own lessons

(continued)

(continued)

Chapter	Layer 1	Layer 2			Layer 3	Layer 4
	Daily life problem/SSI	Knowledge	Skills	Dispositions	Scientific practice	Teaching strategy
12	Environmental problem (industrial pollution in a coastal area)		CAP	SE	Inquiry	Service-learning
13	Energy problem (purchase of Russian gas)		CAI DM		Argumentation	Role-playing
14	Different daily life problems		A		Argumentation	Augmented reality (Zapworks)
15	Different daily life problems for preschool	KA	CAI	EE	Argumentation	Digital storytelling
16	Different socio-scientific issues		A DM		Argumentation	Role-playing, technology (mobile apps)
	Different everyday life situations and socio-scientific issues	KA VS			Argumentation	Multimedia resources (movies and TV series)
	Different everyday life situations and socio-scientific issues			PA	Argumentation	Self-reflection

Note Dimensions for the knowledge domain: *KA* knowledge application; *VS* vision of science
Dimensions for the skill domain: *CAP* comprehensive analysis of the problem; *CAI* critical analysis of information; *A* argumentation; *DM* decision-making; *CS* communicative skills
Dimensions for the dispositions domain: *PA* personal autonomy/metacognition/reflection; *SE* social engagement/activism; *EE* emotional engagement

References

Bailin, S. (2002). Critical thinking and science education. *Science & Education, 11*(4), 361–375. https://doi.org/10.1023/A:1016042608621

Bellocchi, A. (2018). Negative emotional events during science inquiry. In K. Tobin & S. M. Ritchie (Eds.), *Eventful learning: Learner emotions*. Bold Visions in Educational Research (vol. 61, pp. 87–104). Brill.

Bravo, B., & Jiménez-Aleixandre, M. P. (2018). Developing an initial learning progression for the use of evidence in decision-making contexts. *International Journal of Science and Mathematics Education, 16*(4), 619–638. https://doi.org/10.1007/s10763-017-9803-9

Chen, L., & Xiao, S. (2021). Perceptions, challenges and coping strategies of science teachers in teaching socioscientific issues: A systematic review. *Educational Research Review, 32*, 100377. https://doi.org/10.1016/j.edurev.2020.100377

Dwyer, C. P., Hogan, M. J., & Stewart, I. (2014). An integrated critical thinking framework for the 21st century. *Thinking Skills and Creativity, 12*, 43–52. https://doi.org/10.1016/j.tsc.2013.12.004

Ennis, R. H. (1987). A taxonomy of critical thinking dispositions and abilities. In J. B. Baron, & R. J. Sternberg (Eds.), *Teaching thinking skills: Theory and practice.* Freeman.

Evagorou, M., Jiménez-Aleixandre, M. P., & Osborne, J. (2012). Should we kill the grey squirrels? A study exploring students' justifications and decision making. *International Journal of Science Education, 34*(3), 401–428. https://doi.org/10.1080/09500693.2011.619211

Facione, P. A. (1990). *Critical thinking: A statement of expert consensus for purposes of educational assessment and instruction.* California Academic Press.

Franco-Mariscal, A. J., Cano-Iglesias, M. J., España-Ramos, E., & Blanco-López, A. (2025). The ENCIC-CT model for the development of critical thinking. In A. J. Franco-Mariscal (Ed.), *Critical thinking in science education and teacher training* (Chap. 1) (pp. 3–42). Springer. https://doi.org/10.1007/978-3-031-78578-8_1

Halpern, D. F., & Danna, D. S. (2023). *Though and knowledge. An introduction to critical thinking* (Sixth Edition, vol. 2). Routledge.

Henderson, J. B., McNeill, K. L., González-Howard, M., Close, K., & Evans, M. (2018). Key challenges and future directions for educational research on scientific argumentation. *Journal of Research in Science Teaching, 55*(1), 5–18. https://doi.org/10.1002/tea.21412

Jiménez-Aleixandre, M. P., & Puig, B. (2022). Educating critical citizens to face post-truth: The time is now. In B. Puig & M. P. Jiménez-Aleixandre (Eds.), *Critical thinking in biology and environmental education. Facing challenges in a post-truth world* (pp. 3–19). Springer. https://doi.org/10.1007/978-3-030-92006-7

Kolstø, S. D. (2006). Patterns in students' argumentation confronted with a risk-focused socio-scientific issue. *International Journal of Science Education, 28*(14), 1689–1716. https://doi.org/10.1080/09500690600560878

Lazer, D. M. J., Baum, M. A., Benkler, Y., Berinsky, A. J., Greenhill, K. M., Menczer, F., Metzger, M. J., Nyhan, B., Pennycook, G., Rothschild, D., Schudson, M., Sloman, S. A., Sunstein, C. R., Thorson, E. A., Watts, D. J., & Zittrain, J. L. (2018). The science of fake news. Addressing fake news requires a multidisciplinary effort. *Science, 359*, 1094–1096.

National Research Council (NRC). (2012). *A framework for K-12 science education: Practices crosscutting concepts, and core ideas.* National Academy Press.

Next Generation Science Standards (NGSS). (2013). *Next generation science standards: For states, by states.* Achieve, Inc. on behalf of the twenty-six states and partners that collaborated on the NGSS.

Paul, R., & Elder, L. (2005). *A guide for educator to critical thinking competency standards.* Foundation for Critical Thinking.

Rocard, M., Csermely, P., Jorde, D., Lenzen, D., Walberg-Henriksson, H., & Hemmo, V. (2007). *Science education now: A renewed pedagogy for the future of Europe.* European Comission.

Sadler, T. D., & Zeidler, D. (2005). Patterns of informal reasoning in the context of socioscientific decision making. *Journal of Research in Science Teaching, 42*(1), 112–138. https://doi.org/10.1002/tea.20042

Simonneaux, L. (2001). Role-play or debate to promote students' argumentation and justification on an issue in animal transgenesis. *International Journal Science Education, 23*(9), 903–927. https://doi.org/10.1080/09500690010016076

Vieira, R. M., & Tenreiro, C. (2016). Fostering scientific literacy and critical thinking in elementary science education. *International Journal of Science and Mathematics Education, 14*, 659–680. https://doi.org/10.1007/s10763-014-9605-2

Vila, L., Márquez, C., & Oliveras, B. (2023). Una propuesta para el diseño de actividades que desarrollen el pensamiento crítico en el aula de ciencias. [A didactic proposal for the development of critical thinking in the high school science classroom.] *Revista Eureka sobre Enseñanza y Divulgación de las Ciencias, 20*(1), 130201. https://doi.org/10.25267/Rev_Eureka_ensen_divulg_cienc.2023.v20.i1.1302

Zeidler, D. L., Herman, B. C., & Sadler, T. D. (2019). New directions in socioscientific issues research. *Disciplinary and Interdisciplinary Science Education Research, 1*(1), 1–9. https://doi.org/10.1186/s43031-019-0008-7

Correction to: Critical Thinking in Science Education and Teacher Training

Antonio Joaquín Franco-Mariscal

Correction to:
A. J. Franco-Mariscal (ed.), ***Critical Thinking in Science Education and Teacher Training*****, Contemporary Trends and Issues in Science Education 64, https://doi.org/10.1007/978-3-031-78578-8**

In the original version of the book, Volume number has been replaced from 2 to 64. The FM and the cover have been updated with the changes.

The updated version of this book can be found at
https://doi.org/10.1007/978-3-031-78578-8

A. J. Franco-Mariscal (ed.), *Critical Thinking in Science Education and Teacher Training*, Contemporary Trends and Issues in Science Education 64,
https://doi.org/10.1007/978-3-031-78578-8_18

The manufacturer's authorised representative in the EU is Springer Nature Customer Service Centre GmbH, Europaplatz 3, 69115 Heidelberg, Germany. If you have any concerns regarding our products, please contact ProductSafety@springernature.com

Printed and bound by CPI Group (UK) Ltd, Croydon, CR0 4YY

30/09/2026

02225569-0013